COMPUTER INTEGRATED MANUFACTURING

GUIDELINES AND APPLICATIONS FROM INDUSTRIAL LEADERS

The BUSINESS ONE IRWIN/APICS Series in Production Management

Supported by the American Production and Inventory Control Society

OTHER BOOKS PUBLISHED IN THE BUSINESS ONE IRWIN/APICS SERIES IN PRODUCTION MANAGEMENT

Attaining Manufacturing Excellence *Robert W. Hall*
Bills of Materials *Hal Mather*
Production Activity Control *Steven A. Melnyk and Phillip L. Carter*
Manufacturing Planning and Control Systems, Second Edition
Thomas E. Vollmann, William Lee Berry, D. Clay Whybark
The Spirit of Manufacturing Excellence *Ernest Huge*
Strategic Manufacturing: Dynamic New Directions for the 1990s
Patricia E. Moody
Total Quality: An Executive's Guide for the 1990s
The Ernst & Young Quality Improvement Consulting Group
Forecasting Systems for Operations Management
Stephen A. DeLurgio and Carl D. Bhame
Common Sense Manufacturing *James Gardner*
Purchasing Strategies for Total Quality: A Guide to Achieving Continuous Improvement *Greg Hutchins*
Logistics in Manufacturing *Chris Gopal and Gerry Cahill*

COMPUTER INTEGRATED MANUFACTURING

GUIDELINES AND APPLICATIONS FROM INDUSTRIAL LEADERS

Steven A. Melnyk, Ph.D. CPIM
Ram Narasimhan, Ph.D.
Department of Management
Graduate School of Business Administration
Michigan State University
East Lansing, MI

BUSINESS ONE IRWIN
HOMEWOOD, ILLINOIS 60430

NOTICE
This Report was made possible by a research grant from the Education and Research Foundation American Production and Inventory Control Society.

Sponsoring editor: Jeffrey A. Krames
Project editor: Jean Roberts
Production manager: Diane Palmer
Designer: Jeanne M. Rivera
Jacket Designer: Michael Finkelman
Production Services: Caliber Design Planning
Typeface: 10/12 Times Roman
Printer: Arcata Graphics, Kingsport

Library of Congress Cataloging-in-Publication Data

Melnyk, Steven A.

Computer integrated manufacturing: guidelines and applications from industrial leaders / Steven A. Melnyk, Ram Narasimhan.
p. cm.—(The Business One Irwin/APICS series in production management)
Includes bibliographical references.
ISBN 1-55623-538-0
1. Computer integrated manufacturing systems. I. Narasimhan, Ram. II. Title. III. Series.
TS155.6.M45 1991
670'.285—dc20

91–24334

Printed in the United States of America
1 2 3 4 5 6 7 8 9 0 AG-K 8 7 6 5 4 3 2 1

ACKNOWLEDGMENTS

The writing of this book has been a long and hard journey that has taken several years to complete. Its completion is due in large part to the efforts and help of several people. It is to these people that we dedicate this book as a small thank you:

Richard (Dick) T. Smith of Ingersoll Engineering; George Hess of Ingersoll Milling; Chad Frost of Amprotech, Inc.; Larry Barton of Steelcase, Inc.; and Cecil Bradshaw of Oliver Wight and Associates. Our understanding of CIM has been greatly shaped by these people.

Steven A. Melnyk
Ram Narasimhan

TABLE OF CONTENTS

PART 1

PART 1

CIM: MAKING SENSE OF A MANUFACTURING MYSTERY

CHAPTER 1

A TALE OF TWO FARMERS: A CIM PARABLE

THE PARABLE TOLD

Once upon a time, in a land not too far from here, there lived two farmers. These two farmers were neighbors. They lived on two adjacent pieces of property that were almost identical. They were about the same size; they shared the same general terrain; in both, the soil was good, the water plentiful, the sun always shone, and the weather was temperate. In short, the land was good for growing.

Over time the two farmers had become very prosperous. Their produce was always good and it was always in demand. With the money earned from the sales of their produce, both farmers had hired good, conscientious hands to help with the task of working their land.

It was almost inevitable that news of the success of these two farmers would eventually reach farmers in the other vales who were not as successful. In some cases the reason lay in their land. The soil was not as rich. In other cases the problem lay in the climate. In many instances the problem lay in how these other farmers managed their own plots of land. When they heard about the success of the two farmers, many of these farmers decided that they should visit the two farmers so that they could try to learn the secret of their success.

The two farmers prospered. As they prospered, they found themselves visited by farmers from different valleys. Each of these foreign farmers (that is the way that the two farmers referred to those who came from other valleys) had the same request. All they wanted was to see how the two farmers worked their land. They simply wanted to learn from those who really best knew how to farm the land. The more the two farmers prospered, the more farmers came to visit them.

After a time the two farmers began to take great pride in their work. They began to think of themselves as the best farmers in the country. They also began to believe that their ways of managing the land were indeed the best and that these methods could not be improved on. Whenever one of their farmhands approached them with a suggestion for improving yields on the farm, the answer was always the same—no. "That's not the way we've done things here," would reply one farmer. "If it ain't broke, then why fix it," said the other.

The Beginning of the End

Then it happened. At first it took place gradually. The two farmers started to notice that the demand for their crops was falling. When in the past they would sell all that they could harvest, the two farmers now noticed that they were returning with some of their crops. It wasn't that much at first. "The economy," ventured one farmer. "Who knows and who really cares," commented the other farmer. Yet every year it seemed that the amount of crops being returned was getting a little bit larger.

Then one year the situation got worse. The two farmers found themselves with a great deal of produce that no one seemed to want. They first reacted by cutting price or by offering buyer incentives ("two carrots for the price of one"). That seemed to help somewhat but it couldn't stop the slide in sales.

The farmers then reexamined the crops and the methods that they were using to grow the crops. The crops seemed to be as good as ever. The farming methods had not changed. The next step that the farmers took was to tell their hands that they had to work harder. After all, reasoned the two farmers, with all of the success that their farms had enjoyed in the past, it shouldn't come as a surprise that the workers had gotten lax. However, this announcement was not well received. The hands felt that they were already working hard enough. Besides, noted some of the more vocal hands, the farmers should not blame the workers for problems caused by poor management.

The two farmers looked around for other explanations for their problems. They talked to the village people who had bought their produce in the past. The two farmers were surprised to learn that the demand for produce of all types had increased, not decreased. The farmers then asked the villagers whether they knew whose produce they were buying. "Of course," answered many of the villagers. "We are now buying our carrots, potatoes, beans, and such from these farmers living in the nearby valleys. Their harvest is just as good as yours. In fact, it is often better. Their produce is also cheaper and fresher."

The two farmers were shocked when they heard that the same farmers that had once visited them were now raising better crops. The first reaction of the two farmers was to believe that these foreign farmers were somehow cheating. How could they be cheating? According to one farmer, the answer was simple: "They're probably selling their produce for less than it's costing them to raise it. That's the ticket." "You're wrong," said the other farmer. "You know what they're doing? I'll tell you. Either they got low quality seeds or else they got cheap labor. That's the only way."

When the two farmers raised these concerns with the village elders, they were surprised by their response. To the charge that the foreign farmers were selling their crops for less than their cost, the two farmers were told by one of the elders who had visited these farmers, "It ain't so. I've been there and I can tell you that they're selling their goods to us at a fair price. Not only that, they're making a good buck on top of it all."

To the charge that their labor was cheaper or that they were using inferior inputs, the two farmers were again told that they were wrong: "The labor ain't that much cheaper. Furthermore, the seeds that they are using are even better than yours."

One of the elders then turned to the two farmers and said, "You know what your problem is? You've had it your own way for such a long time that you think that you are the best. Well, it looks like that isn't the case now. Those farmers are selling everything they bring to market because they got better and cheaper vegetables and fruits. Why are their vegetables and fruits better and cheaper? Well, as I see it, they work harder and smarter. If you want my advice, I'd see what they were doing and then do the same. Until then, the only thing you're going to get out of me is sympathy."

Both farmers had to admit that the last suggestion was a good one. After they left the village elders, they made arrangements to see several of the more successful foreign farmers.

When they arrived at these farms, they saw several obvious differences. The first was that the farming equipment was much newer than what they had on their farms. Second, the seeds were different. Many of the seeds were newer and were more resistant to such problems as weeds, insects, long periods of drought, and excessive moisture. The hands on these farms seemed to work more diligently. When the two farmers asked the foreign farmers to explain the reason for their sudden success, they heard references to a new way of farming, a way that these other farmers called *Scientific Seeds Farming*, or SSF. One foreign farmer went as far as saying that the only reason he could compete effectively was because his farm used SSF.

Returning home, the two farmers were confused. Here was something that they had never heard of before. They didn't really understand SSF, yet it did seem to account for a good deal of the success now enjoyed by the foreign farmers. SSF seemed to involve seeds and farm equipment. They both decided to visit the farming implement and seeds seller in the village to see if he knew anything about SSF.

The Move to SSF

Of course, the seller had heard about SSF: "SSF, do I know anything about it? Darn right I do. SSF is the newest method for farming. It works. It merges the new technology of agro-engineering with the practices of farming. It has a great payback period. As I like to tell others, if you don't use SSF, then you might as well sell your farm. You know that if you want to have an SSF farm, you got to put your money in better seeds and in better equipment. Unless you have these two things, you really cannot do much with this system. Now if you are interested, I can show you what I've got that just might help you with putting in an SSF system on your farms. Heck, I'll even show you how to use everything."

The two farmers listened to everything that the seller had to tell them. They looked at the various ploughs, spreaders, drills, and harvesting machines. They studied the various types of seeds and listened as the seller described their various properties. It wasn't long before both farmers felt that they had so much information that they simply couldn't absorb anything else. They were even having trouble remembering everything that the agent had just told them.

The two farmers returned home. Each looked at the facts as he knew them. Each thought about how best to correct the worsening situation facing them. Each realized that something had to be done soon since the very survival of their farms was at stake. The two farmers reacted to the crisis facing them in very different ways.

Two Approaches to SSF

The first farmer decided that the key to SSF lay in getting the best farm equipment and the best seeds. He returned to the village implement and seeds seller. The farmer didn't really know what he needed. He did know two things. First, he had to do something to improve operations on his farm—*now*. Second, he had money. If he had to he would get himself the best and newest farm implements and seeds.

When the first farmer entered the store, he approached the seller and told him that he wanted to become a class A SSF user (Class A was the term that the seller used when describing those farms where SSF was used successfully and effectively). "I want to implement SSF and I want to do it now," announced the farmer.

"Well then, lets start by doing it right and looking at the latest and best," said the seller. That day, the seller showed the first farmer the newest seeds and the newest farm equipment. He demonstrated the features of vehicles considered to be the state of the art.

"Consider how envious your neighbors will be when they see this equipment in your fields. Of course, it does cost a few more pieces of gold but look at what you're getting. First of all, you are getting the newest equipment. Nobody else has it. Second, the equipment that you are buying offers a lot of extra features that you wouldn't find in the older equipment. Think of all the uses to which you can put it. Isn't it worth it to spend the extra money?" noted the seller.

The first farmer really didn't know what he would do with those few extra features. He would probably find a use or two for them. In the end he couldn't really disagree with what the seller had told him. It seemed to make sense.

After the first farmer looked at everything that the seller had shown him, he really didn't know what to buy. So he asked the seller to recommend some seeds and the equipment that he thought would be best for the farmer. The seller thought for a moment and pointed out that since he really wasn't familiar with the farmer's crops and needs, the only thing he could do would be to recommend what he thought was the best overall. He did so. The result was a list consisting of the most expensive and most advanced equipment and seeds. The first farmer looked at the list. When he saw the total cost, he was shocked. He knew it would be expensive to improve operations at his farm, but this was unbelievable!

After the initial shock, the farmer decided that there was nothing else to do but buy them. After all, it was only a one-time expense. If it worked, reasoned the first farmer, he could save money in several ways. First, operations would become more efficient. Costs would fall and profits would improve. If operations became more efficient, reasoned the first farmer, the number of workers needed might be reduced. Now here was a way to save money by getting rid of the lazy workers. It all seemed to make sense.

He left the seller with many things to carry to the farm. The first farmer also left the seller much, much poorer.

The approach taken by the second farmer was very different. He knew that he would have to get new equipment and seeds. Yet before he invested any money in making these purchases, he decided that he first had to study how he was farming now.

He looked at the land and the crops he was growing. He talked to his farmhands. He talked to his customers. All of this activity took a great deal of time, but it was very revealing. Among the various things he had learned were the following:

- Much of his fields had been overworked and needed to lie fallow for at least a season.
- Some of the crops that he was growing were not really suitable for his land.
- Some of the crops were not well placed. He had been growing crops requiring dry soil in the clay areas (where it was always wet). He was also growing wet crops in the sandy soil.
- The way he was using his farm help was very inefficient. Often, the men working in the field didn't want to do what they had to do. Instead of weeding the fields (something that they didn't do often), the farmhands would be working on threshing.

- His existing farm equipment was in poor condition. It was out-of-date and poorly maintained.
- He really didn't know what customers he was serving. He was growing too many different crops and not doing a good job of meeting the needs of any one group of customers.
- Each worker had his own way of doing things. There seemed to be no set standards. As a result, the farmer had difficulty in determining whether the problem lay in the methods, the crops, the equipment, the customer, or the land.

The second farmer decided that, before buying any new seeds or equipment, he first had to correct the existing problems.

Correcting the various problems was not an easy task. It took him two seasons to put most of the changes in place. He had dropped certain crops. He rearranged his crops. He improved the use of his farm hands. Weeding was now done regularly. Every day something different was being done.

At the end of the second season, the second farmer visited the village implement and seed seller. Unlike his neighbor, the second farmer knew exactly what he wanted. He was not interested in the newest or the best. He wanted something that worked and did exactly what he needed. He also wanted something that could be easily used by his field hands. To him, the new equipment really didn't replace his workers as much as it made them more effective. It didn't seem to bother him that he wasn't buying the best. "Why buy something that I don't really need or that might not work," said the second farmer. Like the first farmer, the second farmer replied that he too was now implementing SSF.

The Results with SSF Compared

While the second farmer was quietly making changes to his farm, the first farmer was telling everyone about what he was doing. He invited people to see the new equipment that he had bought. He told people that he was now well on the way to becoming a Class A SSF system. He showed the reductions in manpower that had resulted from the introduction of the new equipment. He showed people how the new equipment worked. The people who did visit the first farmer were impressed by the new equipment and the amount of money that he had invested. They marveled at the operation of the new equipment (when it worked, which was not often). "Surely," they said to themselves, "here is an example of how to do SSF right."

In using the new equipment and seeds, the first farmer had not really changed any of his past practices. He had simply replaced the old seeds with the new seeds and the old equipment with the new equipment. He replaced manpower with technology. Everything else stayed the way it had always been.

At the end of three seasons, the two farmers met at the local village market. They discussed their experiences with scientific seeds farming.

To the first farmer, SSF was a failure: "I've spent a fortune on new equipment and seeds. It hasn't been worth it. We really don't want anything to do with this new equipment. Much of the time, it doesn't really work. The seeds aren't really that much better than the old. I'm not selling any more than I used to at these markets. You know, I thought that SSF was supposed to really reduce my costs. Well, it hasn't. If anything, with my investments, my costs are now higher than ever. If you ask me, SSF is a crock. It's just another way for sellers to unload on us seeds and equipment that we don't need."

The second farmer looked at the first farmer. He didn't know what to say. For him SSF did work. The benefits of SSF were not that evident in costs. Instead, SSF seemed to enhance his ability to offer better quality produce. His crops were better than ever. While not as cheap as some of the produce offered by the foreign farmers, it was now in demand. He was making more gold than ever before. His farm hands were happy. His customers were happy. Even the foreign farmers admitted that they couldn't compete successfully with him. The second farmer felt good because he now knew that the crops he was growing and selling really met his customers' needs.

As the two farmers walked away from each other, the second farmer couldn't help but wonder why SSF was a success for him and such an expensive failure for his neighbor. One thought kept reoccurring to him:

If you cannot do the basic things right, all of the technology in the world won't help you.

COMPUTER INTEGRATED MANUFACTURING: AN INTRODUCTION TO THE CASEBOOK

According to the Random House Dictionary, a parable is "a short story designed to convey a truth or moral lesson." The purpose of beginning this book on the principles and practices of effective computer integrated manufacturing with a parable is to identify and expose many of the issues critical to the success and, more important, to the failure of CIM. Present in this parable are the major themes of this book:

- The role of technology in the implementation and operation of computer integrated manufacturing.
- The relationship between automation and the human in the system.

- The minimum prerequisites that management must first ensure are present before they can consider the implementation of a CIM system.
- The definition of CIM.
- The extent to which technology can be used to compensate for or replace ineffective or inefficient management practices.

The rest of this book will try to help the reader resolve the underlying problems of the parable. That is, why was the first farmer unsuccessful with SSF, the parable's version of CIM, while the second farmer was able to succeed?

The case studies presented in this book will demonstrate that technology, by itself, is not enough. That is, the acquisition of new and advanced technology by any firm does not automatically confer on that firm the CIM label. It will also show that CIM is an *enhancement* to the effective and integrated manufacturing planning and control system rather than a replacement for an ineffective system. Finally, it will show that CIM is only effective when it is used as part of a coordinated system's response—a response intended to develop a strategic competitive advantage for the implementing firm.

These lessons will be developed in two ways. The first involves a comprehensive examination of the body of knowledge dealing with computer integrated manufacturing. The second, and the heart of this book, is a detailed examination of the experiences of a selected number of leading users of CIM systems. The companies studied in this book are ones that have been able to successfully solve the CIM puzzle. They are like the second farmer: they have made CIM work for them.

After reading this book, we hope that you, the reader, will also be able to make the CIM concept work for your company. After all, the implementation of CIM is expensive and time-consuming. If done well, its benefits are both long term and significant. If done poorly, it is an expensive and a useless waste of time, manpower, effort, and money. No firm is so rich that it can afford to invest in failure.

The Organization of the Book

The book is organized into three major sections consisting of 11 chapters and 3 appendices. The first section defines the general groundwork for the research study and for examining the concept of computer integrated manufacturing. In Chapter 2, we examine the reasons why managers are now interested in CIM and why there is a critical need for a study such as this one. Chapter 3 develops the conceptual framework used in this book for describing and studying computer integrated manufacturing. This chapter presents definitions for many of the terms and concepts associated with the theory and practice of CIM (e.g., concepts such as automation, flexibility, and integration). In Chapter 4, we summarize the major principles gleaned from the CIM conference and the five case studies contained in this study. These principles are ones that should, we think, apply to all manufacturing firms, irrespective of size, product, or the nature of ownership (private versus public).

This study has examined the experiences of selected firms at two levels. At the first level, seven firms were invited to participate in a CIM conference that was held at Michigan State University. From these seven companies, five were selected and their experiences with CIM studied in greater detail. The experiences of these five companies are documented in the case studies found in this book. The process by which the initial seven firms were identified for the conference and the subsequent five were selected for detailed study is summarized in Chapter 5.

This chapter provides an overview of the seven leading-edge CIM users examined by this study. It describes the process used in identifying the participating firms. It also identifies the major characteristics of each participating firm and those lessons that each firm best illustrated. It concludes with an overview of the five case studies found in this book.

The second section, consisting of Chapters 6 through 10, contains the five detailed case studies. It concludes in Chapter 11 with an epilogue. In this epilogue we revisit the parable presented in this chapter in order to resolve the question of why the first farmer failed while the second succeeded.

The third and final section consists of 3 appendices, which present the reader with the names of the members of the CIM Steering Committee, a glossary of major CIM terms, and a bibliography of major CIM-related articles and references.

CHAPTER 2

COMPUTER INTEGRATED MANUFACTURING AND AMERICAN INDUSTRY: RESOLVING THE ENIGMA OF MANUFACTURING IN THE 1990S

OVERVIEW

This is a book about computer integrated manufacturing, or CIM as it is more commonly known. It is not simply a book about the elements typically found in a CIM system, elements such as group technology, flexible manufacturing systems, automated identification technology (e.g., bar coding), manufacturing cells, and artificial intelligence. These elements, while important, are nothing more than ingredients in CIM. If someone is really interested in these aspects of CIM, there are many other more appropriate books and articles that provide useful information on these topics. Instead, this is a book that tries to understand, explain, and ultimately resolve the CIM mystery. What is the CIM mystery? It can be described in the following terms:

> CIM represents an approach to manufacturing that *should* work. It draws on technology that is now becoming widely available (and increasingly less expensive). It is a feasible solution to many of the problems now facing American manufacturing firms. It represents a solution to a *real* problem facing the typical American manufacturer—that of improving the overall competitiveness of the manufacturing function, specifically, and the firm in general. It is a system that has been shown to work. The benefits reported by firms that have successfully implemented CIM have been significant. Yet in spite of all these facts, the CIM record has been dismal. For most firms the implementation of CIM has become an exercise characterized by high costs, excessive organizational trauma, and low benefits. The mystery is that of trying to explain why a system such as CIM, a system that should work, has not fulfilled its promise.

In this chapter we lay out and discuss the various elements of the background to this CIM mystery. These elements are important, for they present the reader with much of the motivation behind the need for this study. The following is a list of the major elements examined in this chapter:

- Defining the dimensions of the manufacturing problem facing American industry in the 1990s.
- The crisis in American manufacturing.
- New paradigms for new manufacturing realities.
- CIM: old wisdoms versus new wisdoms.
- CIM: resolving the mystery.

These elements not only set the background to the CIM mystery, they also provide much of the rationale underlying the need for the study described in this book.

THE MANUFACTURING CRISIS IN AMERICAN INDUSTRY

The last two decades have seen enormous changes in the state of manufacturing in America. Major business periodicals such as *Business Week, Fortune, The Economist, Forbes,* and the *Harvard Business Review* have all carried articles discussing the events surrounding the change of fortunes in American manufacturing. Numerous books have been written describing the changes that have taken place in North American manufacturing and setting out the changes that have to take place for American manufacturing to become more competitive. Universities, governmental agencies, and consulting corporations have formed commissions and published reports dealing with these changes. Television programs and specials have been written and produced dealing with these changes. Everyone is aware that since 1960 the manufacturing landscape has changed significantly.

American Manufacturing: 1960 versus 1990

In the early 1960s the United States was the major manufacturing power in the world. No country in the world could match its capacity, range of manufacturing facilities, and manufacturing expertise. American gross national product (GNP) was the highest in the world. The American standard of living was the highest in the world. American products were highly prized internationally. In short, the United States was the standard by which every other country judged itself.

By the mid 1980s, a drastic change had occurred. American industry was no longer considered to be the world's leader. That position now had to be shared with countries such as Japan, West Germany, Hong Kong, Taiwan, Singapore, and South Korea. Each of these new "partners" had been successful in aggressively developing an industrial base that was well organized and very efficient. The entire manufacturing landscape had been greatly revised.

Everyone realizes that the industrial position of American industry is no longer as strong and secure as it had been. Evidence of this deterioration can be found in the following data:

• The American productivity rate has been gradually but continuously falling. In the 1970s it averaged 0.6 percent. In the first half of the 1980s it fell to an average of 0.04 percent.[1] In addition, output per head (one indication of industrial efficiency) has been declining (see Table 2-1). One very significant statistic that can be obtained from Table 2-1 is that for the three average measures (i.e., output per head, average rate of industry growth, and average rate of growth of output per head), the United States is *below* the average of developed countries.

• In comparing the trends in annual growth in gross domestic product among seven industrial nations, in the periods 1973 to 1985 and 1981 to 1985, the United States was last in both periods (Table 2-2). Furthermore, a comparison of those 20 countries with the highest gross domestic product per head in 1970 and in 1988 shows that the United States has fallen from being first in 1970 to ninth in 1988 (Table 2-3).

• At the end of 1981 the United States was a creditor nation with a net balance of $140 billion. By the end of 1986 the United States was the world's largest debtor nation with a net debt of $220 billion. This was a swing of $360 billion in five years. Manufactured goods went from an $18.1 billion surplus to a $150 billion deficit from 1980 to 1986.[2]

• The ratio of debt to the GNP has increased by 30 percent in the five years following 1982.

• The prominence of American industry in gross exports of manufactured goods has fallen significantly since 1980. The gross exports of manufactured goods from the main manufacturing countries rose from $134 billion in 1969 to $1,048 billion in 1986 and an estimated $1,177 billion in 1987. The percentage of these exported declined for the United States (by 29 percent), Canada (22 percent), and the United Kingdom (30 percent) while rising by almost 60 percent for Japan.

• The consumption rate per worker has grown three times faster than that of the net output per worker. As a result, American workers are now consuming more than they produce. This trend further accentuates the preceding problem.

---

[1]Richard D. Lamm, "Crisis: The Uncompetitive Society." in *Global Competitiveness: Getting the U.S. Back on Track*, ed. Martin K. Starr (New York: W.W. Norton, 1988).

[2]Richard D. Lamm, "Crisis: The Uncompetitive Society." in *Global Competitiveness: Getting the U.S. Back on Track*, ed. Martin K. Starr, (New York: W.W. Norton, 1988).

• Manufacturing is now becoming a less prominent component of gross domestic product (GDP) in the United States as compared with six other developed

TABLE 2-1
Industrial Efficiency (Developed Countries)

	1988 Industrial Output (American $)		*1965–1988 Percent Average Annual Growth*	
	Total ($ billion)	*Per Head ($)*	*Industry*	*Output per Head*
Developed Countries Average	4,634.40	5,624	3.6	2.7
Australia	72.63	4,394	2.9	1.2
Austria	50.63	663	3.4	3.2
Belgium	46.88	4,726	3.3	3.1
Canada	171.06	6,592	3.4	2.2
Denmark	30.79	6,002	1.8	1.5
Finland	35.50	7,172	3.9	3.5
France	304.95	5,458	3.0	2.4
West Germany	479.69	7,838	2.1	1.9
Greece	15.90	1,588	4.8	4.1
Iceland	1.63	6,520		
Ireland	12.03	3,398		
Italy	286.00	4,979	3.1	2.7
Japan	1,155.41	9,424	7.2	6.2
Luxembourg	2.19	5,919	1.8	1.3
Netherlands	76.48	5,182	2.9	2.1
New Zealand	11.58	3,520		
Norway	28.74	6,843	4.8	4.3
Portugal	16.56	1,591		
Spain	126.60	3,242	4.1	3.2
Sweden	75.17	8,906	2.2	1.8
Switzerland	63.37	9,573		
Turkey	26.07	497	7.0	4.6
United Kingdom	295.00	5,168	1.7	1.5
United States	1,249.54	5,073	2.1	1.1

Source: *The Economist, Book of Vital World Statistics* (New York: Times Books, 1990), p. 70.

TABLE 2-2
Annual Growth in Gross Domestic Product per Employee

Country	*1973–1985*	*1981–1985*
United States	0.4%	1.0
Canada	1.1	1.6
France	2.1	1.7
Japan	2.9	2.9
South Korea	4.9	6.0
United Kingdom	1.6	2.6
West Germany	2.2	2.1

Source: Terry Hill, *Manufacturing Strategy: Text and Cases* (Homewood, Ill.: Richard D. Irwin, 1989), p. 6.

countries. Increasingly, the United States economy is turning to the service sector for the major share of GDP (Table 2-4).

- Investment in research and development (R & D) by American firms is falling behind the levels reported in countries such as Japan. The investments made by major Japanese firms in the research and development of key technologies have increased greatly since the early 1980s. As a percent of sales, the largest Japanese firms are spending about 40 percent more of their revenues on R & D than the largest U.S. companies are.[3] Although the average R & D budget of these Japanese firms is far less than that of the largest American firms, the R & D expenditures in Japanese firms are increasing in both absolute and real terms whereas the R & D budgets for the average large American companies are declining.
- Savings are one of the fundamental building blocks on which investment and economic growth are built. In the United States, as a percentage of gross domestic product, the savings rate is among the lowest of all the developed economic countries (see Table 2-5). Furthermore, the resource gap in the United States has become negative (see Table 2-5). The resource gap is a measure of a country's dependence on the outside world. It is expressed as the difference between exports and imports as a percentage of gross domestic product. A negative figure indicates that a country consumes more in imports than it exports.[4]
- In 1985 there were 77,251 patents issued in the United States. Of these, 43.8 percent went to applicants from other nations. In 1986 that number increased to 45 percent. Furthermore, in the early 1980s the top five patent holding companies

---

[3]James C. Abegglen and George Stalk, Jr. *Kaisah: The Japanese Corporation* (New York: Basic Books, 1985).

[4]*The Economist, Book of Vital World Statistics* (New York: Times Publishing, 1990), p. 36.

TABLE 2-3
GDP per Head: (Top Twenty Nations in 1970 and 1988)

Rank	*1970 (GDP in US $)*	*1988 (GDP in US $)*
1	United States ($4,922)	Switzerland ($27,748)
2	Sweden ($4,149)	Bermuda ($23,793)
3	Canada ($4,107)	Iceland ($23,640)
4	Kuwait ($3,830)	Japan ($23,325)
5	Luxembourg ($3,468)	Norway ($21,724)
6	Switzerland ($3,350)	Finland ($21,156)
7	Denmark ($3,215)	Sweden ($21,155)
8	West Germany ($3,049)	Denmark ($20,988)
9	Norway ($2,887)	United States ($19,815)
10	France ($2,381)	West Germany ($19,743)
11	Bahamas ($2,700)	Canada ($18,834)
12	Belgium ($2,669)	Luxembourg ($18,000)
13	Netherlands ($2,585)	France ($17,004)
14	Australia ($2,520)	Austria ($16,675)
15	Finland ($2,380)	Qatar ($15,909)
16	United Kingdom ($2,209)	UAE ($15,560)
17	Austria ($1,948)	Netherlands ($15,421)
18	Japan ($1,930)	Brunei ($15,417)
19	Italy ($1,1875)	Belgium ($15,394)
20	Israel ($1,830)	United Kingdom ($14,477)

Source: *The Economist, Book of Vital Statistics* (New York: Times Publishing, 1990), p. 40.

were American. In 1988 only one of the top five companies was American. Finally, General Electric has traditionally been responsible for registering the most patents. In 1988 it was fourth, with the top companies (in patents issued) being from Japan. More important, there was evidence that the United States was beginning to lose its position of preeminence in the high-technology arena (see box 1). This has always been one area on which many American manufacturing managers, business executives, and governmental officials have relied on to maintain future industrial growth and to protect America's position as a manufacturing and industrial leader.

TABLE 2-4
Origins of GDP for Seven Industrial Countries

Country	*Agriculture*	*Industry*	*Of which Manufacturing*	*Services*
Canada	3.0	35.0	19.0	62. 0
France	3.6	32.1	22.0	64.3
Japan	2.8	40.4	29.0	56.8
South Korea	10.8	43.2	31.6	46.0
West Germany	1.5	39.7	33.0	58.8
United Kingdom	1.4	35.7	23.7	62.9
United States	2.1	25.6	18.9	72.3

Source: *The Economist, Book of Vital World Statistics* (New York: Times Publishing, 1990), p.36.

TABLE 2-5
An International Comparison of Savings and the Resource Gap

Country	*Savings as Percent GDP*	*Resource Gap Percent*
Canada	26.3	1.5
France	21.1	0.1
West Germany	26.1	5.7
Japan	28.2	0.2
South Korea	38.2	8.5
United Kingdom	16.2	–3.6
United States	13.8	–1.9

Source: *The Economist, Book of Vital World Statistics* (New York: Times Publishing, 1990), p.50.

- A recent study noted that the United States had dropped from second to seventh place among industrial nations in the skill level of its workers.[5]
- After adjusting for inflation, the average American family made less money in 1988 than it had in 1969.
- The Commission on Industrial Competitiveness noted that in the period from 1967 to 1987, 7 out of 10 technologically oriented industries in the United

---

[5]Richard D. Lamm, "Crisis: The Uncompetitive Society," in *Global Competitiveness: Getting the U.S. Back on Track*, ed. Martin K. Starr (New York: W.W. Norton, 1988).

BOX 1
Patent Failure
Although U.S. researchers made key breakthroughs in the new field of superconductivity, America now trails far behind Japan in marketing the revolutionary technology. A new study, by *The Cambridge Report on Superconductivity*, finds that Japanese companies have filed 300 of 354 patent applications since high-temperature superconductivity's discovery in 1987; U.S. organizations have filed 34. Seventy-five of Japan's are for devices such as transistors and motors.

Source: *Newsweek*, March 20, 1989, p. 8

States had lost ground in the world markets. In 1967 only 8 percent of American industry faced foreign competition.

• Many American firms considered in the past to be leaders in their respective fields are either no longer in existence or now exist in much smaller versions. For example, the Bullard Machine Company of Bridgeport, Connecticut (where the vertical lathe was invented), is bankrupt. The Morse Cutting Tool Company of New Bedford, Massachusetts (founded in 1864 by Samuel A. Morse, the inventor of the twist drill), has been sold and is no longer the same major driving force in the tool industry.[6]

In the 1960s "Made in America" was considered to be synonymous with a well-built, quality, low-cost, well-designed product. In contrast, anything that was built in Japan was considered to be cheap and poorly built. If you were only interested in something that you could use a couple of times and then throw away, it was acceptable to buy something made in Japan. There was always something comforting in knowing that the product you had just bought was American built. Names such as General Motors, IBM, Ingersoll Rand, Burgmaster, Sperry, and Kodak were identified with quality, innovation, and good value.

By 1990 this was no longer the case. It is the Japanese products that American consumers now view as being synonymous with quality and good value. The market leaders in many of the products we buy are now Japanese. For example, in stereo equipment the market leaders are Sony, Denon, and Yamaha. In photographic equipment, Kodak is no longer one of the primary manufacturers of quality cameras. Its place has been taken over by such companies as Nikon, Canon, Olympus, and Minolta (all Japanese). One of the most significant recent developments in photography, a simple single-lens-reflex

---

[6]Martin K. Starr and John E. Ullmann, "The Myth of U.S. Industrial Supremacy," in *Global Competitiveness: Getting the U.S. Back on Track*, ed. Martin K. Starr (New York: W.W. Norton, 1988).

camera offering one button shooting combined with autofocusing, was successfully introduced in the Maxxum series of cameras by Minolta of Japan. In cars, companies such as Honda and Toyota consistently provide products considered to be among the best car values on the market. In contrast, it is the American products that are now often viewed by consumers as being cheap and shoddy. Many American manufacturers find themselves in the position of trying to show consumers that their products are as good as or better than those built in Japan.

Decline or Transition: Interpreting the Data

The evidence presented in the preceding section does not paint a flattering picture of the current state of American manufacturing. However, there is still the question of how to most appropriately interpret this evidence. Depending on the assumptions made, there are two different conclusions that can be drawn from this data. Each leads to a very different set of conclusions about the future viability of manufacturing in the American economy and the role of such manufacturing-based systems as computer integrated manufacturing.

The first is pessimistic. It views the evidence as leading to one inescapable conclusion: *American manufacturing has entered a period of decline that will probably result in the American economy becoming primarily service-based in orientation.* The symptoms of this demise are obvious. The decreasing importance of manufacturing and the increasing importance of services in the American economy (Table 2-4) are the results of several long-term, fundamental trends. The American economy is unable to generate the savings necessary to sustain and grow a manufacturing economy. Furthermore, the basis of manufacturing excellence on which American industry has been built has been eroded. The rate of growth of productivity is not keeping up with foreign competition. Innovation and research and development are not generating those advances needed to provide industry with a competitive advantage. Japan has now taken over the technological leadership. Further saddling American manufacturing is a consumer base that is consuming more in imports than it is providing in exports. An adverse balance of payments is limiting the growth of manufacturing. These "facts" point to the fall of American manufacturing. The passing of such past industry and market leaders as Burgmaster is nothing more than a symptom of this decline.[7]

---

[7]For an excellent and insightful discussion of the rise, growth, and fall of a past market leader, the reader is strongly advised to read Max Holland, *When the Machine Stopped* (Boston: Harvard Business School Press, 1989).

What is the answer to the problem of economic deterioration? This first perspective would argue that there are two basic options open to American industry and its management. The first is to turn to the government in hopes of developing an industrial policy similar to those found in such countries as Japan, Sweden, South Korea, and Singapore.

Alternatively, we could view the decline of the manufacturing sector as simply indicating a natural and evolutionary change in economic focus. This is a refocusing of the American economy from being manufacturing-based to one that is primarily service-based (the so-called postindustrial society.[8] Acceptance of this shift in economic focus can be found in such business magazines as *Forbes*[9] and in works presented by the New York Stock Exchange,[10] the Office of the U.S. Trade Representative,[11] and such well-known economists as Gary S. Becker.[12]

In the world of the postindustrial society, manufacturing is no longer the major economic strength of the economy. As a result, there is little need for such new manufacturing developments and systems as CIM, Just-in-Time (JIT), and simultaneous engineering. These developments offer little benefit to the new economic champion, services.

The second interpretation arrives at an entirely different conclusion. American manufacturing is not experiencing a fundamental decline. Rather, the changes that are adversely affecting American manufacturing are simply symptoms of a basic change in the nature of the manufacturing task. Such changes are not uncommon. Manufacturing is always undergoing change. These changes reflect the influence of competitive actions, new technology, and new products. These changes also point to an important paradox in manufacturing: Management is always torn between the need to maintain stability while preparing the production system for the changes that ultimately will be made. When planning for stability, one objective is to minimize the impact of change. However, when planning for change, the need for stability becomes a hindering factor.

According to proponents of this second perspective, manufacturing is not dying but simply changing and evolving. The problems that we see reported in the statistics can be viewed as "growing pains." As is pointed out by Stephen

---

[8]Daniel Bell, *The Coming of Post-Industrial Society: A Venture in Social Forecasting* (New York: Basic Books, 1973).

[9]*Forbes*, April 11, 1983, pp. 146,149.

[10]New York Stock Exchange, *U.S. International Competitive: Perception and Reality* (New York: New York Stock Exchange, August 1984).

[11]Office of the U.S. Trade Representative, *Annual Report of the President of the United States on the Trade Agreement Program*, 1984–1985 (U.S. Gov't. Printing Office).

[12]Gary S. Becker, *Business Week*, January 27, 1986, p. 12.

Cohen and John Zysman,[13] the "postindustrial" economy is a myth; what does and will always matter is manufacturing. In this world, developments such as CIM and JIT are not only relevant but a fundamental part of the process of transition. CIM and JIT are not part of the past or present; rather they are part of the future of manufacturing. This study takes this second view. All of the results presented in this book are predicated on the following assumptions:

- The future of the American economy will not be based primarily on the service sector.
- Manufacturing is the major long-term source of economic growth and well-being.
- Manufacturing is undergoing a major transition—one that requires not simply new tools and technologies but, more important, new management frameworks (paradigms).
- CIM and JIT are important to American manufacturing in the future because they will play a key role in the development of these new required management frameworks.

The view and position taken by this book can be best summarized by Mark Twain, who was told that a newspaper had mistakenly printed his obituary:

The rumors of my death are greatly exaggerated.

TRANSITION IN AMERICAN MANUFACTURING: A HISTORICAL PERSPECTIVE

In 1958 Peter Drucker noted that "American industry has solved the manufacturing problem." The problems that industry is now experiencing indicate that Drucker was wrong. However, while wrong for today (the 1990s), Drucker was right for 1958. The reason is that manufacturing is continually undergoing a process of change and transformation. Most of this change is evolutionary in nature; some are revolutionary.

In reviewing the history of manufacturing in America, Romeyn Everdell divided the changes that have taken place in American manufacturing into six major stages[14]:

---

[13] Stephen S. Cohen and John Zysman, *Manufacturing Matters: The Myth of the Post-Industrial Economy* (New York: Basic Books, 1987).

[14] Romeyn Everdell, "From Lowell to Sunnyvale: Manufacturing in the United States," *Strategic Manufacturing: Dynamic New Directions for the 1990s*, ed. Patricia E. Moody (Homewood, Ill.: Dow Jones-Irwin, 1990) pp. 20–50.

- 1800 to 1850: The Age of the Technical Capitalists
- 1850 to 1890: The Introduction of Mass Production
- 1890 to 1920: Scientific Management in Manufacturing
- 1920 to 1960: The Golden Years of U.S. Manufacturing
- 1960 to 1980: The Decline of the American Factory
- 1980 to ?: Challenge from Overseas

These stages identify responses to various influences and developments: the introduction and use of new or evolving technologies, changing markets, the arrival of new competition, product proliferation, and economic growth. They also reflect two other major influences that always occur whenever any firm or segment develops a very successful approach to the management and use of manufacturing: *model refinement* and *model destruction.*

Model refinement occurs when competitors identify the firm or firms that they consider to be the most successful. These become the models that they use in shaping their own systems. Management studies these "models" and tries to identify any refinements that they can introduce to make their new system just that much better than the models they have studied. Over time this process has had two major effects. First, it has led to more and more refined models. Henry Ford's River Rouge plant could be viewed as the ultimate extension of the assembly process already present in the Swiss watch industry and in the meat processing plants of the American Midwest. Second, it has created a narrowing of differences between systems as the systems start to become more and more alike.

In contrast, model destruction takes a very different approach. Instead of emulation, model destruction tries to provide consumers with an alternative. This alternative is built by studying the best manufacturing systems with the goal of identifying those areas in which these systems are weakest. Firms using this approach try to attack the successful systems by developing new systems that are strongest where the successful systems are weakest.

There are several examples of model destruction in the history of American manufacturing. By 1927 Henry Ford's River Rouge plant was considered to be the ultimate version of the automotive assembly line. Henry Ford had focused on costs as the major component in his corporate strategy. His goal was to continually reduce the price of the Model T Ford thus making it available to more and more consumers. The result was an emphasis on continual process improvements and the elimination of anything that did not reduce costs. For Ford it did not make sense to offer different models every year. Model changeovers created higher costs. For Ford the best approach to options was symbolized by his view that the customer could have any color he wanted as long as it was black. Given Ford's focus on cost reduction, this approach to options was logical. Car options increased costs. If the goal was to reduce cost, options had to be eliminated wherever and whenever possible. The economies of scale present in the River Rouge plant were consistent with this goal of cost reduction.

Henry Ford and the River Rouge plant, however, represented a major obstacle to Alfred P. Sloan and General Motors. General Motors did not have any facilities that could rival the economies of scale present at River Rouge. Furthermore, the company did not have the money necessary to build a similar factory. What Sloan had was a series of divisions and plants that were better at building cars with more options. Sloan understood that the American buying public wanted more than simply cheap cars. Income was rising; more people had cars. What the public wanted now was variety. Here was the major weakness of Ford's system. Ford had abandoned variety to achieve lower production costs.

Sloan responded by developing systems that emphasized variety. His divisions targeted different market segments. Oldsmobile represented the technological leader. Any new innovations or features introduced at General Motors always came out first in Oldsmobiles. Cadillac was the premium car. There was no car better than the Cadillac—a fact recognized even by the employees who worked on the Cadillac lines. Chevrolet was basic, inexpensive transportation that offered few frills. These differences were emphasized by such aspects as pricing, options, and design.

Variety now also came from the introduction of options. Buyers could now chose different color patterns. They were also free to add whatever options they wanted to the basic car model that they were buying. For the freedom of adding these options, consumers were willing to pay a premium.

Finally, Sloan emphasized variety by introducing the model year. Every year, the lines at General Motors were shut down so that they could be changed over to the production of next year's car. Customers could now identify the car models produced in different years.

As a result, Sloan had developed a new manufacturing system that emphasized those areas in which Ford's system was vulnerable. It appealed to those buyers who wanted more than basic transportation and who were willing to pay a premium for variety. When Ford realized the threat that General Motors posed to his position in the marketplace, he found himself trapped by the very manufacturing system that had made him successful in the past. It took Ford almost a year to change over the lines so that he could make the Model T's replacement, the Model A.

Irrespective of the factors causing change, it should always be remembered that one of the major forces propelling the movement from stage to stage is that of *sustainable competitive advantage*. That is, every firm tries to change its manufacturing system so that it can better satisfy the needs of its customer base.

Whenever manufacturing makes the transition from one stage to the next, one of the most important effects is that how the firm is run also fundamentally changes. It is not enough to simply run operations by using existing techniques and tools better. Such an approach emphasizes *refinement*. Instead, what is now needed is a set of new frameworks, or *paradigms*.[15] To develop new paradigms requires a *rethinking* of the manufacturing task. As managers rethink this task, they will develop new tools—tools that are different in content, orientation, and structure from those used in the past.

Paradigms and Manufacturing

Paradigms are critical to all manufacturing firms and to managers. They define the perspectives that managers use in evaluating, operating, and changing the manufacturing task. They establish for managers the underlying belief structure and they set in place those unstated assumptions. They identify what is acceptable and, more important, what is not acceptable. They provide the frameworks that managers use in interpreting the world around them and in understanding the relationships between factors. Paradigms also form the basis on which value structures and beliefs are based.

Although manufacturing managers have recently become aware of the importance of paradigms,[16] scientists have long been aware of the power of paradigms. Their views on paradigms are best stated by James Burke, who noted that when paradigms change, at that moment, the universe changes:

> You are what you know. Fifteenth-century Europeans "knew" that the sky was made of closed concentric crystal spheres, rotating around a central earth and carrying the stars and planets. That "knowledge" structured everything they did and thought, because it told them the truth. Then Galileo's telescope changed the truth.
>
> As a result, a hundred years later everybody "knew" that the universe was open and infinite, working like a giant clock. Architecture, music, literature, science, economics, art, politics—everything changed, mirroring the new world view created by the change in the knowledge.
>
> Today we live according to the latest version of how the universe functions. This view affects our behavior and thought, just as previous versions affected those who lived with them. Like the people of the past, we disregard phenomena which do not fit our view because they are "wrong" or outdated. Like our ancestors, we know the real truth.
>
> At any time in the past, people have held a view of the way the universe works which was for them similarly definitive, whether it was based on myths or research.

---

[15]Dave Garwood and Michael Bane, *Shifting Paradigms: Reshaping the Future of Industry* (Atlanta: R.D. Garwood, 1990).

[16]Dave Garwood and Michael Bane, *Shifting Paradigms: Reshaping the Future of Industry* (Atlanta: R.D. Garwood, 1990).

And at any time, that view they held was sooner or later altered by changes in the body of knowledge.[17]

In short, changes in paradigms cause people to redefine their reality. When a paradigm changes, the old reality and views cease to exist. They are inadequate. In their place there is a new paradigm that defines a new reality. What was important previously becomes inconsequential. What is important now previously was treated as either irrelevant or not as important as some other factors.

The impact of paradigm changes and reality can be best illustrated by returning to an example provided by James Burke.[18]

The Importance of Paradigm Changes: An Example

You awake in an Italian village in the 1500s. It is morning. You look out to the east and see the sun rise. At dusk, as you prepare to go to bed, you look out to the west and see the sun set. How do you explain this phenomenon? The Italy of the 1500s is a society whose views of astronomy are based on the Ptolemaic views. At the heart of these views is the belief that the sun and the other planets revolve around the earth. To you, what you have seen (the rising and setting of the sun) is nothing more than evidence of the sun's movement around the earth.

Now two hundred years have passed. You wake up in the same Italian village. Nothing has changed. You see the sun rise in the east in the morning and set in the west in the evening. The phenomenon has not changed. Yet your understanding of it has changed dramatically. Why? During this time, a Polish monk by the name of Copernicus was born and died. He advanced a very different view of planetary movements. It is a heliocentric view. The sun, not the earth, is the center of the solar system. This new theory, now widely accepted, provides a very different explanation of the events that you have witnessed. The sun has risen and set because of the rotation of the earth.

The point of this example is very simple. In the two instances the facts did not change. The sun rose in the east and set in the west. What changed was our understanding of those "facts." The paradigm of the 1500s has been replaced by a new paradigm. Predictions made in the 1500s are no longer accepted in the 1700s. When we accepted Copernicus's view of astronomy, it could be said that the universe "changed."

Given the preceding comments, the transition between the last two stages (The Decline of the American Factory [1960 to 1980] to The Challenge from

---

[17]James Burke, *The Day the Universe Changed* (Boston: Little, Brown, 1985), p. 9

[18]This example is taken primarily from the PBS television series that accompanied the book written by James Burke.

Overseas) reflects not the introduction of new technology or new products. Rather, it reflects a change in paradigms. It is this shift in paradigms that provides the key to resolve the CIM paradox. However, before we can proceed to resolve this paradox, we must first understand the paradigms that underlie these last two stages.

Managing the Manufacturing Task: 1960 to 1980

The typical manufacturing system found in the United States during this period represented in many ways the ultimate refinement of the trends begun in the 1920s with the introduction of large-scale factories built around the assembly line. This was a system that stressed large-scale production. Large-scale facilities, after all, enabled management to capture the benefits of *economies of scale*. The manufacturing system of this time was capacity based. It tried to increase capacity, since the demand for products was not quite in line with the amount of capacity available.

The system of this time emphasized cost reduction and cost minimization. It also treated cost, quality, lead time, and flexibility as components of a trade-off. You could do well on one dimension (e.g., cost) but not on all the other dimensions. Any improvement that you obtained in one area was always at the cost of corporate performance in the other three areas.

Corporate and Manufacturing Philosophy

The approach taken by managers of this time was based on certain underlying beliefs. The most important of these were as follows:

An Emphasis on Cost Reduction. Cost reduction was seen as an effective strategy for gaining market share and for improving corporate profits. The price was seen as something that could not easily be manipulated. Any increases in price would result in reduced consumer demand. By reducing costs, profits were maximized.

Micromanagement. Many of the solutions implemented during this period reflected a *ceteris paribus* assumption. That is, they treated the problem and solution in isolation and assumed that "all other things were equal." Interactions were often either ignored or their importance minimized. Machines were introduced that by themselves were very efficient. However, the impact of these new items of capacity on overall system balance was often ignored. In short, during this period solutions focused in on finer and finer segmentation of the problem. Ultimately, we became interested in solving one motion exercised by one person at one point in time.

This emphasis on solutions in isolation was reflected in the performance measurement system. Departments in most manufacturing firms of this time were evaluated on their performance alone. The performance measurement system did not evaluate or track situations in which the actions taken by one department would adversely affect the performance of another department. As a result relationships between departments often became antagonistic. Departments were encouraged to become inwardly oriented, separated by strong, invisible barriers.

The Inevitability of Uncertainty and System Variance. A major assumption made at this time was that uncertainty, in the form of quantity, quality, or lead time, was a "fact" of manufacturing. It was unrealistic to expect suppliers to provide shipments that were 100 percent quality and delivered consistently on time. This view was reflected in the use of such statistical quality control concepts as *acceptable quality levels, average outgoing quality*, and *consumer-producer risk*. Purchasing contracts were signed that specified levels of quality less than 100 (e.g., 97). These views on uncertainty were also applied to the manufacturing system.

Scrap, salvage, and rework were accepted aspects of manufacturing life. For example, setup times were often long and variable. In setting up for production, "run-ins" were expected. That is, the setup would be completed and a few parts produced. These would then be checked against tolerances. If acceptable, production began. Otherwise, the run-in parts were scrapped and the setup was readjusted. The process would be repeated until the initial parts could be qualified.

A final source of uncertainty came from the customer and marketing. It was expected that orders would be changed (in terms of quantity, timing, or features) one or more times before the shipment was completed. Master production schedules were seldom stable.

As a result, the manufacturing system of this period relied on buffers in the form of extra capacity, extra production (for scrap), safety stock, and longer than normal lead times.

The Appropriateness of Technical Solutions. Many of the solutions and proposals implemented during this time could best be described as technical in nature. That is, specific solutions such as MRP, CRP, and inspection were often recommended and used. The reason for this approach is relatively simple. The problem that faced manufacturing was well defined. It was one of meeting production schedules and requirements at the lowest possible costs. Given this definition, management could now turn to solutions that were appropriate for this problem. MRP was appropriate because it promised to reduce manufacturing inventories while improving schedule performance.

Traits

Given the assumptions presented in the preceding section, the manufacturing practice of this time can be described as follows:

High Levels of Specialization. Specialization became the order of the day with equipment, skills, and labor classifications experiencing more and more finely defined focus and specialization. This trend for increased specialization was consistent with a manufacturing system that was competing on the basis of lower costs and higher volumes.

Centralized Decision-making and Control. Control of operations moved away from the shop floor and to experts at the middle and upper levels of the organization. These people were expected to make the critical decisions and the people who really knew the process and the shop conditions (i.e., the employees and the first-level supervisors) were expected simply to carry out the dictums of top management.

Systems That Were Inflexible. The combination of the emphasis on specialization, high-volume production, cost minimization, and centralized decision-making created a manufacturing system that had difficulty responding to changes (quantities, lead times, and mix). Customer-driven changes to products other than those that could be manufactured by the existing manufacturing facilities created especially high levels of stress and difficulties. An example of these difficulties can be seen in the inability of the American automotive industry to respond quickly to the need for small, well-built, fuel-efficient cars in the wake of the 1973 oil crisis.

The Acceptance of Scrap, Obsolescence, and Variability. The manufacturing systems were based on the assumption that problems were inevitable. As a result, they were designed to buffer production from the effects of these problems. One of the major effects of this buffering was a heavy reliance on inventories.

Driven by Schedules. Finally, the manufacturing systems were driven by schedules, which were often broad in nature. In many instances the orders had to be out by the end of the month. This fact resulted in so-called hockey stick schedules.

At the start of the month, production is relatively low (Figure 2-1). It ramps relatively slowly until it reaches the last week of the month. Suddenly, everyone realizes that the bulk of the orders for the month have yet to leave the plant.

FIGURE 2-1
Hockey Stick Scheduling (An Example)

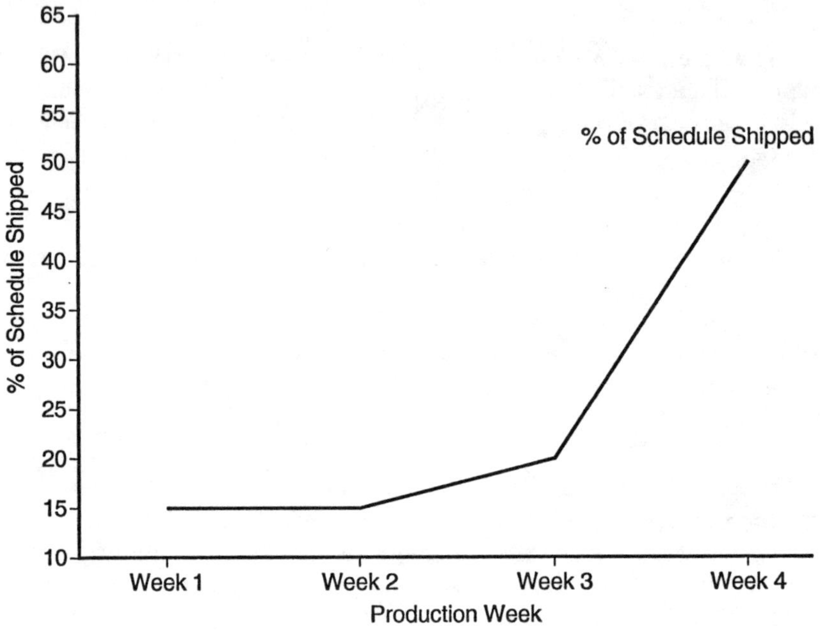

Overtime is ordered. Deviations in quality are allowed so that schedules can be made (it is assumed that the problems can be picked up and corrected in the field). Parts are expedited from suppliers. Shipping costs increase as firms send out shipments at premium rates. When the orders for the month have been completed, activity drops down again (and the cycle repeats itself).

Manufacturing Tools

Managers working in the American manufacturing facility of this period typically relied on a set of highly quantitative tools. The presence and proliferation of these tools represented the impact of several influences. First, they were a result of a growing number of manufacturing consultants and technical societies. These societies often mirrored the increasing specialization and departmentalization developing in most American manufacturing companies.[19] Second, they also reflected

[19] Romeyn Everdell, "From Lowell to Sunnyvale: Manufacturing in the United States," *Strategic Manufacturing: Dynamic New Directions for the 1990s*, ed. Patricia E. Moody (Homewood, Ill.: Dow Jones-Irwin, 1990).

trends within the major business schools. Increasingly, many faculty entering business schools had scientific or mathematics backgrounds. For these people, the quantitative techniques were the most appropriate and suitable ones to teach (as they were often the areas in which the faculty were doing active research).

The following is a partial list of the most significant and widespread of these quantitative techniques:

- Statistical quality control (SPC)
- Facilities layout
- Lot-sizing algorithms (e.g., economic order quantity, part period balancing, periodic order quantity)
- Material requirements planning (MRP)
- Capacity requirements planning (CRP)
- Scheduling procedures (e.g., critical ratio, earliest due date, earliest operation due date)
- Linear programming
- Project scheduling procedures (e.g., PERT/CPM)
- Line of balance (LOB)

Managing the Manufacturing Task: 1980 to ?

By 1980 manyAmerican manufacturers were coming to realize that the face of manufacturing was beginning to change. These changes are best embodied by the Japanese challenge. Many Japanese firms, especially those in the automotive industry, have been able to successfully put together a very different approach to the task of manufacturing and satisfying customer demands. It is an approach that is both innovative and integrative. It is holistic in its orientation.

Unlike the paradigm of American manufacturing in 1960 to 1980, this new approach is driven primarily by value and waste. Value maximization and waste minimization are the new goals, not necessarily cost minimization. It is long term in orientation. It attempts to build proactive manufacturing systems that are constantly refined to increasingly simplify operations. Complexity is viewed as an unnecessary evil and a contributor toward waste.

The results of this new approach are better-built products and more productive employees.[20] Increasingly, this approach is able to create and deliver those products wanted by customers in shorter lead time and at lower costs.

---

[20]Romeyn Everdell, "From Lowell to Sunnyvale: Manufacturing in the United States," *Strategic Manufacturing: Dynamic New Directions for the 1990s*, ed. Patricia E. Moody (Homewood, Ill.: Dow Jones-Irwin, 1990).

This improved performance can be largely explained by recognizing that the Japanese have developed and implemented a new paradigm or framework for viewing the manufacturing task.

Philosophy

Underlying this "new" approach to the manufacturing task now being developed, there are certain critical underlying beliefs that form the philosophy of this new paradigm:

The System Is More than the Sum of the Parts. The system is critical because it forces managers to understand not only how the individual elements work by themselves but how the actions taken by one area can adversely affect the operation of other areas. The major task of any manager is to ensure that all of the elements are working together as a unified whole and that any actions that help one area at the expense of another are avoided.

This tenet brings with it three important implications. The first is that solutions that deal with the entire corporate system will always work; solutions that focus on manufacturing or engineering in isolation will never work effectively. Second, there are only corporate initiatives, not marketing initiatives. Third, if the firm is to be managed effectively, managers must understand the process that links together all of the components of the corporate system.

Simplicity Is to Be Prized; Unnecessary Complexity to Be Avoided. Simple systems are easy to understand, easy to operate, and easy to control. Simple systems are always effective because they create a direct bridge between need and response. They are also the systems that are best able to respond to changes and needs in the shortest possible lead times. Complexity, on the other hand, creates confusion and becomes self-perpetuating. As a result, the firm must always be involved in the task of evaluating all the processes and activities that take place within it with the goal of identifying and eliminating all unnecessary complexity.

Value, Not Cost, Is Critical. Ultimately, the task facing any firm is not to reduce cost. Rather, it is to win new customers and maintain existing customers by continually providing those goods and services that the customers most prize and that generate the highest level of value for customers. By accepting the need to be value-driven, the firm and its management must recognize certain realities. The first is that value is never defined by the firm; it is defined by the customer. As a result, the firm must be able to identify very specifically that group of customers that it is trying to serve. The firm must always be in touch with its customers. Second, value is dynamic. It is always changing. These changes are the result of many factors: competitive actions, governmental regulations, changing customer preferences. As the concept of value changes, the firm must be able to anticipate these changes and be ready to meet the new needs.

The opposite of value is *waste*. Any action that increases the desirability of a good or service in the eyes of the customer adds to value. Conversely, any action that does not directly add value must, by definition, create waste.

As a result, the firm must be continuously involved in the task of changing the manufacturing process so that it maximizes the value generated and minimizes the level of waste created.

People Are Critical to Success. The new paradigm incorporates the belief that most of the improvements in and enhancement of the manufacturing system do not come from outside the firm. Rather, they are the result of actions taken by those people who work with the process. These people, by virtue of their experience with and understanding of the system, are the real experts. As a result, the task for the firm is to develop its people and their skills so that they can better contribute to the continued growth and success of the firm.

This tenet forces management to recognize the importance of training (as a corporate investment), empowerment (giving its people the necessary tools and authority with which to bring about changes), enfranchisement (expecting its people to become the major change agents), and teamwork (recognizing that the group is a more effective problem-solving instrument than the individual working in isolation). This tenet also forces management to recognize that cooperation, not conflict, is the preferred mode of operation and that everyone who works in the firm is fundamentally good and driven by a desire to make the firm better. Everyone, after all, likes to identify with a winner.

While management can plan corporate strategy from the top down, the implementation of these plans and the introduction of any changes must occur from the bottom up (i.e., from the people).

Variance Is the Root of All Evil. Variance is any behavior that causes the manufacturing system to become unpredictable. When systems become unpredictable, performance begins to suffer and costs increase as people start to rely on buffers (lead times, inventories, capacity) to protect themselves from the effects of these variances.

Variance indicates the presence of underlying problems in the system. By identifying these underlying problems and eliminating them, the firm can develop a very effective, simple, and highly predictable system. This is a system in which problems are always immediately visible.

Systems Can Be Changed if Management So Desires. The final tenet views all systems as subject to change. While it is true that all systems are "fixed" in the short term, they are still malleable in the long term. If you do not like what the current system is doing, you can and should change it. Change is always possible if management is willing to invest the necessary time, effort, and resources. As a result, this tenet makes the manufacturing manager proactive with respect to the manufacturing operating environment.

Traits

Given the belief structure described in the preceding section, there is a set of important traits that are present in most firms that are governed by the new paradigm:

A Movement toward the Use of Generalists. Generalists provide the firm with that broad perspective that is required. These are people who are not closely associated with any one group or area. They are driven by problems. Their goal is to uncover problems and to formulate effective system solutions.

Often, such generalists are developed by a training program that exposes the person to different areas, functions, groups, and perspectives. The person is left long enough in one area to understand what that area does and how it contributes to the firm. Yet the person is not left there long enough to develop a close identification with the area.

Decentralized Decision-making and Problem-Solving. Those people closest to the problems are now seen as best able to solve them. The role of management is now to support these people. The result is a change from a situation in which a few "experts" are expected to solve all of the problems to an environment in which many people are solving problems as they occur. With decentralized problem-solving, the objective is to ensure that the time lag from the moment that a problem becomes evident until it is solved is kept as small as possible.

Flattened Organization Structures. One of the major effects of combining decentralized decision-making with an emphasis on system simplification is the creation of a leaner organization. With decentralization and simple systems, there is no longer the same need for the layers of middle management present in other organizations. Since these levels often add no real value, they can be eliminated or greatly reduced. The result is less overhead and less distance between the shop floor and the top of the firm.

Flexible Systems. The new paradigm encourages the presence of more flexible systems. These are systems that are able to adapt rapidly to conditions as they change. Such systems are not held back by the bureaucracy and the complex processes that may otherwise discourage or hinder quick response.

Large bureaucracies and complex processes cause inertia to creep in. Responses and information must now filter through more and more stages. With a flatter organization, top management can become aware of changes taking place either externally (in the market) or internally (on the shop floor). They are now better able to respond to these changes.

Value-driven. The firm recognizes what the customer wants and values from the firm and focuses its energies and capacities on satisfying these needs. It is willing to experience increased costs in the short term if the resulting actions can help the firm better serve its customer base.

Often, the firm and market become closely identified. The firm knows what its customers want; its customers know that the firm will be able to satisfy their needs now and in the future. The firm is described as "owning" the market. It is consistently able to deliver better value for its customers faster than its competitors. It strives to maintain the initiative, thus forcing its competitors into a reactive mode.

The Use of Broad-based Solutions. Typically, the firm now addresses its problems not by relying on specific technologies for solutions (e.g., MRP, CRP). Instead, its solutions often are general in nature and focus on the process. In most instances the first step to solving the problem takes the form of studying the process as it currently exists. This process is documented and the reasons for every activity observed in the process questioned. Those activities that cannot be justified are subsequently eliminated.

To solve the problems, the firm brings together people from different areas into a group. This group can view the problem from a different perspective. The group uses "simple" tools for attacking the problems before them. These include process analysis, cause and effect diagrams, and Pareto analysis.

Elimination of Variance and Waste. The manufacturing processes are now ongoing continual refinement and improvement. Any activity that either does not contribute directly to value or is a source of variance becomes a target for either elimination or revision. The goal is to create processes that are always highly predictable, under control, and create value in the least number of steps.

Manufacturing Tools

The tools of the 1960s and 1970s are not appropriate for the manufacturing environment of the 1980s and 1990s. These new tools are very much reflections of the new philosophies of today. They stress solutions that deal with the system as a whole (not individual areas). They try to avoid the creation of problems instead of coping with problems once they have been created. They aim at bringing the entire system under control. They recognize the interdependences that exist in any manufacturing organization. The following are the most significant and widespread of these new manufacturing tools:

Total Quality Management. A corporate-wide system for establishing quality objectives and for subsequently ensuring that the firm as a whole is able to satisfy and maintain these objectives. Total quality management embodies traditional quality control procedures such as statistical process control. However, these procedures are just one component of this approach.

Kaizen. The process of continual improvement in all aspects of the firm. This is an activity that involves everyone, managers and workers alike.[21]

Design for Manufacturing (DFM) and Design for Assembly (DFA). This approach is also known as simultaneous engineering. Its use reflects management's recognition that during the design phases of product development, about 5 to 10 percent of the costs are incurred. Yet the activities that take place during these stages influence and often fix up to 80 percent of the resulting manufacturing costs. As a result, better products that are subject to fewer problems can be built by better spending our time during the design stage.

One way of eliminating or reducing problems encountered at this stage is to bring together all the areas that could be affected by the proposed product. The representatives from these areas would then be free to review the design and to suggest improvements and modifications in the product's design while the costs of change are still relatively low. The result is not the best product but rather a better product more quickly produced.

Just-in-Time Manufacturing. This corporate-wide approach is aimed at improving the strategic position of the firm by continuously uncovering and eliminating all sources of corporate waste.

Strategic Alliances and Value-adding Partnerships. This last tool serves to improve corporate performance by recognizing the importance of the value chain. That is, the ability of any firm to produce a high-quality product in a timely and cost-efficient manner is not simply determined by how well the firm is able to manage its own internal operations. Instead, it is greatly influenced by the performance of its suppliers.

---

[21]Masaaki Imai, *Kaizen: The Key to Japan's Competitive Success* (New York: Random House Business Division, 1986).

These programs encourage closer relationships between the buying organization and its suppliers.[22] In part, the closer relationships are the result of a reduced supplier base. They are also a product of greater sharing of information and greater supplier involvement in the design and problem-solving activities of the buying organization. They also encourage joint efforts aimed at improving value (rather than at reducing costs). As a result the buying organization tries to make the supplier as reliable and predictable as the internal system.

Each of these tools is often supported by some strong quantitative instruments. However, in each case, the devices present are not the distinguishing features of the tools described. Rather, each of these tools shares the same concern with developing system-wide, simple, effective solutions. Any quantitative devices present are simply ways of getting to an end. They are not the ends in themselves.

EXPLAINING THE CIM PARADOX

All of the pieces are now in place that are necessary to resolve the CIM mystery that began this chapter: why have firms not generally realized the potential benefits present in CIM systems? The answer is not technological. Rather, it lies with the paradigms that managers used when implementing CIM systems and in developing CIM-based strategies.

CIM can be viewed in one of two different ways. At one level CIM can be viewed in terms of its computer-based technologies (e.g., CAD, CAM, robotics, flexible manufacturing systems, and artificial intelligence and expert systems). At this level CIM is nothing more than a series of techniques that can be used to either replace or enhance the current manufacturing and engineering capabilities of the firm. In this view of CIM the emphasis is on the "C" and the "M" of CIM.

Alternatively, the emphasis can be placed on the "I" of CIM. That is, CIM can be viewed as a systematic, strategic approach to improving the overall competitive position of the firm by using, as its vehicle, computer-based technology. Under this second perspective, technology is the means to an end, but it does not define the end.

The first perspective is consistent with the manufacturing paradigm of 1960 to 1980. The second reflects the manufacturing paradigm now being developed. Each perspective, however, brings with it very different implications and consequences for the manufacturing firm.

---

[22]Keki R. Bhote, *Strategic Supply Management: A Blueprint for Revitalizing the Manufacturer-Supplier Partnership* (New York: American Management Association, 1989).

As will be shown in this book, the differences between success and failure in CIM reflect the approach taken by the firm. Those firms that have succeeded with CIM have regarded it as a strategic solution to a corporate system problem. Those managers approached the task of evaluating and implementing CIM with a paradigm that is more consistent with the manufacturing world of the 1990s than with the manufacturing world of the 1960s and 1970s. In contrast, those firms that failed with CIM did so because they regarded CIM as a technical problem to a strictly manufacturing problem. They failed to realize that CIM, if successful, can affect the operation of all the departments in the firm. They also failed to realize that CIM, to be successful, requires the cooperation of many areas and groups outside of manufacturing and engineering (the primary target group for most CIM implementations).

CIM is successful only when viewed from a *macro* or system perspective; CIM fails when treated within a *micro* or very myopic manufacturing or technology perspective.

CONCLUDING COMMENTS

This chapter has painted the setting against which any realistic study of computer integrated manufacturing must be examined. CIM has been and continues to be a manufacturing mystery. Its introduction was hailed as offering manufacturers an opportunity to enhance operations by introducing the advantages of new and sophisticated technology into the various manufacturing and engineering activities. However, this bright promise has developed a hollow ring to it for many firms: many rushed into the implementation of CIM only to end up with expensive, highly automated systems that were less effective than the manually driven systems that they replaced. As a result of these difficulties, many manufacturing people have begun to view the CIM concept with a great deal of distrust and suspicion.

In this chapter we have shown the reader that the resolution to this mystery lies not in CIM itself but rather in the paradigms or frameworks that managers have used in formulating, implementing, and operating CIM strategies. We have also set the stage for CIM. American industry is now going through a period of transition. The old methods and paradigms are no longer adequate. The costs of using the old systems and approaches can be expected to take the form of the continued deterioration of America's manufacturing position. To survive in this new manufacturing world, new paradigms must be formulated and implemented. These new paradigms favor approaches such as CIM because they are broad-based, systems-oriented, and value-driven.

As a result, it is now time to reexamine the CIM concept. The developments of the last 10 years have created a situation in which we view CIM as a concept *whose time has now come*. However, if we are to make the greatest use of CIM, we must first understand what the CIM system and concept is. That task is the focus of the next chapter.

CHAPTER 3

COMPUTER INTEGRATED MANUFACTURING: THE FRAMEWORK

As was mentioned previously, this is a book about the principles, practices, and theory of computer integrated manufacturing. Studying CIM is not an easy task. Although it is a relatively new concept (the term was coined by Harrington in 1973[1]), CIM has received a great deal of attention from production and inventory control managers, consultants, and researchers. Along with just-in-time (JIT) manufacturing, total quality control (TQC), and world class manufacturing (WCM), CIM represents an important path that American manufacturing can follow to reestablish and improve its competitiveness in domestic and global markets. The extent to which CIM can positively influence manufacturing and corporate competitiveness has been demonstrated by the experiences of such leading proponents of CIM as Frost of Grand Rapids, Michigan; Ingersoll Milling of Rockford, Illinois; Allen-Bradley of Milwaukee, Wisconsin; and Cone Drive of Traverse City, Michigan.

However, in spite of the attention devoted to this concept, CIM still remains a manufacturing mystery. Computer-aided technology is critical to the implementation and operation of any CIM system. Yet, the differences between CIM and automation are not clear. It is also an ill-defined concept. There is a great deal of disagreement about what CIM is and, more important, what it is not.

Mention the term CIM to a production and inventory control (PIC) manager, consultant, or researcher and you will inevitably receive a wide range of responses and views. To some, CIM is nothing less than the full use of computer-aided technology. CIM is viewed as almost synonymous with the use of machine vision, automated handling systems, robotics, and flexible manufacturing systems (a few examples of computer-aided technology). To others, CIM is primarily an information-based system. It represents one way of managing, structuring, and representing the corporate database. Finally, to still others CIM

[1]Joseph Harrington, Jr., *Computer Integrated Manufacturing* (New York: Industrial Press, 1973).

is primarily a strategic system. It draws on computer technology to help the firm better meet the needs of its marketplace.

Given the wide diversity of views on CIM, any investigation into the concept must first acquaint the reader with the view of the CIM concept on which it is premised. This is the primary purpose of this chapter, in which we will examine and summarize various views of CIM. Drawing on these views and the results of the research project summarized in this book, we will present our view of computer integrated manufacturing. In defining CIM close attention will be paid to the relationship between corporate strategy and CIM. We will also introduce the four types of CIM. The differentiation introduced by categorizing CIM installations into one of these four types is important, for it can help the reader understand why it is possible to have two similar firms with very different CIM implementations. Finally, this chapter will examine in broad terms the different paths available to managers interested in implementing a CIM system in their firms.

THE CIM PROBLEM: BRIGHT PROMISES, HARSH REALITIES

To many of its proponents (e.g., Gunn and Clark), CIM promises a world in which manufacturing has been transformed through the use of technology into a truly competitive weapon. Indicative of this are the following benefits directly attributed to CIM and summarized by James Clark[2]:

- 5 to 20 percent reduction in personnel costs.
- 15 to 30 percent reduction in engineering design costs.
- 30 to 60 percent reduction in overall lead times.
- 30 to 60 percent reduction in work-in-process.
- 40 to 70 percent gain in overall production.
- 200 to 300 percent gain in capital equipment time.
- 200 to 500 percent gain in product quality.
- 300 to 3500 percent gain in engineering productivity.

These benefits have been documented in such American firms as Allen-Bradley, Ingersoll Milling, Cone Drive, Frost, and Continental Can. For these and other firms, CIM has not simply meant better products delivered faster and at lower cost; it has helped each firm develop and maintain a *sustainable com-*

---

[2]James T. Clark, "CIM Planning," *APICS 31st Annual Conference Proceedings*, Las Vegas, October 17–21, 1988, pp. 401–407.

petitive advantage in its market. It has helped each firm become a market leader in innovation and, in many cases, become a dominant force in its market.

However, behind these bright promises offered by CIM there is the harsh reality that few firms will succeed in implementing CIM. For every CIM success that is reported, there are many failures that go unreported. In the absence of studies reporting reliable statistics about the chances of success in implementing CIM, it may be safely assumed that the success rate for CIM is far less than that for a material planning and scheduling system such as MRP. Generally, the success rate for firms attempting to implement MRP has been between 5 percent and 20 percent. In a 1981 study of MRP implementation and practice,[3] it was observed that only 9 percent of the companies surveyed classified themselves as Class A users, while about 62 percent were either Class C or D. Since it is a much broader system that places greater demands on the firm in terms of prerequisites,[4] CIM should be far more difficult to implement than MRP.

Even more daunting is the realization that being a CIM success in the past does not necessarily guarantee success with CIM either now or in the future. For example, consider what took place in IBM's Lexington, Kentucky, plant. This plant, which produces typewriters, keyboards, and printers for IBM, has long been considered to be one of the major success stories in CIM. It was a plant where, through the use of advanced technology and better product design, IBM was able to compete head-to-head with major foreign producers of these same products. But by 1990 IBM announced that its Lexington plant was no longer consistent with IBM's future strategic direction since the Lexington plant had not performed as well as had been expected.

There are two other illustrations of this nature that were uncovered in this study. The first involved a major manufacturer located in the Midwest. Representatives of this firm were approached to determine if they would be interested in participating. The representatives told the authors that, although they still consulted in the area of CIM, their own system was no longer operating. One manager from the company went as far as stating that the firm considered its own system to have been a failure.

In a second instance the authors had identified a small midwestern manufacturer that had participated in various CIM studies and workshops in the mid

[3]John C. Anderson, Roger G. Schroeder, Sharon E. Tupy, and Edna M. White, *Material Requirements Planning: A Study of Implementation and Practice* (Falls Church, Va.: American Production and Inventory Control Society, 1981).

[4]Examples of these prerequisite requirements are presented in the next chapter, which summarizes the various principles for the successful implementation and use of CIM.

1980s. After talking with various representatives from the company, the authors learned that

- The firm was no longer actively pursuing CIM.
- The CIM implementation team had been disbanded.
- Although the people in the firm had been proud of their success with CIM in the past, top management had mandated that no one in the firm was even to talk about CIM. The president had even gone so far as to remove all mention of CIM from the company reports.

Finally, in a world enamored with speed, lead time, quality, and just-in-time manufacturing, the glamour frequently associated with CIM in the past is starting to fade. Many managers are trying to cope with the fact that CIM offers the promise of large gains and also the potential for losses that are as large or in many cases even larger. We are now starting to see CIM as a double-edged sword for manufacturing.

BARRIERS TO CIM: THE MANAGEMENT OBSTACLES

Many managers do not experience the trials and tribulations of undertaking a CIM implementation. They do not have to worry about the double-edged sword of CIM. The reason is that they cannot even come to accept CIM as a viable alternative for improving operations in the corporate system. They see CIM as a system that is surrounded by insurmountable barriers.

There are several major obstacles to the acceptance of CIM as a practical management solution. The following are some of those most frequently encountered (summarized in Figure 3-1):

CIM Is an Expensive Undertaking That Can Only Be Done by Those with the Deep Pockets to Afford It. Most managers see CIM in terms of its technology. This technology is inevitably expensive and not easy to implement. It requires adequate resources for testing and debugging the systems. In addition, these investments are seen as ones that must be made up front. For many firms these costs represent a significant barrier. A good example of this obstacle was encountered by the management of Frost when they were exploring the suitability of CIM for their company. The management listened to a presentation in which one system supplier described the equipment that would be needed initially to automate their manufacturing floor. The required investment was in excess of $12 million dollars. The president remarked that the system being proposed was "nice" but far too expensive for Frost. After all, Frost's annual sales were less than $12 million.

FIGURE 3-1
Management Barriers to Acceptance of CIM

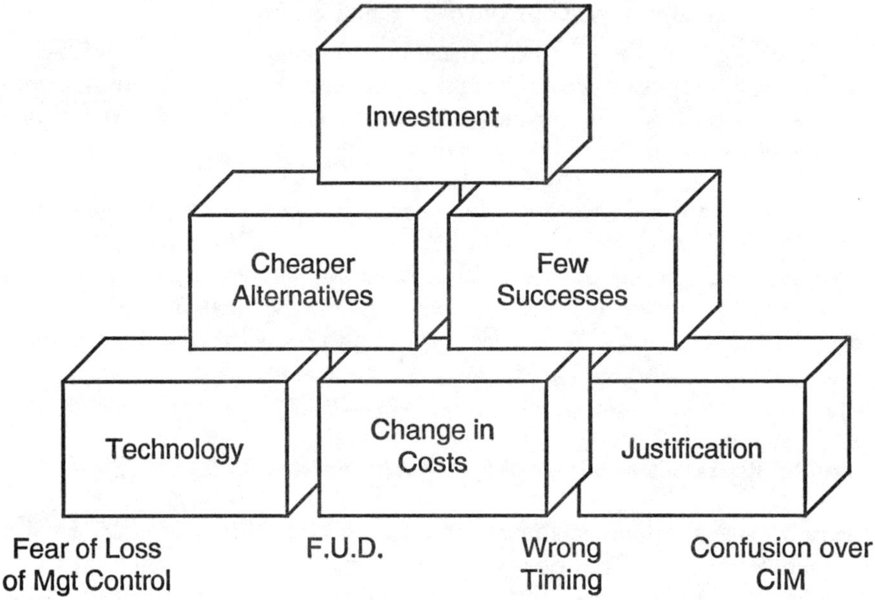

The Implementation of CIM is a Long-term Undertaking. When most managers look at the experiences of firms that have been successful with CIM, they see endeavors that took many years to complete. For a company that is faced *now* by dropping sales, falling margins, and rapidly eroding profits, there is little attraction in examining a system that may require four or five years before it yields any benefits.

There Are Cheaper Alternatives to CIM Available. For many managers who are interested in improving manufacturing performance and enhancing corporate competitiveness, there are other alternatives available. These alternatives include system simplification, time-based competition, competing on quality, and just-in-time manufacturing. What makes these alternatives so appealing is that they appear to require less investment. Consider, as an example, just-in-time.

One trait frequently associated with all just-in-time systems is setup reduction.[5] In most cases significant reductions in setup (up to 75 percent) can be

[5]Edward J. Hay, *The Just-in-Time Breakthrough* (New York: John Wiley & Sons, 1988).

obtained by investing in a video camera and monitor and having the workers record and observe their own setup activities. Such a payback for a relatively low investment (often between $1,000 and $1,500) is both impressive and attractive. The alternatives, by emphasizing working smarter, seem to be less risky. They also appear to be more promising. Why should a manager spend over a million dollars for a flexible manufacturing system when similar results can be obtained by investing in low-technology equipment that is laid out in a manufacturing cell and is manned by skilled, cross-trained employees?

There Are Relatively Few Success Stories. As was pointed out previously, the concept of CIM has been around since the early 1970s. During this period, many firms across the country have attempted to take advantage of the potential benefits of CIM. Firms have tried to implement technology such as flexible manufacturing systems, robotics, MRP, NC machines, AGVS, and CAD systems. However, very few of these attempts have succeeded. The success stories in CIM are both few and well publicized (thanks in part to the LEAD awards made annually by CASA/SME). However, in general the success rate has not kept pace with the time, money, and effort that have been expended on CIM and CIM-related technologies. As a result, many managers view any investment in CIM as being large and high risk. Such a combination does not help to overcome management resistance to CIM.

High-Technology Barriers. For most managers technology and CIM are closely intertwined. The discussion of new technology leads to a discussion of CIM; an examination of CIM is seen as requiring a thorough understanding of the technology on which it is based. Most managers, however, are intimidated by technology. There are several factors that contribute to this sense of intimidation.

First, most discussions of new technology are surrounded by *jargon*. Jargon is useful because it often becomes the *shorthand* of discussion. It enables people to communicate quickly and efficiently. However, jargon can often be a formidable barrier to clear communication. To someone trying to get more information about a new development or system, the presence of jargon tends to frustrate.

Second, the sheer volume of information involving new technology is also intimidating. The number of articles dealing with technology increases with every passing month. The manager interested in learning about technology is presented with the problem of keeping pace with this overwhelming tide of information. There is always the hidden fear that by not examining all of the relevant literature, the manager will overlook some critical new information or development.

To better understand the sheer volume of information pertaining to technology and CIM, consider the following. One quick method of obtaining information about technology and CIM is to access the databases stored at *Dialog*

Information Services[6] and available through the *Knowledge Index*. Many of the major business journals are abstracted and stored by Dialog. There are two databases that would be of most interest to someone interested in learning about CIM or new technology. These are *ABI/FORM* (BUSI1)[7] and the *Trade and Industry Index* (BUSI2).

Each database could be searched by several key words. For example, a search of each database using four key words, *CIM, automation, technology*, and *FMS*, for articles published between 1985 and 1989 yields the results summarized in Tables 3-1 and 3-2.

Consider the situation faced by a manager interested in learning about automation. In 1989 that manager would have had to read about 4,437 articles (1,354 from *ABI/Form* and 3,083 from the *Trade and Industry Index*).

Third, every manager must deal with the problem of picking the "right" technology. That is, when evaluating current developments, managers are faced by a critical decision: should they pick technology that is currently available and that might work, or should they wait for the introduction of newer and better technology? After all, if they buy something today, it might be made obsolete by a development announced tomorrow. The result is a form of "paralysis of analysis"—a situation in which the firm never invests in new technology for fear of making an expensive mistake.

The Fear of Change in the Cost Structure. Some managers see any investments in CIM as changing the fundamental cost structure of the firm. For many managers when the technology required to make CIM operational is introduced, the anticipated result is that the firm's fixed costs are increased as a percentage of overall costs, while variable costs fall as a percentage. This increase in fixed costs occurs for several reasons. First, the acquisition of capital equipment increases the fixed-cost base of the firm. Second, to service and run the new equipment, a different type of work force is required. This is a work force that is very skilled and is able to maintain, program, run, and correct problems as they occur. It is also a work force that tends to be salaried or paid on the basis of a fixed number of hours per week (e.g., 40 or 45). It tends to be more fixed than variable in nature (again increasing the fixed-cost component).

Increases in the fixed-cost component tend to be viewed negatively by most managers. This perception is attributable primarily to two major factors. First, fixed costs tend to be reflected in the overhead or burden rate. Reducing or managing overhead rates is often more difficult than reducing or managing vari-

---

[6]Dialog Information Services, 3460 Hillview Avenue, Palo Alto, CA 94304.

[7]This name is the one under which the database is stored and accessed when using the *Knowledge Index*.

TABLE 3-1
Breakdown of Articles by Key Words and Year (ABI/FORM)

Publication Year	*CIM*	*Automation*	*FMS*	*Technology*
1985	94	1,720	32	3,756
1986	157	1,872	56	4,286
1987	132	1,326	33	4,149
1988	190	1,414	50	5,098
1989	214	1,354	33	4,936

TABLE 3-2
Breakdown of Articles by Key Words and Year (Trade and Industry Index)

Publication Year	*CIM*	*Automation*	*FMS*	*Technology*
1985	40	2,604	88	4,340
1986	60	2,762	54	3,823
1987	88	2,598	32	4,596
1988	78	2,732	30	5,468
1989	85	3,083	49	6,902

able costs. Second, as the overhead rates increase, the break-even point for operations in the firm is increased. This, in turn, reduces the perceived flexibility of the firm.

Difficulties in Justifying the Investments in CIM. A major obstacle encountered by any CIM implementation is that of justifying the need to invest the time, effort, money, and resources needed by CIM. Every investment must pass through some form of formal justification procedure. There must exist definite benefits attributable to the proposed project that can offset the anticipated costs (within a reasonable period of time). Most of the commonly used justification procedures (e.g., payback, net present value, discounted cash flows) are appropriate when dealing with a project in which the benefits are well defined, quantitative in nature (i.e., can be measured in dollars), and occur within a relatively short time period (i.e., frequently less than three years). Many CIM projects seldom satisfy these requirements. Most CIM projects are long-term undertakings (often requiring more than three years for completion). Their benefits tend to vary from firm to firm. They are also difficult to identify in advance.

The benefits include both quantitative and qualitative elements. The impact of CIM on inventories, for example, must be determined as well as its impact on quality, lead times, and enhanced system predictability. It is extremely difficult to assign accurate dollar values to these latter elements. Finally, it may take a relatively long time for all of the benefits resulting from the implementation and use of the new CIM system to become apparent. As a result, developing a convincing justification has become a significant hurdle for most potential CIM projects. Many CIM implementations that should have taken place never saw the light of day because they could not be adequately justified.

The Fear of Loss of Management Control. For many managers (especially those of small to medium-sized firms), CIM represents a potential loss of management control. These managers tend to understand the business of building and delivering their firm's products. They tend to have "hands-on" experience. They have been with their companies for a long period of time. Often, their authority in their firms is based on their knowledge of the firm, its production processes, and its products.

CIM represents a major challenge to their position and authority within the firm. By introducing automation and technology, CIM often redefines the production process. It changes how the firm plans, schedules, executes, and tracks production. Much of the skills and knowledge possessed by the current management may no longer be pertinent to this new CIM-run environment. These managers may feel that they no longer understand operations in this climate.

In addition, managers find that they are now faced with a new set of *experts.* Typically, these experts are people needed to program the equipment, maintain it, or design parts for production. These people, not the managers, are perceived as having the necessary knowledge of the system and its operation. Managers might choose to view them as a threat to their authority and position within the firm. In short, CIM changes the "hands-on" orientation of the firm to that of an "arms-length" relationship.

Increased Resistance from the Rest of the Firm. Management is not the only group that may experience increased uncertainty as a result of a possible decision to implement CIM. Employees may also be adversely disposed toward CIM. There are several reasons for this negative attitude.

First of all, CIM is change. With any change it is normal for people to experience *FUD* (fear, uncertainty, and dread). CIM will require changes in the way that people carry out their tasks. CIM must be explained and understood by people within the firm. There are often numerous misconceptions about what CIM is and is not.

Second, employees may see CIM as part of larger *hidden agenda* that is being implemented by management. This hidden agenda may be perceived as including replacing all or a portion of the current work force, breaking a union, or reducing the number of employees (a result of CIM increasing each employee's productivity). The employees' objection to CIM may be an objection to this perceived hidden agenda. Finally, some employees resist change because they do not want to change.

The Desire for Stability. Many managers and employees like stability. Stability leads to a sense of comfort and security. People have a chance to develop routines or patterns with which they can feel comfortable. CIM represents instability along two dimensions.

First, there is the change that takes place when CIM is introduced. Second, there is the ongoing change that accompanies CIM. Part of this change is due to advances in technology; other components of this change result from users learning about CIM and uncovering new capabilities. As a result people must now learn to live with change. Learning to live with change is a major undertaking for some people—an undertaking that they may not wish to begin.

The Fear of Developing External Dependencies. As was noted previously, CIM has a very technical side associated with it. Some managers may choose to deal with this side of CIM by hiring consultants who are seen as being knowledgeable in this area or by relying on a well-established and respected supplier. In either case, there is a hidden danger. Since management is not familiar with the technical side of CIM, it may have to rely on the insights and knowledge of these external experts. Some may fear that they may be put in the situation of doing not what they think is best for the firm but rather what their experts tell them is best for the firm.

The Timing Is "Wrong." There is a potential paradox that some firms may encounter when evaluating the CIM decision. The best time to implement CIM should be when business is good. The reason is that there are enough resources to fund the necessary investments. There is less pressure to rush into CIM (and to make unnecessary mistakes). However, few firms turn to CIM when business is good. When business is good, there is little incentive to change. The attitude of many toward CIM can be best summarized by the old saying,

If it isn't broke, don't fix it.

For others, when business is good, there is a strong pressure to build products for customers. The firm may not be able to release those people essential to implement CIM because they are needed for day-to-day operations.

In contrast, many firms seriously consider CIM *only* when business is going poorly. CIM is then seen as necessary to survival. People are anxious to have a CIM system in place and generating the benefits typically associated with it. The problem is that this is often the worst time to implement CIM.

When the firm is in trouble, it may not have the resources needed for making CIM a success. Banks, for example, are very hesitant to lend to firms that are experiencing difficulties. Furthermore, the firm may not have the luxury of time. When operations are not going well (or are deteriorating), the firm needs a solution that will have an immediate positive impact on operations. CIM may be viewed as being too long-term a project. In either instance, the timing of the decision is "wrong."

Confusion over What CIM Is Exactly. The final management obstacle to CIM is that of confusion. This confusion can take one of three forms. First, there is confusion over what CIM is and what it is not. Second, there is confusion over whether the firm has in place all of the prerequisites demanded by a successful CIM implementation. Third, there is confusion regarding whether CIM can address the problems and concerns facing the firm. This confusion over CIM is not unique to managers. It is a confusion that is repeatedly echoed in the literature.

A review of these obstacles points to one critical factor: many of these obstacles are not directly attributable to CIM. Rather, they reflect confusion—confusion both within the firm and confusion about what CIM is and what it is not. It is a confusion present in how managers and employees react to CIM. As a result, before we examine the principles and practices of CIM in operation, we must first present the reader with a framework for understanding the CIM system. A first step for developing such a framework is to examine how others have defined CIM in the past.

COMPUTER INTEGRATED MANUFACTURING: ALTERNATIVE PERSPECTIVES

One of the basic obstacles to developing an understanding of CIM is that of trying to define what CIM is. Writers and managers have been trying to define the concept of *computer integrated manufacturing* since 1974. Beginning with Joseph Harrington, who first coined the term, numerous definitions have been offered. Unfortunately, these definitions have often resulted in more, not less, confusion. People have disagreed over what CIM is and what CIM is not. As a result, it is important to recognize that at present *there is no universal agreement* about CIM. Instead, there is a wide spectrum of views about how different people perceive CIM.

In part, this spectrum reflects the development of CIM. The first CIM systems implemented in the early 1970s dealt primarily with the manufacturing side of the firm. They focused on the application of computer-aided technology (e.g., numerical control and CAD) to the task of building products. However, since that time CIM began to evolve. It began to broaden its focus. Included in CIM were issues of information management, integration with other areas, and fit with corporate strategy. These issues have influenced the way in which writers have tried to describe the concept of CIM to others.

Although there are certain differences, these various views of CIM tend to fall into four distinct categories (see Figure 3-2):

- CIM as automation.
- CIM as manufacturing integration.
- CIM as information system.
- CIM as strategic integration.

It is useful to examine each view of CIM. This assessment should not be regarded as comprehensive. Not every definition for CIM will be examined. Instead, what we will do is to select definitions that are representative of the four categories. As a result the review will be both selective and limited.

CIM as Automation

Of the four categories, this tends to describe the initial views of CIM. According to proponents of this view, the test for the presence or absence of CIM is relatively simple: is computer-aided technology being applied to the manufacturing and engineering areas of the firm? If it is, that system can be classified as being a CIM system. It is in some ways a myopic view, for it limits the development and use of CIM primarily to the manufacturing and engineerings areas of the firm. There are several examples of views that fall into this category. The first is that offered by Harrington.

Harrington saw computer integrated manufacturing as being a very limited system in which automation and computer technology were applied to only manufacturing. It was not a corporate system but rather a manufacturing system.

This view has been echoed by other writers such as Joseph Aiello. In a 1985 presentation, Aiello described CIM as being the integration of factory automation and information.[8] As such, it represented the latest stage in a continuing

---

[8]Joseph L. Aiello, "CIM: The Factory of the Future...NOW," *Computer Integrated Manufacturing and Flexible Manufacturing Systems Seminar Proceedings*, New Orleans, La., April 24–26, 1985, pp. 41–43.

FIGURE 3-2
Alternative Views of CIM

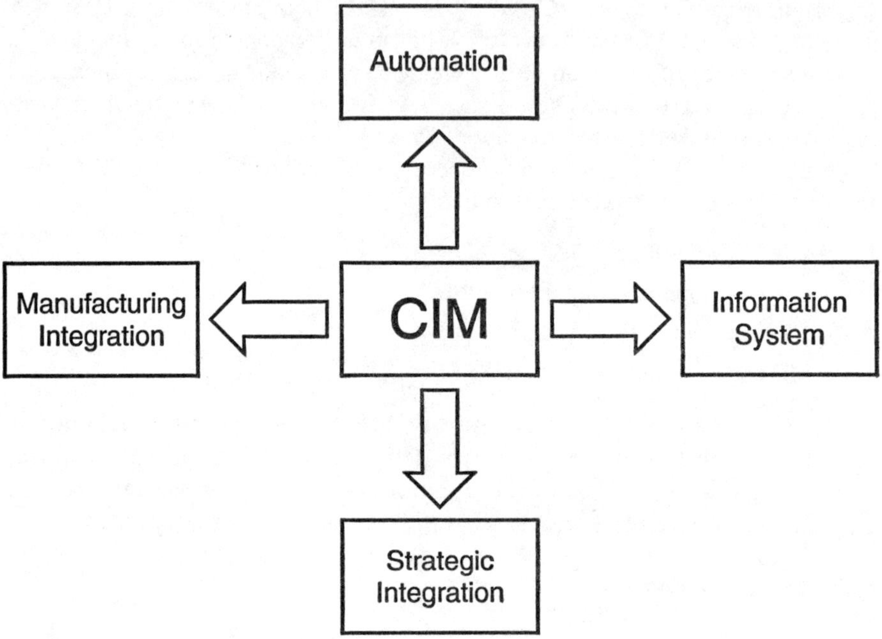

evolution of manufacturing systems. Typically, the following were identified as being major elements in the CIM umbrella:

- Robotics
- Computer-aided design
- Computer-aided manufacturing
- Flexible manufacturing systems.
- Automated storage and retrieval systems.
- MRP II
- Data-collection equipment
- Automated test equipment.
- Distributed systems
- Transportation systems

Speaking at the same conference, R. Miller presented a similar view of CIM. To him, CIM was "any use of computer technology and systems to support the manufacturing process from product definition through design and manufacturing to the delivery of the finished product in the customer's hands."[9] A successful CIM, under this view, consisted of a number of critical and mutually supportive elements:

- Computer-aided design
- Flexible automation
- An automated shop environment.
- Committed vendors
- Lot sizes of 1.
- Continuous flow
- The handling of uncertainty with capacity, not inventory.
- The minimum distance between workstations.
- Materials stocked at points of usage.
- Kitting eliminated
- Components packaging that varies by workstation.
- The lot tracking of components.

Paul Ranky presented his views of CIM in a book that was published in 1986. To Ranky, the central trait of CIM lay in its extensive use of computer technology and automation:

> Computer Integrated Manufacturing (CIM) is concerned with providing computer assistance, control and high level integrated automation at all levels of the manufacturing industries, by linking islands of automation into a distributed processing system. The technology applied in CIM makes intensive use of distributed computer networks and data processing techniques, Artificial Intelligence and Data Base Management Systems.[10]

A final definition of CIM that stresses the use of automation comes from Paul Adler. In 1988 Dr. Adler published an article dealing with the task of managing flexible automation.[11] He described CIM as being characterized by the convergence of three types of automation:

---

[9]R. Miller, "Factory with a Future...No Pushover," *Computer Integrated Manufacturing and Flexible Manufacturing Systems Seminar Proceedings,* New Orleans, La., April 24–26, 1985, pp. 146–161.

[10]Paul G. Ranky, *Computer Integrated Manufacturing* (Englewood Cliffs, N.J.: Prentice-Hall International, 1986).

[11]Paul S. Adler, "Managing Flexible Manufacturing," *California Management Review,* Vol. 30 (1988), No. 3 (Spring), pp. 34–56.

- Design automation
- Manufacturing automation
- Administrative (or control) automation.

The value of these linkages and the new automation technologies is most apparent in their impact on cost, quality, and time.

As is evident from these definitions, the key to CIM is *automation*. In many ways the CIM equation can be stated as follows:

CIM = automation

CIM as Manufacturing Integration

Definitions that fall under this second category differ from the automation perspective in that they see CIM as a vehicle by which a firm can better integrate the operation of the manufacturing function. The function of CIM is not to automate the operation of the various components of the manufacturing system. Automation is often present in a CIM system. However, automation by itself is not a sufficient condition for determining the absence or presence of CIM. Instead, for CIM to be present the use of computer-aided technology must be present in the planning system and on the shop floor, and all aspects of the manufacturing system must work together as a seamless whole. Often, the primary means of integration is through the flow and use of information. However, the primary focus is on the integration of the manufacturing system.

"Islands of automation," which are acceptable under the automation perspective, are unacceptable by those who define CIM as a form of manufacturing integration. There are several examples of definitions that fall into this category.

Ann Majchrzak, in examining the human side of factory automation, defined CIM as follows:

> The concept of integration, or computer-integrated manufacturing (CIM), is one of the most technologically sophisticated gains over conventional equipment. CIM is a closed-loop feedback system, with prime inputs as product requirements and product concepts, prime outputs as finished products, and processes as a combination of software and hardware that incorporates functions of product design, production planning, production control, and production processes.[12]

This definition is based largely on the views offered by Schaffer.[13]

[12]Ann Majchrzak, *The Human Side of Factory Automation* (San Francisco: Jossey-Bass, 1988).

[13]G. Schaffer, "Implementing CIM," *American Machinist*, August 1981, pp. 71–93.

Similar views have been offered by writers such as Voss,[14] Hegland,[15] and John Arnold and Hank Johansson. For example, Arnold and Johansson were succinct in their view of CIM when they said that CIM could be defined as "*the application of computers and computer technology to integrate the manufacturing process.*"[16]

However, some writers have tried to expand the view of CIM outside the manufacturing environment. They have tried to show that CIM is not simply concerned with integrating the manufacturing and engineering functions. Instead, CIM also tries to integrate manufacturing and nonmanufacturing elements. This broadening of the integration focus is most evident in the definitions of CIM provided by writers such as William Thurwachter:

> CIM is the logical organization of engineering, production and marketing and supporting functions into a computer integrated system. Functional areas of the enterprise (e.g., design, analysis, planning, purchasing, cost accounting, inventory control and distribution) are integrated with the factory floor functions (e.g., materials handling and management with direct control and monitoring of all process operations) through computers.[17]

In definitions belonging to this category, the focus is primarily on integration. In the third category of definitions, the notion of integration is still very evident. However, the focus now shifts to the vehicle of integration, which is *information.*

CIM as Information System

For a third major group, CIM is a system that is organized around the computerized flow of information and data. Information and its management and use (often through computerized methods) is viewed as the hallmark of all successful and effective CIM systems. It is also this emphasis on information that helps to differentiate a CIM system from simple factory or system automation. This view of CIM is one that emphasizes information flows, databases, and supporting computer technology.

---

[14]C. A. Voss, *Managing New Manufacturing Technologies* (East Lansing, Mich.: Operations Management Association, 1986).

[15]D.E. Hegland, "CIM—The Foundation of Factory Automation," *Production Engineering*, May 1985, pp. 36–42.

[16]John C. Arnold and Hank J. Johansson, "Establishing a Framework for the Analysis and Justification of Computer Integrated Manufacturing," *Computer Integrated Manufacturing and Flexible Manufacturing Systems Seminar Proceedings*, New Orleans, April 24–26, 1985, pp. 58–66.

[17]William A. Thurwachter, "FM$ vs. FMc," *Computer Integrated Manufacturing Flexible Systems Seminar Proceedings*, New Orleans, April 24–26, 1985, pp. 179–188.

This view of CIM was presented by John Ettlie in his study of advanced manufacturing technology and CIM. Ettlie saw CIM as "the integration of design, manufacturing and business functions through computer technology so that information is sent where it is needed and the manufacturing process moves from raw material to finished production without disruption."[18] Similar views were presented by the Manufacturing Studies Board[19] and by Stevens.[20]

Again, the definitions and views of CIM that fall into this category are primarily tactical in nature. That is, they focus on the application of technology. Strategy, when present, often plays a minor role in the definition.

CIM as Strategic Integration

The fourth and last major view of CIM is the broadest of the four categories. The focus is not primarily on technology, integration, or information. Those are important aspects, but they do not get to the heart of CIM. Rather, technology, integration, and information are seen as vehicles. They are means to an end, nothing more. The critical trait found in definitions belonging to this last category is the emphasis placed on CIM as a *strategic system*. That is, what makes a CIM system unique is that it involves the use of computer-aided technology in support of specific strategic objectives. Whereas automation is driven by the simple application of technology to a specific problem, CIM is driven by strategy. It is a concrete expression of how the firm competes in the marketplace.

The best example of a definition belonging to this category is found in the views of George Sibbald, who defined CIM as "the computer-assisted integration of the production process and operations control functions needed for optimal deployment and productive use of the minimum manufacturing and management resources required to meet *specific strategic business objectives*[21] (italics the authors')."

With this last group of definitions, the confusion surrounding CIM can be better understood. Some of the definitions take very different views. These differences are greatest between those who view CIM as automation and those belonging to the other three categories. In other cases the differences are more of degree and focus. However, there is one point that must be stressed again: there is no single, universally accepted view of CIM.

---

[18] John Ettlie, *Taking Charge of Manufacturing* (San Francisco: Jossey-Bass, 1988).

[19] Manufacturing Studies Board, *Toward a New Era in U.S. Manufacuring: The Need for a National Vision* (Washington, D.C.: National Academy Press, 1986).

[20] Larry Stevens, "The CIM Connection: Reaching Out to the Shop Floor," *ComputerWorld*, Vol. 22, No. 22 (May 30, 1988), pp. 45–52.

[21] George W. Sibbald, *CIM Today* (Menlo Park, Calif.: SRI International, March, 1986).

COMPUTER INTEGRATED MANUFACTURING: THE CONCEPT DEFINED

The view of computer integrated manufacturing taken in this study parallels very closely the corporate or strategic perspective on CIM. That is, we define CIM as a broad-based system that ultimately is strategically driven, or as *a corporate system that, by drawing on computer-aided technology, is designed to help top management organize and direct the company to focus on a market with the goal of developing and maintaining a* sustainable *competitive advantage.*

This definition highlights certain traits that are central to all effective CIM systems. There are four such traits:

- CIM is a corporate system.
- CIM is a strategic system.
- CIM is a means of organizing the firm and mobilizing its resources.
- The goal of CIM is the development of a sustainable competitive advantage.

Each of these four traits is significant, for each helps to differentiate CIM from simple automation. The major difference lies not in the technology but in how it is used and how it is viewed by management. With CIM, management creates through strategy the vision of what the company should be. It uses CIM as the vehicle of realizing this vision. In turn, the technology that CIM draws on in achieving this vision makes available to the firm new strategic opportunities.

CIM as a Corporate System

The first thing that any CIM definition should try to do is to differentiate between the application of such focused computer-aided systems as CAD/CAM and CIM. The difference between these two systems does not lie in their use of technology but rather, in the extent to which the systems are narrow or broad in their scope.

Ultimately, any firm can survive only by better serving the needs of its customers. This task can never be achieved by manufacturing and engineering by themselves. Instead, it requires the involvement of all areas of the firm, from sales to order entry to shipping.

CIM as a Strategic System

Our view of CIM is very explicit on one critical point: the objective of any CIM implementation project can never be the development of a CIM system. Firms should never implement CIM simply to have CIM systems. Instead, CIM should be regarded as being nothing more than a vehicle. It is a means to an end. CIM does not set or formulate the goals that the firm will achieve. Rather, it helps the firm achieve its objective(s). This aspect can be better illustrated by returning to the analogy of a car.

A car offers the driver certain capabilities. The car, however, is a tool in the hands of the driver. It is the driver who determines where the car will go and its speed. More important, it is the driver who initially determined the capabilities (and implicitly the limitations) of the car during its acquisition.

If CIM is the car, the driver is *corporate strategy.* It is corporate strategy that sets the directions that drive CIM. To understand CIM it is first necessary to understand corporate strategy and its role.

In other words, corporate strategy determines the *what* of the company. It identifies the goals to be achieved by the firm. One way of achieving these goals (i.e., the *how)* involves the use of CIM. For CIM to be effective and truly supportive of corporate objectives, an important requirement must first be satisfied: there must be a well understood and widely communicated corporate strategy in place prior to the consideration of CIM. This strategy can be formally developed and implemented (the desired state for strategy) or it can be informally formulated and communicated (the result of a clear understanding on everyone's part of what it is that the firm does and how it serves its customers in the market). CIM without corporate strategy is like a car without a driver.

Corporate Strategy: An Overview

Understanding the concept of corporate strategy is not an easy task. One major hindrance is that it has been used so loosely by some business writers that it has lost much of its meaning. It has come to represent everything from the broad guiding principles that management uses in formulating and communicating goals, plans, and policies to the very detailed implementation plans that are carried out within the firm.

In this study our view of strategy is very similar to that presented by Robert Hayes and Steven Wheelwright in their 1984 book on manufacturing strategy.[22] When we talk about the concept of corporate strategy, we refer to the process by which the firm establishes its purpose(s), sets directions, identifies the target market segment(s), determines the time horizon for achieving these objectives, develops plans, mobilizes and organizes corporate resources, and takes major action (both to implement the plans and for corrective actions should problems arise) with the goal of achieving a distinctive competitive advantage. To this

---

[22]Robert H. Hayes and Steven C. Wheelwright, *Restoring Our Competitive Edge* (New York: John Wiley & Sons, 1984), chap. 2.

end there are five traits common to the use and practice of strategy that are critical when dealing with CIM:

The Development of a Purpose That Is Internally and Externally Consistent. Corporate strategy begins by identifying the purpose of the firm. The purpose is critical for two major reasons. First, it identifies how the firm views itself. For example, some firms such as Steelcase of Grand Rapids, Michigan, see themselves as risk reducers. That is, they are suppliers of high-quality products that are:

- Consistently delivered on time.
- Better suited to the customers' needs than any products provided by their competitors.
- Supported by top-notch customer service at *every* step in the procurement process.

Such firms do not compete on cost. Instead, they compete on providing better value to their customers. To these firms, lowest cost does not necessarily equal highest value.

Second, the purpose provides the firm with a bond—both within the firm and between the firm and its customers. For this bond to be effective, the definition of the purpose should be one that is consistent both *internally* and *externally*.

The term "internal consistency" has two dimensions. The first dimension refers to the extent that the purpose reflects how the people within the firm view themselves and their unique competencies in the marketplace. Often, this internal view has been called the company philosophy. The company philosophy can be defined as "*the set of guiding principles, driving forces, and ingrained attitudes that help communicate goals, plans, and policies to all employees and that are reinforced through conscious and subconscious behavior at all levels of the organization.*"[23] Another view of the company philosophy may be that of a *set of common values.*

The second dimension of internal consistency is that of resource consistency. This dimension deals with the extent to which the purpose contained within the strategy is consistent with the corporate and manufacturing capabilities of the firm. When this dimension of internal consistency is not present (i.e., when the purpose cannot be supported by the manufacturing process of the firm), the purpose will not be achieved.

Terry Hill has recently developed product profiling,[24] an interesting procedure for evaluating this second dimension of internal consistency. This proce-

---

[23]Hayes and Wheelwright, *Restoring Our Competitive Edge*, p. 25.

[24]Terry Hill, *Manufacturing Strategy: Text and Cases* (Homewood, Ill.: Irwin, 1989).

dure provides a graphic representation of the degree to which strategic demands and desires are supported by *current* manufacturing capabilities.[25]

The concept of external consistency involves the extent to which what the firm is willing to deliver really matches the demands of the market. To evaluate this fit requires that the firm be able to identify its specific group of customers. For example, the customer who buys primarily on cost and is interested in acquiring a desk chair will not find a chair manufactured by a firm such as Steelcase as particularly attractive.

When dealing with external consistency and corporate purpose, the management of the firm must recognize the potential presence of three different types of customers:

- Customers to whom the firm thinks that it sells.
- Customers who actually buy the firm's products.
- Customers to whom the firm should be selling.

External consistency is strongest when these three types of customers are one and the same.

When there is a problem with consistency, management has two basic strategic options on which it can draw. First, it can change the market segment that it is targeting. Alternatively, it can either change its purpose (to better fit the perceived needs of its targeted market) or invest resources in the firm to develop the necessary capabilities. In either case, resources must be committed and change must take place.

The Definition of the Relevant Customer Base and the Identification of How the Firm Proposes to Generate Value for These Targeted Customers. The corporate strategy identifies to the rest of the firm the target market. This definition can be stated in geographic terms (i.e., location), relevant traits (e.g., income level, family structure), size (i.e., the number of potential customers), or the degree of homogeneity.[26] It also defines the concept of *value* for the members of the firm. That is, the strategy should describe the general guidelines that the firm intends to follow in satisfying the perceived needs and requirements of the target market (ideally at the lowest cost to both the customer and the firm).

---

[25]Those interested in learning more about product profiling should see chapter 4 of the previously cited Hill book on manufacturing strategy.

[26]There is a critical trade-off present between market size and the degree of homogeneity. These two are typically negatively correlated. That is, as the size of the target market grows, the degree of homogeneity falls.

The Identification of the Time Horizon for the Achievement of Strategic Objectives. Corporate strategy specifies the time horizon over which the objectives and changes needed to achieve these objectives must take place as well as the time period over which the effect of these changes should become apparent.

The Establishment of the Blueprint for Resource Allocation. By setting the overall goals for the firm and the time period within which to achieve these goals, the corporate strategy largely lays out the requirements that must be satisfied by any *appropriate* investment or system. It is these requirements that must be considered when evaluating a system such as CIM or when justifying the investment in CIM and CIM-related technology.

The Building of the Organizational Structure. The term "infrastructure" describes the organizational frameworks, reporting relationships, performance measurement systems, staffing and recruiting practices and policies, and training and education guidelines that are in place. In short, it describes the environment found in the firm.

A critical element of the structure is the *performance measurement system.* This is one of the feedback mechanisms that firms develop to help manage day-to-day operations. It defines, often in very detailed or operational terms, what people within the firm are expected to do well. These expectations are reflections of how the firm tries to compete in the market. They also establish time horizons for acceptable performance. They determine whether the employees go for actions that result in immediate paybacks or are willing to undertake longer-term projects that will generate significant benefits that take longer to become evident. Finally, the performance measurement system influences the reaction of its employees to qualitative and quantitative benefits. That is, if managers are measured on how dollars are saved per period rather than on improvements in quality or reductions in late or early shipments, they will tend to do things that make them look better on that quantitative dimension.

Strategy has a major impact on the structure. It determines, for example, whether the organization is structured along functional lines or product line. It also determines whether the firm is guided by short-term actions or by a longer-term perspective. The structure ideally should be a reflection of a corporate strategy. Strategy and structure together form the setting into which CIM is introduced.

Corporate Strategy and CIM

Why, then, is corporate strategy so critical to the success of CIM? Simply stated, corporate strategy shapes the setting into which the CIM system is introduced. It establishes the justification for the implementation of CIM. It sets out the objectives and measures against which the new CIM system will be measured. Strategy provides answers to the questions of *who* (i.e., customers, market segment), *what* (product or service delivered by the firm), *when* (time horizon), and *where* (specific areas of the firm to be involved). Within this context CIM is simply part of the *how*.

Strategy is the element that is always present in CIM systems and almost always absent in automation projects. It is the element that differentiates automation from computer integrated manufacturing.

Automation versus Computer Integrated Manufacturing

Automation and CIM are distinct from each other. While they are based on the same or similar technology, their orientation is different. Consider first automation (see the Glossary for a discussion of the various definitions of automation). The major objective of any automation project is the application of technology (be it mechanical, electrical, electronic, or computer based) to the operation and control of production systems. Typically, automation is introduced as a means of breaking a bottleneck, improving the efficiency of a specific segment of the production system (e.g., using a robot in place of a worker for an operation that takes place in a hot or dangerous environment), or increasing the capabilities of the process (e.g., introducing an automated paint line to enhance the flexibility of painting and to expand the paint mix now supported).

Irrespective of the reasons for its use, automation is *myopic* in its approach. For example, we have an area in which we need increased speed and greater accuracy of repetition. We define the requirements for an "acceptable" solution very narrowly. The requirements are often set out in technical terms that are applicable to the situation currently being evaluated. For example, in the situation where accuracy of repetition is desired, the most acceptable solution would be the technology that is the fastest in movement speed and offers the greatest accuracy in terms of repetition of movement. Automation solutions seldom examine the impact of the proposed solution on the rest of the production system. The question of how the new technology fits in with the rest of the system (or even whether it can be successfully and efficiently integrated) is not a primary consideration. Automation solutions rarely consider the question of whether the proposed solution is consistent with or supportive of corporate strategy.

It is the myopic nature of automation that is the major reason for the emergence of so-called islands of automation. With an island of automation, we have a situation in which the performance of one element of the production system is optimized. However, from the perspective of the larger system, the performance of the entire system has either not improved significantly or may have deteriorated. This deterioration in performance may occur for several reasons:

Integration Problems. The larger system may be experiencing problems in integrating the resulting solutions. In some cases these integration problems may be primarily information based. This is the case when the new automated system generates information in a form that is not readily usable by the rest of the system.[27] In other cases the new automated system may require the use of a different unit of analysis or measurement. For example, the introduction of a flexible manufacturing system may require that we think, plan, and schedule in terms of part families rather than individual parts. However, the planning system may be organized in terms of separate and individual parts.

Bottleneck Problems. Automating one part of a manufacturing system can result in an increase in the capacity of that section. This increase in capacity may now uncover bottlenecks elsewhere in the system. This situation is most critical if the system was previously balanced. The ensuing change in capacity now results in a situation in which the rest of the system is essentially a bottleneck. Automation does not always consider the issue of balance in system capacity when introducing technology into the system.

A Change in Resource Requirements or Specifications. Automation may create problems with the adequacy of the resources being used. Resources that were previously acceptable may no longer be considered acceptable. The resources affected can include people, materials, and information. Workers' skills may have to be *upgraded*. They may have to be retrained so that they are familiar with the equipment and can run (and maintain) it. They may have to learn how to program the equipment. Materials have to meet higher standards and satisfy closer tolerances. Information requirements may have to be redefined. Bills of material may have to be maintained in terms of part families and individual parts. Information must now be far more accurate. Inventory accuracy of 95 percent, for example, may now have to be considered the minimum floor (with the desired goal being 99 percent plus). Manufacturing information must also be complete and current. The firm may have to invest significant time, effort, and resources in upgrading the resources so that they better fit the requirements of the new technology.

Problems with Strategic Consistency. Automation may introduce new capabilities that are at odds with corporate strategy. Consider, for example, the introduction of a flexible manufacturing system. Such a system can reduce setup time (thereby contributing to a reduction in the overall manufacturing lead time). It can also give the firm a new capability. By reducing setup, flexibility is

---

[27]These problems with information format compatibility are one of the major reasons behind the development and growth of such information protocols as MAP (manufacturing automation protocol) and TOP (technical and office protocol)

enhanced. The manufacturing system can now produce smaller order quantities efficiently. This capability may be wasted if the firm competes primarily on large orders. Under certain conditions, this new capability may create problems. Managers, as they become aware of this new capability, may go out and solicit business that takes advantage of it. The firm is now going out for both small- and large-quantity orders. These types of orders do not compete on the same set of criteria. Ultimately, the result is a loss of focus and the creation of confusion over what the firm is really selling its customers.

In short, automation is a procedure that is problem driven and that attempts to improve performance from a *local* perspective. In contrast, CIM is driven first and foremost by strategy. The problems that it addresses are seldom defined in terms of bottlenecks or the need for process substitution (i.e., the replacement of one type of process or procedure by another). Instead CIM tries to enhance the overall effectiveness of the entire system and to develop for the firm a distinctive competitive advantage. It tries to improve performance from a strategic or *global* perspective.

CIM as a Means of Organizing the Firm and Mobilizing Its Resources

Every firm draws on four assets to design and build its goods and services and to compete in the marketplace. As summarized in Figure 3-3, these are:

- *People.* All of the personnel employed by the firm. Included under this resource are all people engaged in all corporate activities (direct, indirect, administrative, and support).
- *Information.* All of the information that the firm draws on for building products, processing orders, improving efficiencies, responding to customer requests, designing and testing new products, and meeting schedules (to name a few).
- *Processes.* The systems (formal and informal) that the firm has developed for planning and controlling all flows through the firm. These flows consist primarily of either information or product.
- *Resources.* This is a catch-all term that describes all of the remaining resources that the firm can draw on to meet schedules, manage day-to-day operations, and respond to change. Resources can be either physical (machines, plants, locations) or financial (lines of credit).

These four assets are similar to ingredients found in a recipe. By themselves they have no form. They offer potential. By themselves they do not make the firm distinctly different from its competitors in the eyes of its customers. To effectively

FIGURE 3-3
Four Primary Company Assets

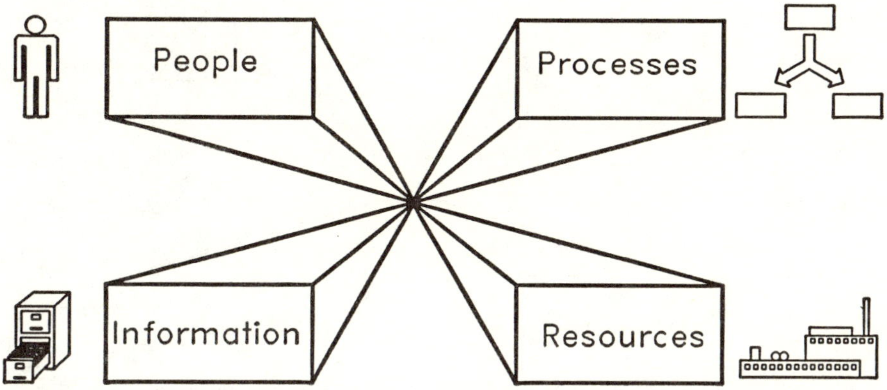

organize the assets in support of corporate strategic objectives, the firm draws on three major agents: *technology, organization* (i.e., how the firm arranges people and functions as well as the resulting interrelationships), and *strategy* (see Figure 3-4). The process of applying these three agents to the four assets describes the implementation and use of computer integrated manufacturing.[28]

CIM and Sustainable Competitive Advantage

Ultimately, the application of technology, organization, and strategy to people, processes, information, and resources is driven by one key objective: the generation and maintenance of a competitive advantage that is *sustainable* (the fourth trait). These two traits must be present if the firm is to survive in the long term. A competitive advantage describes any action that the firm takes that is designed to differentiate the firm favorably from its competitors. However, competitive advantages that are not sustainable offer the firm little long-term relief. If the advantage is not sustainable, competitors can quickly take actions to counter or offset the advantage. In the end, resources have been allocated and consumed but there has been no long-term change in the relative position of the firm.

---

[28]This process will be described in greater detail later on in this chapter, since it is the basis of the CIM framework used throughout this book.

FIGURE 3-4
Major Agents of the Firm

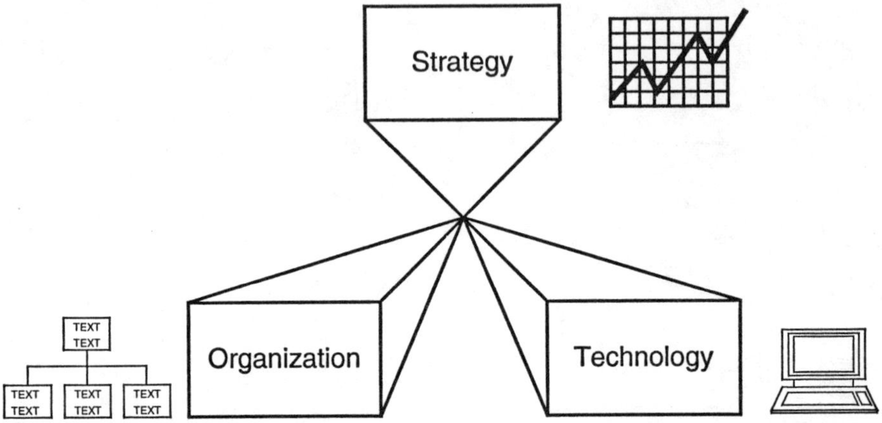

A sustainable competitive advantage occurs whenever one firm has developed such a dominant position in its targeted market that it effectively "owns" the market. The advantage offered by the actions taken are long term. Competitors cannot easily or quickly respond to actions taken by the firm. They have difficulty in countering moves taken and in dislodging the firm from its leadership position. Its position in the marketplace is both well entrenched and well protected. Often the firm is able to surround itself with high barriers to entry. These barriers can take many forms:

- Reputation
- Constant innovation (both in the product and process).
- Market share
- Capital base (minimum investments needed in plant and equipment).
- Well-established distribution system.

A firm that has developed a sustainable competitive advantage tends to be an aggressive competitor that continuously pushes to enhance its position in the marketplace by improving performance in such areas as cost, lead time, quality, risk, and flexibility. Finally, it has successfully put its competitors into a *reactive* mode—one in which they are constantly responding to initiatives taken by the market leader, yet are never able to mount a successful initiative of their own.

Ultimately, a sustainable competitive advantage occurs whenever one firm is able to develop and implement a system that ensures that its products represent (ideally) the highest level of value for its customers. The pairing of sustainable competitive advantage and value is critical. It forms one of the elements of the CIM framework.

Aside from the companies examined in this study, there are other examples of firms that have been successful in generating such sustainable competitive advantages (not necessarily through the use of CIM). These include Hewlett-Packard (laser printers), Motorola (pagers), and Union Special (industrial sewing machines).

THE CONCEPTUAL FOUNDATIONS OF CIM

The model of CIM used in this book incorporates three major dimensions. The first dimension deals with the elements of a CIM system and the manner in which they interact to create computer integrated manufacturing. In other words, the first dimension deals with the *structure* (i.e., elements) found in all successful CIM systems. The second dimension is a *process* view of CIM. We can envision CIM systems as passing through a series of major stages from past to finish. The second dimension describes these stages and their sequence.

The problem with the first two dimensions is that they result in very generic frameworks. They do not really help the reader differentiate between the different types of CIM systems that exist now. To differentiate between the various CIM systems, we must examine the nature of the "fit" between the various CIM systems and corporate strategy. The result of this examination is a *strategic* framework that identifies not one general CIM system but rather four distinct CIM types. The third dimension deals with these CIM types. To understand CIM we must explore and understand all three of these dimensions.

A Structural View of Computer Integrated Manufacturing

The CIM framework used by this book is somewhat generic in that it views CIM as the result of the application of the agents described previously (i.e., strategy, organization, and technology) to the firm's four assets (i.e., people, information, systems, and processes) with the goal of achieving and maintaining a sustainable competitive advantage. At first, there seems to be little in this framework to differentiate from other systems such as just-in-time-manufacturing. However, the difference between CIM and other competitive approaches lies in the nature of the primary change agent. In a CIM system the primary change agent is *technology* (see Figure 3-5).

The term "change agent" is used to describe the element that is used to bring about change in the firm. It is the primary means to a desired goal. This change can occur anywhere in the firm—from the shop floor to how the people of the firm communicate. In short, all CIM systems rely on technology to bring about

FIGURE 3-5
CIM: A STRUCTURAL FRAMEWORK

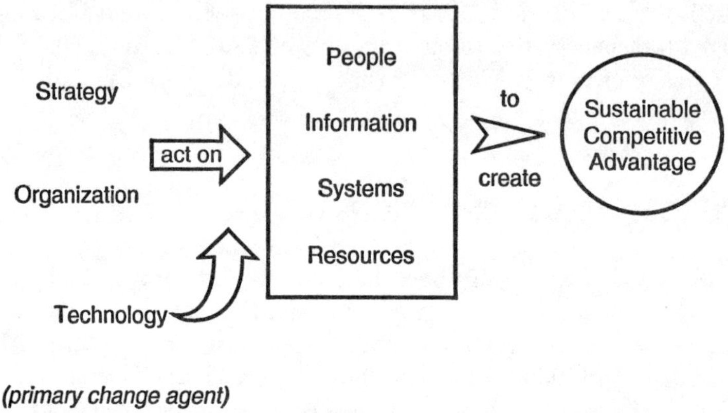

change. However, technology by itself is not enough. It needs to be supported by (and to be consistent with) the other two agents, strategy and organization. However, before we can explore the relationship between these three elements, it is first necessary to develop a better understanding of the concept of technology.

Technology

Technology, especially computer-based technology, is the major change agent in any CIM system. However, the concept of technology is often difficult to discuss because it is surrounded by a large number of different perspectives and, in many cases, misconceptions.

To some, technology is closely tied up with such areas as basic research, applied research, engineering, research and development, and innovation. It is embodied in machines, computers, robots, and fixtures and pallets. It is simply applied knowledge in its most basic form. However, this view of technology is far too narrow. It ignores the dynamics that are present when a new technology is introduced into an existing environment. It ignores the links between technology, strategy, and organization.

In this book we take a very broad, *holistic* view of technology. Our views of technology closely follow those presented by Lowell Steele.[29] Although various definitions of technology can be found in the Glossary, we view technology as *the system (consisting of knowledge and physical methods) by which the firm satisfies its strategic needs and desires.* Technology draws together (as well as

[29]Lowell W. Steele, *Managing Technology: The Strategic View* (New York: McGraw-Hill, 1989).

draws on) the physical and information or knowledge components. It represents a means to an end that can be applied at every stage of the process used by the firm in conceiving and bringing a product to market:

- Basic research
- Applied research
- Development
- Design engineering
- Production engineering
- Manufacturing engineering
- Quality control
- Fabrication
- Assembly
- Application engineering
- Physical distribution
- Product service

Ultimately, all technology affects the firm and its processes in one of two ways. First, it can create new products or capabilities that previously were not present (or were available at much higher cost or longer lead time). Alternatively, it can provide the firm with a substitute (or replacement) for existing products or processes. Finally, all technology affects either the product (what the firm offers its customers), the processes (the means by which the product is delivered), or both. While technology provides the means to an end, technology by itself cannot significantly change the future prospects for the firm.

For technology to have a significant effect on the firm, it must be coupled with strategy and organization. Strategy provides technology with a direction; it also provides technology with needed justification, a rationale for its existence, top-level support, and legitimacy. Organization, on the other hand, affects the environment in which people (the ultimate users of all technology) operate and in which the technology must exist. Organization determines how people are arranged, to whom they report, how positions are staffed, how people are trained, and how the employees are measured and evaluated.

Technology, strategy, and organization are linked together in the form of a dynamic equilibrium. Each element must be consistent with the others; each element must be supported by the other elements. As each component is altered, the change can and will affect the operation and shape of the other two components. The introduction of new technology, by changing the product or the process, can provide the firm with new capabilities. These new capabilities can give rise to a new set of strategic advantages (thus resulting in a change in corporate strategy). Furthermore, taking advantage of these new capabilities may require a new organizational structure (e.g., a shift from a functionally organized firm to a product-oriented organization). It is the presence of this dynamic equilibrium that gives rise to the dynamic nature of CIM systems.[30]

Types of Technology

There are a number of different dimensions that are present when trying to understand technology. The first dimension pertains to the *types of technology*. However we define technology, all technology belongs to one of three major types as defined by Steele.[31] These are as follows:

Product Technology. The first type embodies the use of technology in support of product design activities. In other words, it entails all activities involved in the conceptualization, layout, evaluation, revision, maintenance, storage, and retrieval of a product design. Included in product design are such activities as

Product planning. The task of product conceptualization. Involved in this task is the identification of customer needs and the formulation of appropriate product specifications.

Product engineering. The process of converting the specifications developed during product planning into a design that (1) satisfies requirements, (2) is feasible (can be manufactured using the firm's manufacturing facilities), and (3) is efficient (can be built at an acceptable cost both to the intended customer and the firm).

Application Engineering. The activity of customizing existing product designs to better satisfy the needs of specific customers.

Service Engineering. The process of developing systems and procedures to support field installation, maintenance, and repair.[32]

Manufacturing Technology. The second type encompasses the use of technology in the process of taking the design created by *product technology* and implementing it in the manufacturing planning and control system. Included as part of this technology are the activities of developing and acquiring necessary equipment and tooling, implementing and using manufacturing planning and scheduling systems (such as MRP), identifying appropriate materials (and

---

[30]Discussions of the dynamic nature of CIM are found in Chapter 4 as well as in the various case studies.

[31]Steele, *Managing Technology.*

[32]Steele, *Managing Technology*, p. 16.

their supply sources), developing standards (costs, bills of material, routings, labor profiles), specifying and implementing appropriate material handling systems, instituting quality control systems, and developing suitable maintenance systems.

A distinction that is often applied when discussing manufacturing technology is between *hard automation* and *soft automation.* This discussion is most pertinent when examining the problem of designing the production system. Hard automation describes the use of specialized machines and machine lines to manufacture and assemble items. Typically, hard automation involves dedicated equipment, in that each machine or line can only do one function or task or a very limited range of functions and tasks. Hard automation is most appropriate when dealing with mature products that are built in large quantities and that compete primarily on price. In contrast, with soft automation the goal is to develop a manufacturing system that is flexibile (i.e., where competition is not on price but on the response speed of the system). That is, with soft automation the goal is to build a system that can respond quickly to changes in product mix or product design or features. This flexibility is obtained by incorporating flexible, reprogrammable machines. These machines can be reprogrammed quickly to do a variety of tasks or functions.

Information Technology. The third type of technology describes the collective means used to assemble and electronically store, transmit, process, and retrieve words, numbers, images, and sounds, as well as the electronic means to control machines of all kinds, from everyday appliances to vast automated factories.[33] Included as part of information technology are such developments as *expert systems, decision support systems, database management systems*, and *corporate connectivity systems* (e.g., local area networks).

These three types of technology are found in almost all firms. They are closely associated with the process of designing, delivering, and maintaining an acceptable product in the marketplace. They also draw in varying degrees on both physical and information means.

Radical versus Local Technology

Technology is not only differentiated by its application (i.e., to the design of the product, to the production of the item, or to the management of information). It is differentiated by the magnitude of its impact on the firm. This impact can be described using two major dimensions. The first is the *prerequisite impact.* This dimension deals with the preconditions that the technology requires first be in place for it to operate effectively. The second dimension is the *operation*

---

[33]Marc S. Gerstein, *The Technology Connection* (Reading, Mass.: Addison-Wesley, 1987).

impact, which describes the extent of the effect of the technology on operations and practices elsewhere in the firm.

Both of these impacts can be envisioned as spectrums. Each of the two spectrums is anchored at one end by technologies that are completely *local* in impact. Such a technology is one that is essentially invisible to the rest of the firm. It makes no demands on the rest of the firm in terms of prerequisites; its impact, once implemented, is confined primarily to the area where it was introduced. An example of such a technology can be found in the introduction of a pick-and-place robot.

The introduction of such a robot is often a decision of substitution—replacing one method of production (a person managing a workstation) by another method of operation. This substitution makes no demands on the rest of the firm. It does not require a change in the manufacturing database; it may not require a change in inventory or purchasing practices. Furthermore, it has no real significant impact on the operation of the rest of the firm. The types of products manufactured are not altered; the scheduling practices of the manufacturing system are not changed; marketing has no real reason to be aware of the presence of this robot. As a result, when a technology is purely local, for all practical purposes the rest of the firm is unaware of the change or its presence.

At the other end of each spectrum are technologies that are *radical.* Such technologies in nature have system-wide effects. They make demands on the rest of the firm in terms of the level of employee training, the minimum skill levels for the employees, database accuracy, the structure of the database, and other such conditions that must be satisfied before the firm can attempt to implement the new technology. Their impact on operations may also be system-wide. They affect how orders are planned, scheduled, and executed. They can also affect how costs are tracked and standards determined. They may also affect how marketing sells to the marketplace (or what it sells). Finally, they may even affect strategy by generating new capabilities that have significant strategic impact.

Whereas a pick-and-place robot can be considered to be an example of a local technology, a flexible manufacturing system or a CAD/CAM system tends to lie toward the radical extreme of the two spectrums. In general, the more radical the technology, the greater the concern about the fit between the technology and the organizational environment of the firm. Furthermore, the more radical a technology, the more likely it is that the technology is tied to strategy (in an iterative fashion).

Finally, the task of justifying new technology becomes more difficult as the technology becomes more radical. Often when the technology being considered is local, the benefits are often quantitative and gained in the short term. Furthermore, the experiences of other firms with similar or identical operations can be used in identifying the costs and benefits. As a result, cost justification procedures such as discounted cash flows, payback, and internal rate of return are appropriate.

However, technology that is radical in nature is more difficult to justify for several reasons. First, many of the potential benefits tend to be more qualitative and therefore more difficult to assess (for a discussion pertaining to CAM systems, see Edosomwan[34]). Second, the exact nature of the impact varies from firm to firm. How radical technology affects the firm and its operation depends on such factors as the quality and condition of the database, the nature of operations within the firm, the effectiveness of the formal planning and control system, the nature of the organizational structures, and the degree to which integration is present. No two firms are identical when measured on all of these dimensions. As a result, the experiences of one firm with a specific radical technology may not be appropriate to another firm considering the same technology. Finally, the time horizon for realizing the benefits offered by the new technology tends to be longer. People must learn about the new technology; they must become familiar with and feel comfortable with its operations. They must learn to identify the new capabilities offered by the technology. These activities take time. The same justification procedures that work well for a local technology will not work well for a radical technology.

Technology: Concluding Comments

Technology can affect the firm and its manufacturing process in several ways. Its impact can be limited to certain areas, such as product design or manufacturing. Alternatively, it can make significant demands on the rest of the firm in terms of prerequisites, and its impact, once it is up and running, can be felt throughout the firm. As a result it is now possible to understand why technology is such a difficult concept to understand. Technology can take various forms ranging from something that is very limited and very local to an entity that affects all of the activities (product design, manufacturing, and information) in very radical ways. In short, technology is complex and dynamic—traits that it, in turn, bestows on the resulting CIM system.

A Process View of Computer Integrated Manufacturing

The process view of CIM deals with the stages through which most CIM implementations pass and their sequence. It identifies where in the implementation process management focuses on technology, strategy, and organization. It presents conditions that must be achieved before any further development can take

---

[34]Johnson Aimie Edosomwan, "Understanding Computer-Aided Manufacturing Impact on Total Productivity," *Computers and Industrial Engineering*, Vol 12 (1987), No. 4, pp. 283–290.

place. In short, it adds the elements of dynamics to the structural model presented in the previous stage.

The process framework of CIM breaks up the implementation process into four major stages (see Figure 3-6):

- Focus
- Integration
- Simplification
- Automation

Focus

Focus is the first stage through which all CIM implementations must pass. During the focus stage the implementation is driven by two major considerations. The first is to ensure that the necessary building blocks are in place. In general, these building blocks, which form the foundation of CIM, are (1) a manufacturing or corporate database; (2) a well-developed, formal corporate

FIGURE 3-6
CIM: A PROCESS FRAMEWORK

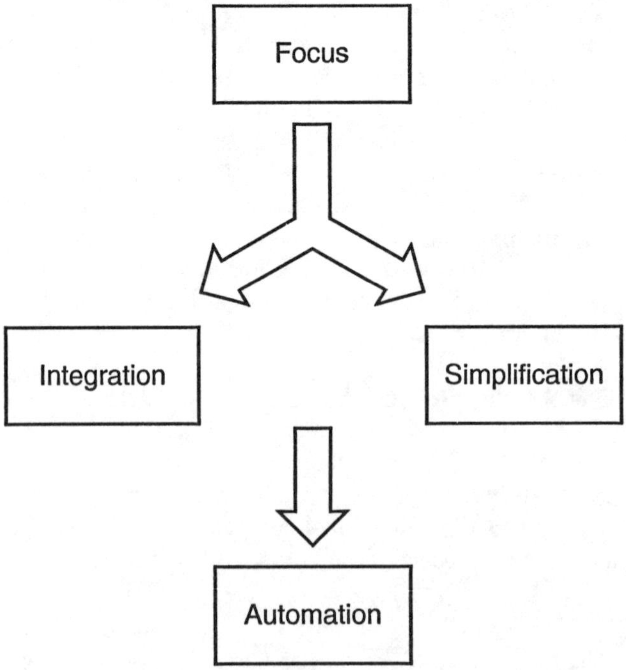

strategy; and, (3) a formal manufacturing and corporate planning and control system. Each block is critical to the success of CIM. The database provides the information which most formal systems need. The strategy provides direction. The planning and control system provides the procedure by which the firm manages its assets (specifically people and resources) to achieve a set of corporate objectives (often contained and communicated within such plans as the production plan, the sales plan, and the master production schedule). Not only must these blocks be in place, they must also be operating effectively and efficiently. This requires that management correct any problems and deficiencies in the current manual system.

The second consideration is that of strategic focus. The firm must know what its target market is and how the firm intends to serve this segment. The firm must understand how it intends to compete in the marketplace and how it intends to generate and maintain a sustainable competitive advantage.

Simplification

The next two stages of the process framework, simplification and integration, appropriately occur simultaneously and after the focus stage. Simplification describes the process by which the current systems and procedures are reexamined with the goal of identifying and eliminating

- Redundant steps
- Unnecessary steps (i.e., those that add activity without adding any corresponding value to the product).
- Steps or activities that could be combined.
- Steps or activities associated with inventories, delays, rework, and inspections.

Simplification attempts to reduce the complexity of the existing process. Simplification is applied primarily to the processes by which products are designed, production is planned and initiated, and information communicated. Eliminating unnecessary or extraneous steps results in a more efficient system that is also easier to automate and computerize.

Integration

Integration is one term that many researchers and managers freely use but rarely define. As a result a great deal of confusion surrounds this term. In this book our view of integration is relatively straightforward. We view integration as simply *the coordination and cooperation of systems, individuals, or organizational units within functional areas, between functional areas, and between firms.*

The concept of integration is complicated by two critical dimensions: (1) the application of integration and (2) the types of integration. The first dimension, the application of integration, describes how integration is implemented and supported. There are two levels to this first dimension. The first level deals with the structure of integration. This level refers to formal and informal devices and structures that management has introduced to encourage integration. Examples of these devices can be found in matrix organizations, cross-functional committees, and the presence of a corporate network for the communication of ideas and comments. These devices are present to provide a means of encouraging coordination and cooperation. They attempt to facilitate such cooperation. Their presence alone does not necessarily result in integration, however. As a result we must consider the second level, the practice of integration. This level is the acid test in that it refers to the practice or *tradition* of working together that is present. Considered within this level is not only the practice but also the attitudes of the people toward working together and joint problem-solving. It is in this dimension that we consider such factors as the degree to which people feel comfortable working together and the extent to which people from one area are willing to solicit information from other areas when faced by a problem or decision.

The second dimension of integration deals with the different types of integration that can be found in the firm. There are four basic types that we will consider and work with in this book:

Functional integration. This integration refers to the extent to which the functional area is integrated internally. Within the context of CIM, functional integration applies to the manufacturing area. It deals with the linkages and feedback loops that exist between planning, scheduling, and execution on the floor. In short, functional integration describes a manufacturing function that is internally integrated and in which all the major components are in "synch."

Organizational integration. This type of integration deals with links and coordination between the different functional areas of the firm (e.g., between marketing and manufacturing or between manufacturing and accounting).

Strategic integration. The third type of integration examines the extent to which the processes of formulating, implementing, and revising strategic objectives are coordinated with the activities of the functional area. With strategic integration, we allow for the presence of closed loops between the functional areas and the strategic formulation system. That is, the strategic system generates the objectives that drive the activities of the functional areas. However, the activities and problems of the functional areas can influence the objectives and their formulation.

Interorganizational (channel) integration. The fourth and final type of integration examines the linkages that exist external to the firm.

Specifically, this type of integration considers two different directions that this type of integration can take. The first is forward interorganizational integration. This deals with the coordination of activities between the firm and its customers. In contrast, backward interorganizational integration, the second direction, focuses on the integration between the firm and its suppliers. Both directions must be considered.

Like simplification, the major objective of integration is to improve the overall effectiveness and efficiency of the current manual system. Without integration, costs tend to increase. Why? There are several reasons. Without integration, redundancy of information and organizations cannot be minimized. Individual areas tend to duplicate activities taking place elsewhere in the firm because they cannot get timely access. Lead times are also longer, since decision-making tends to be sequential. Each area examines a problem or makes a decision without access to information or insights available from other areas. When they have arrived at a decision, the decision is then passed to the next area where the process is repeated. Often this process is expensive and time-consuming. These expenses are most evident in those situations in which a problem has been identified that requires action by a different department. The reason is that the entire process must be repeated.[35] Furthermore, decision-making is myopic in that people often are unaware of how their actions affect other areas in the firm. Integration attempts to overcome and eliminate these and other similar problems by ensuring that cooperation between areas is present.

Automation

Automation is the last stage of CIM. It occurs after the other stages have been completed. It is applied to those areas where focus, simplification, and integration have not eliminated all of the problems or sufficiently improved the overall effectiveness and efficiency of the corporate system. The position of automation relatively late in the process emphasizes a major trait shared by all CIM systems: success with CIM is greatly enhanced by automation; success, however, is not *solely* dependent on automation. As was pointed out by Ingersoll Engineers:

Technology is good, but it should come last.[36]

---

[35]This factor is one of the major reasons for the development and growth of simultaneous engineering systems.

[36]Ingersoll Engineers, "'The Factory of the Future' Is a Myth" (Rockford, Ill.: Ingersoll Engineers Incorporated, 1987, Publication No. IEI 713-687-5M-TR).

Comments on the Process Framework

The process model offers some interesting implications concerning the implementation of CIM. The focus of the first three stages is not technology. Rather, the first three stages are devoted to the task of preparing for the introduction of CIM. This emphasis on laying the groundwork for CIM is critical, for this is where most firms will do the greatest amount of work and spend the most amount of time. This factor also emphasizes the observation previously made that technology, by itself, is not enough. It cannot correct problems created by a manual system not working effectively and not linked with corporate strategy. The importance of this groundwork to the successful introduction and use of computer-based technology is a point that is repeatedly noted in this book. It is found in the various case studies as well as in the principles summarized in Chapter 5.

A Strategic View of Computer Integrated Manufacturing: Understanding the Four Basic CIM Types

As was noted previously, the prior two frameworks, while helping us understand the elements and structure common to CIM systems, are somewhat generic in nature. They are present in all CIM systems. However, all CIM systems are not the same. This trait is very evident both from a review of the CIM-related literature and from the case studies presented in this book. These differences in the type of CIM system cannot be explained by the two frameworks presented so far. To explain these differences, we must introduce a third CIM framework: a framework that is strategically based. Central to this CIM framework are two tenets. The first is that there are four distinct types of CIM systems. The second is that these differences are reflections of the strategic stance taken by the firm. The four CIM types exist because each is consistent with and supportive of a specific strategic stance taken by the firm.

The Four CIM Types

CIM systems can be found in a number of different forms. In some firms the CIM systems are narrowly focused, in that they are found primarily in the manufacturing or engineering areas. They are characterized by the extensive use of shop floor automation. In other firms the CIM systems are very broadly focused. They encompass not only the firm (and all of its operating areas) but also draw in (by means of electronic communications and other such devices) the suppliers and, occasionally, the major customers. Both extremes are CIM systems. Given corporate needs, each type of CIM system may be very appropriate. Yet they do not represent the same type of CIM system.

These differences reflect not only the extent of the application of technology but, more important, the extent of integration. It is integration that is the basis on which management tailors its CIM system to the market. Integration can be viewed as a response to problems involving product or process design and definition. That is, in general, the less clearly defined the product or the higher the risk in developing the product, the more likely is a firm to build into the organization all four types of integration (for a discussion of the four types of integration, see the section titled "A Process View of Computer Integrated Manufacturing").

How management combines the four classes of integration (functional, organizational, strategic, and interorganizational channel) determines the specific type of CIM system present. These four classes can be found in any of four distinct types of CIM system (see Table 3-3).

There are two observations about this framework that should be made. First, the framework is presented in a form that identifies all the combinations of integration that are possible. However, based on the experiences of the authors with CIM, there are certain combinations that are more likely to occur. For example, a Type III CIM system is most likely to consist of organizational, functional, and strategic integration. In other words, it is likely to be a firm that is tightly integrated *internally*. These "most likely" combinations are indicated in Table 3-3 as shaded blocks.

Second, all four CIM types have one specific class of integration in common: functional integration. This feature reflects a basic perspective of the study. It is

TABLE 3-3
The Four Types of CIM Systems: (The Framework)

CIM Type	*Integration*			
	Functional	*Organizational*	*Strategic*	*Channel*
Type I	x			
Type II	x	x		
	x		x	
	x			x
Type III	x	x	x	
	x	x		x
	x		x	x
Type IV	x	x	x	x

NOTE: x denotes the presence of the specific integration class.

the premise of this model as well as the entire study that all CIM systems, as a minimum, must be *functionally integrated.* That is, the process of designing, planning, scheduling, and building a product must be coordinated and linked (thus resulting in a closed-loop system). This view of integration treats both manufacturing and engineering as part of the same functional group. Any system that lacks functional integration is not regarded as being a CIM system.

Exploring the Framework

These four CIM types define a CIM spectrum—a spectrum that is anchored at one end by the Type I CIM system and at the other end by the Type IV CIM system. Within this spectrum the types are arranged in order of increasing complexity and sophistication (with Type I as the least complex):

TYPE I—TYPE II—TYPE III—TYPE IV

Each type of CIM system is appropriate for certain conditions. These differences can be best understood by briefly examining each CIM type.

Type I CIM. A Type I CIM system is narrowly focused, the primary interest being automating and improving the overall effectiveness of the manufacturing or engineering function. Such an approach to CIM is appropriate if the firm is faced by a mature market in which the products are standard and the basis for competition is both well established and widely known (both in the marketplace and within the firm). That is, there is little or no uncertainty regarding the product and its configuration. There are no real product differences between competitors. The criteria on which orders are won and lost are known. Typically, these criteria deal primarily with cost. As a result, success in this market comes to the firm that is able to do the best job in these areas.

These traits reduce the need for the various types of integration. Since the product is standard and the basis of competition well known, organizational integration is not critical. There is no real need, for example, for marketing and manufacturing to work together. Marketing acts primarily as an order taker. Similarly, there is no pressure on developing strategic integration. Top management knows what the firm has to do to win orders; manufacturing knows what it must do well. Forward channel integration is not required. Product definitions are already present. As a result, it is not necessary for the firm to work closely with its customers to derive such specifications. Similarly, backward channel integration is not needed.

In this environment the major focus is on making manufacturing more effective and efficient. It must be integrated; the manufacturing planning and control process must be automated wherever possible. As a result, a relatively focused and straightforward CIM system is adequate. Such a CIM system is found in Continental Can.

Type II CIM. A Type II CIM system tends to be one in which the focus of competition shifts from price and cost reduction to other dimensions such as availability of capacity or changes in system capabilities. Typically, in a Type II CIM system, the major product is still mature in nature. However, there is now a need for some cooperation and integration. For example, if the firm is involved in capacity selling, marketing must know what manufacturing's current capacity levels are and the exact type of capacity that is available. This latter aspect is critical as not all capacity is the same (a point repeatedly noted by Hill[51]). In this setting, marketing acts as an interface between the customer and manufacturing. As a result, organizational integration represents a useful and needed form of integration. The benefits of having this type of integration do outweigh the costs.

In general, it is the experience of the authors that the Type II CIM system tends to be transitory in nature. Most firms move either to a Type III or a Type I.

Type III CIM. Most Type III CIM systems tend to be ones in which the firm is internally tightly integrated. However, there is very little integration present between the firm and either its customers or its suppliers. In Type III CIM systems, the firm competes on such dimensions as quality, total cost (acquisition, operating, and maintenance costs), delivery speed, and flexibility. In these firms product mix is greater and subject to more change. In general, there are several circumstances in which a Type III CIM is the most appropriate type of CIM system.

The first circumstance involves those firms that face mass consumer markets. These firm must compete by continuously generating new and better products for their customers. The process of designing and bringing these products to market requires functional, organizational, and strategic integration. However, it is now difficult to justify channel integration. With so many different customers in the market, management is faced by a situation in which it is too expensive to build links with the various customers. As a result, the firm tries to maintain a general awareness of customer wants and needs. In this situation backward integration may also not be warranted because the firm may generate purchasing requirements that can readily be filled from existing sources.

The second instance in which a Type III CIM system is appropriate occurs when the firm primarily sells *capability* to its customers. A capability buy occurs when a buying organization meets a specific problem that requires facilities, skills, technology, and expertise that the firm currently lacks. Capability buys usually involve products and services that are not standard. There is no "off the shelf" solution. Instead, the buying organization must secure the services of a firm that is able to provide this needed expertise.

---

[51]Terry Hill, *Manufacturing Strategy: Text & Cases* (Homewood, Il: Irwin, 1989).

In a capability buy the buying organization can describe its needs in sufficient detail for the supplying organization to act. Once the needs of the solution have been set down, the buying organization then turns the problem over to the supplying firm. This firm is now responsible for developing a product or service that can satisfy these requirements. Under a capability buy, contact between the supplier and the buyer is usually limited. Frequently, these contacts are governed by a time frame and a set of milestones.

In a sense, with a capability buy the buying organization gives the problem to the supplier. The supplier, in turn, tosses back a solution when done. One of the best examples of capability selling can be found in the Frost case study found in this book.

Finally, a Type III CIM system may be appropriate where a smaller CIM system is working with a much larger system. The CIM system may be in the role of either a supplier or a buyer. Under ideal conditions, the CIM system would build channel integration with the larger system. However, channel integration often requires that the parties involved change practices or procedures to facilitate the process of integration. The smaller firm may be unable to make its larger partner accept the need for any change. The larger partner may not even see the need for any change or for integration. When this situation takes place, a Type III system develops in place of a Type IV system.

Type IV CIM. In contrast to the preceding three types, a Type IV system is the most comprehensive and complex of all CIM system types. Its complexity is a response to the demands placed on the firm. All four classes of integration are needed because *all* are needed in getting the product to the market. Such a system is appropriate when the firm is dealing with a product that is in the very early stages of its product life cycle (i.e., introduction). Organizational integration is needed, since marketing, for example, must know manufacturing's capabilities and systems since it is these capabilities and systems that it will be selling to prospective customers. Strategic integration is needed because top management and manufacturing must work closely together. As manufacturing develops new capabilities and strengths, top management must be able to incorporate these new abilities into future strategic plans. Channel integration is needed because it is only through the continuous flow of information and contact between the buying organization, the CIM firm, and its suppliers that the final form of the product can evolve. As a result, a Type IV CIM attempts to develop a close partnership with both its suppliers and its customers.

Furthermore, a goal of a Type IV CIM system is not simply to provide its customers with a product. Rather, it is to provide them with access to a process that can take an ill-formed or incompletely specified product concept and convert it into an actual product that is operational (i.e., it works and fulfills the function for which it was intended) and is feasible (i.e., it can be built at a reasonable and fair cost to both the customer and the firm). In other words, what the firm is selling with its CIM system is *process.*

Four CIM Types: Managerial Implications

The four CIM types form a CIM spectrum that is defined in terms of the level of complexity and integration present. This spectrum ranges from the simplest (Type I) to the most complex (Type IV). The spectrum does not describe a path of progress with Type I systems evolving into Type IV systems. Rather, it describes how certain CIM systems are consistent with certain strategic stances taken by the firm.

The four CIM types have certain important managerial implications for managers interested in building effective CIM systems. The following are the most important:

1. *At each CIM type, what we sell our customers changes.* How the firm competes differs for each CIM type. A Type I CIM system is appropriate for a mature market and for products that are commodity-like. The marketplace has defined what constitutes an acceptable product. It has identified the basis on which firms compete. The manufacturing task is often well defined. Under such conditions the task for the CIM firm is to manufacture the products demanded by the customer faster and cheaper. The resulting CIM must be focused to support these objectives. Because the basis of competition is well defined and the emphasis on cost so high, it does not make sense to require any type of integration other than functional. The decision to build the other types of integration also runs counter to the corporate objective of reducing overhead and flattening the organizational structure. In a mature market the emphasis is on cost reduction and the elimination of any unnecessary overhead. Adding any form of integration other than functional would add to overhead without adding any significant offsetting benefits. Why integrate manufacturing and marketing if the basis for competition and for winning orders is well known? Under such conditions marketing needs only act as an order taker. Once the order is received, it is passed on to manufacturing for scheduling. In a Type I CIM system, the firm is primarily a *product* seller.

In a Type II CIM system, the product being sold is often standard but characterized by a great deal of variability in terms of end item options. To cope with this variability, the CIM firm shifts from being a product to a *capacity* seller.

As was previously noted, a Type III firm is often a *capability* seller.

Finally, a Type IV CIM system is appropriate for products that are primarily engineered to order and for which the traits are not well defined in advance. Under such conditions the firm sells *process.* It provides its customers with access to a delivery system that is designed to reduce the risks associated with the design and delivery of such an ill-defined product. The delivery system is company-wide and embodies all of the activities needed to receive, process, design, revise, manufacture, test, and deliver the product. As a result, the task for a Type IV CIM system is to get the customer to buy into this delivery system. This is most often done through partnerships. Under partnerships or joint ventures the customer becomes aware of the strengths and abilities of the Type IV CIM system.

The differences between the different CIM system types and what they sell can be summarized as follows:

CIM Type	*What Is "Sold"*
Type I	Product
Type II	Capacity
Type III	Capability
Type IV	Process

2. *The interaction between the CIM system and corporate strategy changes from Type I to Type IV.* In a Type I CIM system, strategy drives the CIM effort. There is no feedback link. The strategic planning process is fairly rigid. The firm competes on price and lead time; the CIM system must provide a structure that enables the firm to achieve these objectives. Any new capabilities that may develop in the CIM system are irrelevant if they do not support the objectives.

At the other end of the spectrum, in a Type IV CIM system strategic planning must be flexible. Corporate strategy and the CIM system must operate in a *closed-loop* environment. Corporate strategy may begin by driving the CIM system. However, over time the CIM system develops new capabilities. The new capabilities may enable the firm to explore different markets. Once top management becomes aware of these new talents, they may decide to change the firm's strategic direction.

3. *Changing CIM types is an expensive undertaking.* At each type of CIM, management must invest in building certain types of integration. Building each type of integration requires providing a proper infrastructure. It also takes time, effort, and resources (money and people). Once the firm attains a certain type of CIM, management is now faced by a problem: changing the type of CIM (in either direction) takes time, effort, and resources. If the firm needs to add strategic integration, the structure for such integration must be built and the appropriate management practices (e.g., including manufacturing's input into the strategic planning process) encouraged. Attitudes must be changed. All of these activities make change expensive.

If the firm picks a CIM type with too much integration that is unnecessary (e.g., the firm has a Type III CIM when a Type I would be appropriate), there are also costs to be incurred. The infrastructures supporting the unneeded forms of integration (along with their associated overhead costs) have to be identified and eliminated. Management practices and attitudes have to be changed. Again, these changes require time, effort, and resources.

These considerations demand that management carefully consider which type of CIM is most appropriate. Picking the wrong CIM type subjects the firm to two penalties: the costs of implementing the wrong CIM type and the costs of moving to the right type. Under certain conditions the costs of building the wrong CIM system may consume all available resources in the short term and prevent the move to the right CIM type.

4. *The cost and difficulty of implementation increases nonlinearly as the firm moves from a Type I to a Type IV CIM system.* Generally, a Type I CIM system is the easiest to implement, since it confines the implementation process to the fewest areas in the firm (i.e., manufacturing and engineering). The other areas of the firm are not directly affected. However, in moving to each higher CIM type, more types of integration are introduced that now affect these other areas. Management must now overcome not only resistance in manufacturing to the changes demanded by CIM; it must also overcome resistance from the rest of the firm. It must convince areas such as marketing that the changes are both required and beneficial. Management may have to change how these other areas carry out activities such as order processing or master scheduling.

The resistance to change may be greatest when dealing with channel integration. Management must convince a supplier or a customer to change how they place or process orders. This may not be possible. Frost, of Walker, Michigan, is a small firm with a Type III system that is strongly integrated internally. Being small has limited its ability to influence its suppliers and customers. It lacks the volume to give it the necessary clout.

5. *The corporate culture changes from Type I to Type IV.* This relates to the previous observation. In a Type I system, given that the changes are confined to one area, the corporate culture is not affected greatly. However, in implementing Type III or Type IV CIM systems, all areas of the firm will be influenced by the presence of CIM. To have these changes accepted requires a drastic revision in the corporate culture and organizational structure. A hierarchical, functionally organized firm, which is feasible in a Type I system, is not appropriate for a Type III or Type IV system. Instead, what is required is a flexible project- or product-oriented structure that emphasizes the extensive use of networks.

Achieving the changes required by a Type III or a Type IV system requires a different change agent from that found in a Type I system. In a Type I system, it may be adequate for the CIM system to be championed by a vice-president of manufacturing. In a Type III or IV system, the change agent must have greater power to achieve the required transformation. Often, success occurs when the change agent is either the owner or president. This is the case for such firms as Frost, Ingersoll Milling, and Cone Drive.

6. *In moving from a Type I to a Type IV CIM system, the nature of investment (and justification) changes.* With a Type I CIM system, the firm tends to invest primarily in hard automation (technology, computer systems). As the firm considers the other types, it must recognize that it will be investing more and more in a mixture of hard automation (technology) and soft automation (organizational change, interfacing mechanisms). As a result, it is easier to justify investments in a Type I setting. The benefits are often confined to one area and are better defined. The interval between investment and payback is often relatively short. Procedures such as discounted cash flows, payback, and internal rates of return are appropriate. However, with the other CIM types the nature of the benefits changes. They occur across the firm. The time needed for the benefits to become apparent is longer (often because the users have to become familiar with the new system and its capabilities). The benefits are both quantitative (cost reductions, inventory reductions) and qualitative (improved flexibility, better designed products, risk reduction). It is now more difficult to justify these types of investment using the more traditional approaches alone. Justification must now be done using a mixture of traditional procedures and *gut feel* (i.e., are the investments necessary from a strategic perspective?). Firms such as Steelcase have recognized the need for a test other than the traditional payback. To Steelcase management, the first hurdle for any investment is a simple one: is it needed for Steelcase to develop a strong competitive position in the long run in its designated market?

7. *The nature of the relationship between the firm and its customers in a Type IV system differs from that found in a Type I system.* In a Type I CIM system, the relationship between the customer and the CIM firm is well defined. The customer identifies the product, the quantity, and the timing of delivery. The firm, in turn, acts as an order taker. It accepts the order and then fills it, subject to the conditions laid down by the customer. Such a well-defined relationship is appropriate because of the nature of the product (mature, commodity-like).

Such a relationship is not appropriate in Type IV CIM. Channel integration (most commonly found in this system) is present because the product to be built is not initially defined. Its ultimate form may be set but its features are often vague. The design of the product evolves over time as a result of the interaction of the customer, the CIM firm, and the CIM firm's suppliers. Each brings a specific form of expertise needed in designing and building a feasible product that is cost effective. As a result, a Type IV CIM emphasizes partnering and alliance building. Furthermore, these partnerships tend to be long term in nature. A customer that finds a good CIM firm with which to work is often reluctant to change because of all the reeducation of a new supplier that would have to take place.

8. *Desirable manufacturing objectives in a Type I CIM system differ from those found in a Type IV system.* The differences between the various types is clearest when looking at the desirable manufacturing objectives. In a Type I system, the major objectives are closely related to cost reduction and involve such areas as capacity utilization and the reduction of scrap, salvage, rework, and other forms of waste. In CIM systems located at the other end of the CIM spectrum, there is emphasis on such areas as speed to market, reduction in risk, control of system variability, and success in exploiting new technology.

9. *It takes longer to implement a Type IV CIM system than it takes to implement a Type I system.* The more complex the CIM type, the more resources that have to be mobilized and the greater the number of barriers that have to be overcome. All of these take time. For example, it took Frost over four years to lay down most of the foundations of its Type III CIM system. Continental Can, in contrast, required less than three years.

10. *The type of barriers to entry encountered in a Type IV system are often hidden, whereas those found in a Type I system are very obvious.* In a Type I CIM system the barriers to entry are often associated with capital investment and technology. As we proceed toward a Type IV system, these barriers change. The firm cannot depend on technology alone to match or surpass the competitive advantage possessed by a Type IV system. New barriers are now introduced in the form of a well-developed and integrated delivery system that is well matched on an ongoing basis to its customers' needs. The CIM system has changed from being *product* based to being *process* based. The competitor must convince the customer that it has a better delivery system. This is not an easy task, for it requires that the customer leave something that he or she knows and trusts for something that is very much uncertain.

CONCLUDING COMMENTS

Computer integrated manufacturing is a complex concept. It is used frequently by writers and managers, often to denote very different concepts or systems. As a result, the study of CIM has often been surrounded by confusion. However, such confusion cannot be tolerated if we are to truly understand what CIM is and what it is not. Developing that understanding has been the primary objective of this chapter.

In this chapter we began by noting that CIM is a double-edged sword. It has the potential to change significantly any firm and to make it a true world-class competitor. However, this same CIM system can also become an expensive and debilitating failure.

We have tried to understand why managers are often reluctant to even consider CIM. We have tried to show the diversity of views regarding what CIM is.

To avoid the confusion that surrounds CIM, we have presented this book's framework for CIM.

The framework presented in this book is unusual in that it is not really one framework but three separate but related frameworks. Why three frameworks? The answer is relatively simple: one framework is not enough for such a complex concept. The structural framework identifies those elements that are common to all CIM systems. The process framework examines the sequence of events that CIM systems tend to follow. Finally, the strategic framework attempts to explain the diversity in observed CIM systems by relating how each of four different CIM types are consistent with certain strategic stances taken by the firm. Together, these three frameworks give a fuller and more accurate account of what CIM is and what it is not.

Having now established how the concept of computer integrated manufacturing is used in this book, we can proceed with the task of understanding what makes CIM work and what makes CIM fail. The rest of this book is devoted to that task.

CHAPTER 4

THE PRINCIPLES OF AN EFFECTIVE CIM SYSTEM

OVERVIEW

One of the major obstacles facing any person trying to draw a set of generalizable principles and lessons from a study of so-called leading edge CIM users is that of trying to separate the variability present from the underlying lessons that are common to the systems studied. This variability is a result of two major factors. The first is inherent in the very nature of the concept of computer integrated manufacturing. As was pointed out in the preceding chapter, CIM is intrinsically plastic in nature. The exact shape and nature of the CIM system is influenced by such factors as the strategic direction being followed by the firm and the requirements of both the customer being served and the product being delivered.

Given this inherent level of variability, it might be reasonable for the casual reader to question whether it is possible to coax any general set of principles from a study of firms considered to be leaders in the theory and practice of CIM. However, it is possible. An extensive study of the experiences of the participating firms has shown that there are certain lessons or principles underlying the effective implementation, development, and management of CIM that are common to most, if not all, of the participating firms.

The principles presented in this chapter provide a checklist to be used not only when reviewing the experiences of the participating firms but also when assessing any CIM system, either actual or potential.

These principles are not exhaustive. Rather, they represent one attempt to understand and identify those factors that affect the development and use of the CIM concept in firms. It should always be remembered that CIM is inherently dynamic in nature. As a result, as new technology becomes available and is used and as firms learn to take advantage of these new developments, the CIM system will change. It will continue to evolve. As it evolves, the principles and practices associated with effective CIM systems will also change.

The principles are divided into four major categories: (1) general guiding principles, (2) prerequisite principles, (3) operating principles, and (4) system principles. The principles are presented in two parts where possible. In the first part, each is discussed and its importance to the development of an effective

CIM system is examined. In the second part, the principle is illustrated by examples taken from the experiences of one or more of the participating firms. It is not the intent of the authors to examine the experiences of every participating firm that illustrate the principle under discussion. Instead, the primary purpose of introducing these examples is simply to demonstrate the principle in action.

GENERAL GUIDING PRINCIPLES

General guiding principles are those that are relevant to anyone interested or involved in the design, implementation, development, or maintenance or enhancement of a computer integrated manufacturing system. This category of principles applies to all manufacturing systems, regardless of the manufacturing process and its characteristics (i.e., job shop as compared with repetitive or continuous-flow processes), the marketing environment (engineer-to-order [ETO], make-to-order [MTO], assemble-to-order [ATO], make-to-stock [MTS]), the size of the firm (small to large) or the nature of ownership (privately versus publicly held). These principles are basic to all effective CIM systems.

One reason for the importance of these principles is that they are central to the development of a realistic understanding of the CIM concept and realistic expectations of CIM.

1. Computer-based technology should be regarded as a magnifying glass.

The computer-based technology that is an integral element of the CIM concept, when implemented, is never neutral. That is, it will affect the operation of the firm in one or more ways. In some cases it substitutes one type of process (such as a pick-and-place robot) for another (such as a person). In other cases it complements an existing process. CAD allows designers to quickly uncover and make use of the underlying geometry of a given design. It can also enhance an existing process. Automated identification technologies such as bar coding can improve the accuracy with which information is gathered on the shop floor by eliminating operator input error.

Computer-based technologies can influence the way in which physical flows are managed. That is, they can affect the manner in which a product or resources (e.g., tooling, material, labor) is moved, stored, or processed. They also affect the information flows.

Often, the implementation of the technologies underlying CIM results in the replacement of the current manual system by an automated system. This change, when it takes place, uncovers flaws hidden in the current system and exposes the firm to the problems created by these flaws. As a result, these technologies tend to have a magnifying effect. They tend to make an ineffective or marginal system even worse. The reason lies in the nature of the manual system being replaced.

In many firms ineffective or marginal manual systems are often supported by an active informal system and by various buffers. The informal system is a result of the actions of users who are aware of the shortcomings of the current system and who take the appropriate corrective actions. For example, an operator knows that the bill of material for one of the items that he is building is incomplete. Certain critical parts are missing. When pulling the items for the order, he automatically gets the necessary missing components.

Similarly, the manual system is often supported by buffers in the form of excess capacity, high inventory stocks, high scrap allowances, and overstated lead times. These buffers protect the manufacturing system.

When a computer-based technology is introduced, these supporting elements are taken away. As a result, these technologies expose the underlying problems and force management to address them. For example, incomplete bills will now result in missed completion schedules and expediting.

People (both management and shop floor personnel) will often tend to blame the technology for the problems. They do not always recognize that the problem is with the underlying planning and control system rather than with the technology. As the blame is placed on CIM, the credibility of the system often suffers.

In general, this principle can be stated as follows:

CIM acts as a magnifying glass. It makes the good system much better; it makes the poor system much worse.

2. CIM and its associated computer-based technology should be used to enhance the capability of people in the system rather than to replace the human.

The introduction of technology seldom reduces the total number of people required by the manufacturing planning and control system. It often changes the mix of employees. While the number of direct employees is reduced, the number of support people is increased. This increase in support personnel is needed to satisfy the increased requirements for preventive maintenance, machine programming, and other similar activities. When there is a reduction in the number of people required, it is often a result of such external factors as increased competition or market changes.

Instead, CIM and computer-based technology should be used to enhance the capabilities of the user. To achieve this objective, it is important to recognize that humans are well suited for certain tasks and poorly suited for others. People are best used as problem solvers. That is, once a problem has been identified, people tend to be good at identifying the best solution and at implementing this solution quickly and effectively. However, people do not do well when involved in repetitive, extremely exacting, monotonous work (e.g., bookkeeping tasks). Shop floor operators do not do well in recording data. Frequently, the result is varying precision and numerous data-entry errors.

Computer-based technology is most effective when it is used to free the person from such repetitive tasks. It should be used for activities such as data recording; monitoring, tracking, and locating resources such as tooling and material; and transferring information (e.g., from a central database to the shop floor). By taking over these tasks, the computer frees the user to concentrate on the task and to work on improving operations. It also simplifies the operation of the system by taking over responsibility for those various time-consuming and menial tasks.

There is a final reason that the computer should be used to enhance the capabilities of the human in the system. In general, in most manufacturing systems labor costs make up a small (and increasingly shrinking) percentage of total costs. Often, labor costs amount to no more than 10 percent. Even if all labor costs could be eliminated, the impact on total costs would not be that great. By enhancing human capabilities, the person is now freed to attack the other 90 percent of total costs.

This principle is reflected in several of the firms' experiences with CIM. In one firm, engineering was computerized because engineers had become the constraining resource (the bottleneck). CAD was an attempt to make them more productive. In another firm, the data entry and verification process was computerized because the existing manual system was too error prone and time-consuming for the operators. It was also a task universally detested by the shop floor personnel. Computerization left the operators with more time for problem-solving. Decision-making was enhanced by computerization, since the computer system now flagged problems, identified potential alternatives, and gave users a *simulation* capability for evaluating potential systems.

Examples of this principle in operation were most evident in the experiences of Cone Drive, Steelcase, Ingersoll Milling, and Allen-Bradley.

3. CIM systems should be seen as people systems.

People play an important role in the CIM environment. Ultimately, it is the people who determine the success or failure of the CIM system. People are responsible for implementing CIM. They are also responsible for its use, maintenance, and any subsequent enhancements made to it. People, especially the operators on the shop floor, can make CIM more effective by suggesting or adding new features or by using the system in ways not initially envisioned by the planners. As a result, when implementing CIM, the challenge facing management is that of having the people develop a sense of ownership over the CIM system. They should view the CIM system as their system, not as something that management has burdened them with. Developing ownership requires that management devote enough resources toward adequate training and education. It also requires that management be aware of the full costs of CIM and be prepared to share this information with its work force.[1] Trust must precede the move to CIM.

4. Recognize that there* will *be resistance to the implementation of CIM and be prepared to cope with the four forms of 'lags' as soon as they appear.

CIM often represents a dramatic change not only in the way the firm does business but also in the way its people carry out their activities and functions organize themselves. There is resistance to such change. This resistance occurs for one of two reasons. The first is the fear of change. Change represents uncertainty. People are not sure how the new system will affect their tasks. So some may even see the change demanded by CIM as being a danger to them.

The second is the presence of *lags*. That is, the system, in its current form, may not be ready to keep pace with the change being introduced. The participants focused in on this explanation and identified four major lags: people lags, organizational lags, information lags, and strategic lags.

People lags exist because people (often shop floor) are not ready for the new system. They understand their tasks well and typically see themselves as being good at what they do. They realize that the current formal system is inadequate. They know exactly where the deficiencies of the current system are most pressing. The informal system that they have built up over time corrects these deficiencies and lets them build the product. They do not understand the need for a change. They resist CIM because it is seen as making things more difficult. The term people apply to CIM is that of "more red tape." Management must rely on education and training to overcome this form of lag. People must know why the old ways of doing things are no longer acceptable. People must understand why they must change. People must see CIM as making life easier, not harder.

Organizational lags exist because the current structures are not adequate for the requirements of CIM. The linkage between critical functional groups such as marketing and production is often missing or reactive (i.e., present in response to problems). The organization is weak on integration. It often emphasizes functional perspectives at the expense of a company-wide perspective. Inmany instances the existing organizational structures lack the flexibility to accommodate the new requirements imposed by CIM.

---

[1]This aspect of CIM will be discussed in greater detail later on in this chapter.

To overcome organizational lags, management must be prepared to work on building in the missing element of integration. It must also be ready to change existing organizational structures if and when needed. Functional perspectives must be replaced by a product orientation. Organizations must be built to support the delivery of the product, rather than the provision of one activity such as engineering. Simultaneous engineering is an example of such a new organizational structure. Management must recognize that CIM may require the introduction of fluid rather than rigid organizational structures.

Informational lags exist because the existing information system cannot support the requirements of the new CIM system. Data demanded by CIM may either not be present in a form that is readily accessible or it may be missing entirely. For example, CIM may require information structured in terms of part families. The existing database may be structured around individual parts. It treats each part as if it were unique. Often, informational lags indicate underlying problems with the structure and content of the existing information database. For example, the existing information database may be inadequate because of such problems as poor accuracy, incompleteness, long reporting and recording delays (resulting in data that is not current), and excessive duplication (the same item is reported in several different functional databases).

To overcome informational lags, management must be prepared to invest in an extensive restructuring of its existing corporate database. It must also be prepared to recognize that information management procedures that were adequate twenty years ago may not be adequate today.

Finally, strategic lags exist because top management may be unable to accurately assess the impact of the new technologies embodied in CIM on the firm's competitive capabilities. It does not know how to best take advantage of the new technology. Often, after the technology is implemented and management has gained experience with it, management then formulates an "appropriate" strategy. The problem with this approach is that it is reactive. It creates a situation in which technology drives strategy, rather than the other way around.

Overcoming strategic lags requires that top management invest the time and effort to understand the new technologies in terms of their general functions and implications for the firm. They do not have to understand the specific technical details of the new technologies. Instead, they must have a broad understanding of the technologies. They must continually ask the question, What functions do these new technologies provide? The question of how these technologies work can be left for subordinates. Top management must also be able to assess the risks, benefits, and costs that underlie these new technologies. This is a critical requirement, since the decision to fund the implementation of these technologies and to support their use will ultimately come from top management.

PREREQUISITE PRINCIPLES

Prerequisite principles are simply those that identify requirements that must first be satisfied by the firm (either at the manufacturing or corporate level) before management can even begin to consider the implementation of CIM. These principles describe the foundation on which any effective or successful CIM system is built. Ignoring one or more of these principles can hinder the CIM implementation effort or adversely affect the probability of success.

Of the four categories of principles, this one contained the majority of principles. This observation emphasizes one basic conclusion:

All of the technology in the world cannot compensate for a system that is ill prepared for the changes required by CIM.

5. A good understanding of the current system and its strengths and weaknesses is a required starting point for the move to CIM.

The implementation of CIM is an expensive proposition. It places significant demands on *all* of the resources of the firm (i.e., time, money, and manpower). It requires investment in hardware, software, training, debugging, and implementation. It also requires significant changes in management practice, corporate organizational structures, information databases, and worker skills. Because of these factors, there is often a considerable amount of resistance to it. Overcoming this resistance requires that the firm, as a whole, see a real need to change. As a result, the move to CIM frequently does not occur in a vacuum. It is often triggered by the need to improve the operation of the current system.

In some cases, as with Frost, Continental Can, and Cone Drive, CIM was a corporate response to a need to correct a deteriorating manufacturing position. The management in these companies recognized that unless something was done and done quickly, the long-term success of their firms was in jeopardy. Furthermore, radical changes were perceived as being necessary if management was to effectively correct these problems.

In other cases CIM was a corporate response to a financial problem. For DEC, CIM was selected as being the most appropriate method of improving the financial returns generated within the firm.

In the case of Steelcase, CIM was not a method of remedying poor manufacturing performance or improving financial returns. Steelcase was and still is the market leader in the office systems market. Instead, management turned to CIM as a method of maintaining Steelcase's dominant position in the marketplace now and, more important, into the future. It was to be the vehicle by which the firm was better able to serve its customers. It was also to be the integrating mechanism by which Steelcase's suppliers were to be linked to Steelcase and Steelcase, in turn, was to be linked to its customers.

Deciding to implement CIM is an important decision. However, it is not enough by itself. Rather, it is the first step in the CIM implementation process. Once management and employees recognize the need to change current operations, they must decide on the exact form and structure that CIM will take and what technology (or technologies) to use.[2]

Determining the most appropriate form of CIM for the firm requires that management first undertake a close and detailed evaluation of the current system. This analysis should identify the strengths and weaknesses of the current system. That is, the analysis should explicitly identify what the system can do well and, more important, where the current system is deficient. This self-assessment should include such areas as

- The manufacturing planning and control system.[3]
- The corporate planning and control system.
- The corporate database.
- The manufacturing database.
- The governing policies and procedures.

One of the important outputs of this self-assessment should be a determination of whether the firm is ready to even consider the change to CIM. There are instances in which the firm cannot consider implementing CIM because the current system is neither complete (i.e., contains all of the above elements) nor effective (i.e., is able to generate feasible plans that can be carried out on the shop floor). These deficiencies must first be corrected. Two of the participating firms provided good illustrations of this principle in action. These were Continental Can and Cone Drive.

Continental Can had made the decision to implement CIM in its Chicago facilities. The necessary funds had been set aside; management had made the required commitment to being a CIM system. However, when carrying out this self-assessment (at the request of the company's consultants), management arrived at an important conclusion. The current system was inadequate. There was no effective master scheduling; capacity planning was nonexistent; the corporate database was in poor condition. It was then decided that implementing CIM given the status of the current system would simply result

---

[2]This aspect of CIM is discussed in greater detail in Chapter 3. The four CIM types are presented in that chapter.

[3]This aspect of the self-assessment is discussed in greater detail later on in this chapter.

in a situation in which the same mistakes would be repeated. The only difference would be that these mistakes would now be made in "nanoseconds" rather than in days and hours.

As a result of this review, the building and implementation of an effective planning and control system became a top priority for management. It preceded any investment in CIM technology. The experiences of Cone Drive were somewhat different.

The management at Cone Drive used the period preceding the implementation to review their existing business and to identify those changes needed to effectively support their business. As a result of this review, changes took place. Many of the management practices that had long been accepted as normal were challenged and altered because they were not consistent with how the firm *should* be doing business. In general, management found this self-evaluation useful, for it gave them the opportunity to question practices that had long been considered as simply "how we do business here." All practices were examined. These included not simply those that took place on the shop floor but also practices in engineering and in the office.

6. Sound CIM systems are built on a foundation of an effective business and strategic planning process.

There are many different computer-based developments on which management can draw when developing its CIM system. Some of the most significant include such technologies as robotics, machine vision, group technology, flexible manufacturing systems, automated identification (e.g., bar coding, optical character recognition), computer-aided design (CAD), and computer-aided process planning (CAPP). Each technology offers the user a different mixture of capabilities and limitations. How these various technologies are combined can greatly influence (both positively and negatively) the manufacturing capabilities of the firm. Ultimately, the mixture of technology will affect the firm's competitive stance. The problem facing management is that of what capabilities (and, hence, what technologies) are most appropriate given the needs and requirements of the firm. The management in each of the participating firms was guided in this task by the presence of a corporate-wide strategic plan. More importantly, however, these plans were the result of a formal, on-going strategic planning process.

Given a set of well-defined and clearly understood strategic objectives, the management in each of the participating firms translated these objectives into those manufacturing traits (e.g., flexibility in terms of being able to accommodate various combinations of options) needed to support these goals. Management then went looking for the technology (or technologies) that were closest in offering these capabilities.

The key to success in this process of matching technology with strategic needs is a formal business and strategic planning process. The participating firms recognized the strong need for such a process. Each noted that such a process *must* precede the move to CIM. Where none existed, the necessary process was first put in place. Various reasons were cited for putting so much importance on such a process:

- The process drives the development of the CIM.
- The strategic process identifies the corporate objectives and defines the desired manufacturing capabilities needed to support them.
- The strategic process, as well as fostering an understanding of the reasons for these traits, ensures that there is corporate-wide agreement over what capabilities are important.
- The strategic process helps managers to assign relative rankings to the various traits. As a result, it helps to minimize the possibility that the firm might choose technology that is not appropriate or inadequate. It also helps minimize the possibility that technology is bought for the sake of buying the latest or best technology (without any consideration for how it might be used within the firm). This trait has been suitably described as "technolust."
- Such a process avoids the problems created by situations in which technology is first bought and implemented and then management struggles with the task of how to best take advantage of the resulting new (and often conflicting) capabilities.

To be effective, the participants agreed that the business and strategic planning process had to embody the following characteristics:

- It has to be formal.
- It has to involve *all* of the groups (line and support) needed to carry out the resulting corporate objectives. This means that groups such as engineering, purchasing, manufacturing, and quality assurance—groups excluded from the strategic planning process in many companies—have to be included and that they must play an *active* part in the process of identifying, setting, and refining corporate objectives.
- The process has to be ongoing. That is, all of the strategic plans and objectives must be reviewed on a regular basis to determine whether they are still valid and feasible in light of changing market conditions and different competitive responses. At a minimum, all plans have to be reviewed on an annual basis. Some of the companies, such as Steelcase, review their plans on a quarterly basis.
- The process has to ensure that the objectives identified are realistic, clearly

stated in meaningful terms (dollars for finance, units sold for marketing, units produced or standard hours required for production), and clearly communicated to the entire firm.

• The process has to ensure that the objectives identified in the plans are attainable given current resource limitations.

• The process should seek to build for the firm a distinctive competitive edge. Ultimately, such competitive advantages result in market advantage and corporate survival.

• The process should ensure that the plans are integrated. That is, the plans should be examined in terms of their impact on manufacturing, marketing, accounting, finance, engineering, and the other functional groups. Potential problems or conflicts should be identified and resolved in advance instead of after the plans have been implemented.

• The process should ensure that the objectives identified subsequently become inputs into the formal manufacturing planning and control (MPC) process. The objectives laid out in the business plan should be reflected in the resulting production plans, master production schedules, capacity plans, and order priorities.

In discussing the strategic and corporate planning process, most of the participants identified three factors as being at the heart of such planning. These were *quality, cost*, and *lead times*. Every firm, it was pointed out, competed on these three elements. Changing the mixture of these three elements resulted in different types of competitive advantage. As a result, the strategic planning process could be reduced to a process by which the management of a company decided how to mix and blend these three elements to generate unique and marketable advantages for the firm.

The participants strongly agreed that unless a firm knew its strategic direction and posture, the movement to CIM could be very expensive, time-consuming, and frustrating. Without the guidance of a strategic plan, the ultimate fate of the CIM system was a simple one: failure. As one of the speakers put it:

CIM is not a substitution for sound business and strategic planning.

This principle was best illustrated by the experiences of Steelcase. Steelcase had spent a great deal of time, effort, and money in building a sophisticated CAD/CAM system for designing certain parts. This project was ultimately abandoned when management realized that the new capabilities generated by this system were ones that could not be used directly by the firm. They were not traits that were likely to be appreciated by the customer; their impact on improved customer service was at best minimal; the system was not easy to use. As a result, the project was scrapped. It did not fit the strategic goals of Steelcase.

7. When looking for solutions to strategic problems, don't look for CIM solutions alone.

The first step in the CIM journey is to identify the problems to solve and also to establish those manufacturing objectives that are in concert with the corporate-wide strategic objective. The next critical step is to determine what choices would offer the most appropriate means of achieving these objectives. There are many potential paths available to the firm. These include just-in-time manufacturing, total quality control, world-class manufacturing, design for manufacturability and computer integrated manufacturing. Each brings with it different strengths and weaknesses. Each can create different manufacturing systems. CIM is just one possible path. There is no assurance that it is the best.

The participants agreed that management should never limit its options from the outset by settling on CIM. Such an approach may put blinders on the firm and may leave them ignorant of the benefits offered by other approaches. Rather, the firm should explore alternative approaches. CIM should be selected because it is the path that *best* meets the manufacturing and strategic needs of the firm.

In five of the participating firms a wide-ranging search for alternatives was undertaken. The purpose of this search was to uncover the most appropriate choice(s). In no instance did management begin this search by focusing on CIM solutions (e.g., group technology, flexible manufacturing systems) to the exclusion of other procedures. Ultimately, CIM was selected because it was the most appropriate solution.

8. Never buy more technology than is needed to attain the firm's manufacturing and strategic objectives.

One major temptation frequently encountered when acquiring new technology such as equipment or software is that of buying more than is really needed. Often, this investment takes the form of the purchase of *state of the art* technology (i.e., technology that represents the latest or most advanced available on the market) or technology that offers more features or capabilities than the user currently needs. In many cases, these purchases make sense, since the marginal cost may not be that great. State of the art technology or additional features can often be obtained for a *few dollars more*. In addition, some managers are attracted to the acquisition of advanced technology because the presence of such equipment seems to imply that the user is both highly advanced and highly sophisticated. Finally, other managers are drawn to such highly advanced equipment primarily because of the technical features alone (technolust).

According to the participants, this temptation should be resisted. A manager from one of the participating firms described this "leading edge" technology as "bleeding edge" (where the bleeding is done by the user). In many cases the advantages offered by such technology are little more than false economy; in other cases such technology may increase overall costs, not decrease them.

First, such technology is often far more expensive. These expenses come from a variety of different sources. The initial acquisition cost is higher. Often, since the equipment is not as easy to use, training time may be longer and costs higher. Maintenance tends to be more difficult and the resulting maintenance-related costs higher.

Second, because the investments in state of the art equipment are larger, they expose the firm to higher risks. Management must recognize the possibility that the firm may acquire technology that is *inappropriate* (i.e., does not fit with the needs of the company). Initially, the technology may have appeared to be appropriate. However, this view may change after the technology has been implemented and management has gained experience with its use and learned about its capabilities and limitations. It may then learn that the technology is not right. If such investments are large, the ability of the firm to invest subsequently in more appropriate technology may be severely constrained. The resources needed for the new investments may be tied up in this "wrong" technology. As a result, the firm may be forced to live with the consequences for several years.

Third, state of the art equipment or software can expose the firm to a far higher level of potential risk. It may not be thoroughly debugged and tested. Breakdowns may occur. The firm might find itself in the position of fighting two fires at once: that of trying to meet the needs of its customer by shipping product out the door and of trying to repair the breakdowns. Such a position puts the firm at a competitive disadvantage (especially if it is trying to offer its customers lower risk).

Fourth, and closely associated with the preceding problem, is the lack of a history. Standard technology often comes with a well-recorded history. A manager interested in doing *predictive maintenance* (i.e., identifying how long a machine will operate before developing problems and then intervening in the operation of the machine through maintenance activities before this point is reached) can draw on this information in setting up a maintenance scheme. However, with "leading edge" technology, such information might not be available. As a result, the firm must be prepared to gather such data. Until such data are gathered, breakdown maintenance rather than predictive maintenance must be accepted as the standard way of dealing with potential equipment problems.

Fifth, buying more capabilities than are needed to meet the firm's needs may cause manufacturing personnel to "invent" new problems. That is, people may try to find problems to which the new but not yet used capabilities can be applied. As a result, managerial time and effort are wasted in trying to discover such problems.

Finally, when there is a mix of equipment (i.e., some state of the art, the rest not), there is a tendency noted by the participants for this mix to change over time in favor of the more expensive equipment. Management must take explicit steps to curb this tendency. This trait of new technology can be better understood by examining the experiences of Cone Drive.

The management at Cone Drive use monochrome IBM PCs as the basic vehicle for data tracking, recording, and transfer. They decided to experiment with a new color display offered by IBM. An appropriate work center was selected and the monitor placed there. All the other work centers still used their monochrome monitors. Soon, management noticed that every one of the workers wanted to have the new color display. The operators claimed that they could not do their work adequately with the old monochrome displays. The managers who reviewed these requests could not identify what additional features offered by the new displays could help the operators do their work any better. As a result, management decided on a simple, cost-effective, and direct approach: they removed the color monitor. The demand for new monitors suddenly evaporated.

Underlying this discussion of the potential risks present in advanced or state of the art technology is a simple yet major concern. The technology, be it soft or hard, must ultimately be *appropriate*. Appropriate technology is that technology that best fits with or satisfies management's requirements. A company that offers a product line that has a limited range of options and is built in small batches may find that a flexible manufacturing system may offer it capabilities that it can never really use. Changing the product to use this newfound flexibility may inadvertently change not only the products offered but also how the firm competes and the companies against which it competes. When this occurs, it can be said that the technology is driving corporate strategy, rather than strategy driving the technology or automation selection.

The identification of appropriate technology begins with the firm identifying needed functions and capabilities. The definition of these functions and capabilities may be based on strategic objectives (e.g., reducing design lead time or improving product option flexibility while maintaining or reducing product cost). This definition can also be based on specific operational problems (e.g., managing a task that exposes personnel to such adverse operating conditions as excessive heat or toxicity). Once this list of requirements has been drawn up, the next step is to evaluate the competing technologies. The most appropriate technology is that procedure or system that best meets the requirements previously set down.

The process of evaluating technology should be done by a committee that involves all of the groups involved in the process of acquiring and using the new technology. This group should not only include users (e.g., shop floor personnel, engineers) but also groups such as accounting, finance, and data processing. In addition, it should include any person or group charged with integrating the new technology into the existing corporate framework.

By defining the needed capabilities and functions first, management can now ensure that whatever technology is selected is truly appropriate. It is also in a better position to avoid the problems created by technology acquisition based on technolust.

9. Obsolescence in a CIM system is defined by corporate and strategic considerations.

One important and persistent problem facing any manager involved in the implementation of new technology is that of determining the effective life of the technology. That is, the manager must identify when it is necessary to replace one form of technology by another. The first inclination is to base the determination of this life on technological considerations. The existing technology is considered adequate until new technology arrives. At that point the existing technology becomes obsolete. To the participants, this practice is identified as the *technological obsolescence*. However, this form of obsolescence is not the same as *functional obsolescence*.

Functional obsolescence views a technology as viable as long as it best fulfills a given function. A technology becomes obsolete once it can no longer carry out the required functions at the lowest costs. The functions that the technology is required to carry out are defined not by technical considerations but by the needs of the firm. In practice, this principle means that the life of new equipment in some cases will be relatively short, whereas in others, it will be relatively long. This principle is best illustrated by comparing the experiences of two of the participants, Cone Drive and Ingersoll Milling.

At Cone Drive ordinary 8088-based IBM personal computers are used on the shop floor even though there are now better and faster industrial-quality computers available. The reason that these computers are still used is that they still offer a cost-effective means of fulfilling a definite need. They are used for storing, displaying, and entering data. To management, the marginal benefits to be gained from the introduction of newer industrial-quality computers are not enough to justify their use. From a corporate and strategic perspective, they offer little real comparative advantage.

At Ingersoll Milling the maximum effective life of any piece of equipment is five years. Unless the manager can justify their continued use, all equipment that is five years old is automatically replaced by the newest technology. The reason for this replacement policy is strategic. Ingersoll Milling sells its customer its experience with the latest technology. The replacement policy is consistent with this strategic stance.

10. Simplify the current system before *introducing any form of automation.*

Having an integrated formal manufacturing planning and control system, while necessary, is not a sufficient condition for the successful implementation of CIM. It is necessary that the current system be critically examined so that management can identify any and all opportunities for system simplification. Included in such simplification are the identification and elimination of unnecessary reports, the reduction of multiple-page reports to single-page format, and the flagging and elimination of any excessive or redundant steps at either the planning or shop floor stages.

There are several reasons for emphasizing system simplification. First, by simplifying the current system in advance of the implementation of CIM, the result is a simpler and easier-to-use manual system. Such systems are easier to computerize as compared with larger and very complex systems. The costs of computerizing simple systems are significantly lower than those of computerizing very complex systems. One of the participants observed that the costs of computerizing systems increases nonlinearly (i.e., exponentially) in response to complexity. It is also easier to maintain and manage a simple system. Training costs are far lower for a simple system. It is easier to train people to use such systems (often, in part, because such systems are more transparent to the users and subsequently easier to understand and apply). In addition, the total lead time required to implement a simple system is far less than that required by a complex system.

Another reason for requiring system simplification is that the period preceding the implementation of CIM gives the firm an important opportunity to examine and understand the entire process that has been developed to deliver a product (or products). Often, there are elements of this process that at one point in time served an important purpose. However, over time, the need for these elements may disappear. In other cases, some of the elements found in the process may have been introduced as a "quick fix." There may be better ways of achieving the same goals. As a result, management now has a chance to develop a more effective and efficient process for delivering products to its customers. This principle was evident in the experiences of companies such as Continental Can, Cone Drive, Ingersoll Milling, Frost, and Steelcase.

11. Before introducing change, first standardize on the "best" practices available.

The implementation of CIM involves the introduction of change. However, unless there is a fixed basis of comparison, most firms are unable to evaluate the effectiveness of any changes that they have introduced. The reason lies in the variability present in the underlying system. Over time, people introduce their own unique "improvements" in the operation of tasks and similar activities. The problem with these improvements is that since they are unique to specific workers and tasks, they introduce *variance.* It is this variance that complicates the evaluation of any subsequent change such as those introduced by CIM.

To reduce these problems, the first step is to evaluate the practices present at all stages of the process. The practices that are identified as being most efficient or effective for any activity or task must then be made into the *standards* for those tasks. Everyone carrying out those tasks must use these new procedures. There can be no exceptions. If a new and better method of doing the task can be identified, then that method becomes the new standard and the process is repeated. By following this procedure, the level of variance present in the system is reduced. It is now easier to evaluate the effectiveness of any CIM-related change.

This principle is best illustrated by the experiences of Frost of Grand Rapids, Michigan. At Frost, management noticed that everyone seemed to have their own ways of doing things. This carried an unacceptably high level of system variability. To eliminate this variability, tasks were standardized. To the criticism that this approach stifled worker creativity, management replied that the workers were free to identify, develop, and suggest better ways of doing things. If these new procedures were found to be better, they would become the standard for everyone to use. By using this approach Frost benefited in two ways. First, the entire system became not only more efficient but, more important, the system became more predictable and easier to control. Second, whenever a new CIM-related procedure or technology was introduced, the benefits could be identified much more quickly and easily. People no longer had to account for the differences between procedures present between operators. These differences could affect the observed results through an *interaction* effect.

12. A successful CIM system is built on an integrated corporate database.

Any computerized system such as CIM is built around the management and control of two major flows: product and information. Of these two flows, information is the more critical one. It is information that often serves as a major integrating vehicle. It helps bind the CIM system by keeping the various groups aware of developments taking place elsewhere within the firm. However, few managers understand or appreciate the critical importance of information to the system of the firm. As a result, the information control systems (and databases) of many non-CIM users often exhibit the following shortcomings:

Fragmentation. Often, each of the maior functional groups (e.g., top management, accounting, production and inventory control, finance, marketing, engineering) develops its own internal and separate database of information. As a result a problem facing a user in need of a certain item of information is that of determining where (in what specific database) that item is currently located. Such activities are time-consuming and add nothing to value.

Redundancy. The same information often appears in more than one area. For example, engineering may store information about product structure. Slmilar information may be found in the databases controlled by production and inventory control and accounting.

Incompleteness. Not all of the relevant information is available in the databases.

Inaccuracy. The information found in the various databases may not be current or accurate. As a result, one of the major questions facing the user is that of first determining whether the data that they are using is really correct.

Inconsistency. Information systems tend to evolve over time. As a result, there is the danger that as they evolve, they begin to lose consistency with how the firm does business. For example, the manufacturing database treats each item (end item, component) as being distinct and separate. In order to simplify the design and manufacturing processes, management would like to take advantage of the capabilities offered by a database organized around *part families.* The database, in its current form, is not consistent with this objective.

The use of different units. Each group in the firm uses different units of measure. Finance thinks in terms of dollars; engineering may view a product in terms of its design features or components; marketing may think in terms of dollars; and production uses hours of production. These orientations are reflected in the resulting structure of the various databases. A major problem created as a result is that these databases cannot "talk" to each other. Manufacturing faces the problem of being unable to translate actions and decisions taken by groups such as marketing or finance into their implications for manufacturing.

These problems create a situation in which information becomes a corporate bottleneck. Often, a great deal of time is spent on sorting out the information-related aspects of a problem (i.e., in answering such questions as Where is the information? and Whose information is right?). A strong informal system is developed for coping with these problems. Finally, when information is bad, the amount of slack needed (in terms of inventory, lead time, and excess capacity) to ensure that the manufacturing system can deliver a product to its customers is high. To the participants, such a situation is unacceptable. Successful CIM systems are built on good data. Without good data, CIM cannot and will not succeed. As one of the participants put it:

GIGO[4]

Critical to the successful management and control of information is an integrated corporate database. For such a database to be adequate to the demands of CIM, it must satisfy the following requirements:

It must be corporate-wide. There can only be one database that is used by everyone in the firm. No other databases are allowed.

There is no data redundancy. Information about each item in the database is only found in one area. Multiple listings of the same data in different areas or parts of the database are not permitted. Allowing such practices is costly and a threat to data integrity.

It must be complete. All of the relevant information needed to plan and execute must be contained in the corporate database.

It must be accurate. The corporate data must have high database accuracy. The absolute lower limit on accuracy is 95 percent with 99 percent and higher being attainable and realistic goals.

It must be current. The corporate database is an image written in information of the current status of the entire corporate system. It must reflect what is currently happening as well as what has happened.

It must offer its users a conversion capability. Given that the various areas in the firm use different units of measure, the database must facilitate the translation of one unit into another. For example, the database should enable the user to restate decisions made in dollars into the resulting implications as stated in hours or units. This requirement ensures that everyone in the firm can use the database to evaluate the effects of decisions made elsewhere in the firm on their functional areas.

There should be a common format for entering and recording information. There should never be two formats for entering information: e.g., one for accounting and one for manufacturing. The presence of diverse formats always ensures that neither form will be correctly or consistently used. The practice of having more than one record format is inefficient, confusing, and a potential threat to data integrity.

It must be accessible to everyone who has a legitimate need. There must be control over access to the database. Such control is required to ensure the

---

[4]Garbage in, garbage out.

integrity of the underlying data. This control should never inhibit access to the data by anyone with a real need, however.

By ensuring the presence of an effective corporate database, management can begin the process of reducing lead times and costs and improving quality and efficiencies. It can begin to eliminate all the buffers previously built up to protect itself from the effects of bad information. The firm can begin to trade off information against assets.

All of the participating firms had developed such databases. The form and structure of these databases differed from firm to firm. At Ingersoll Milling, the maintenance of a corporate database was a top management mandate. There was one unifying structure. The database resided on the mainframe. Concern over data integrity was so high that management was extremely reluctant to use microcomputers in the system because they were seen as a potential threat to the integrity of the corporate database.

In contrast, Allen-Bradley has taken a different approach. Unlike Ingersoll Milling, which has only one central location in Rockford, Allen-Bradley has several different facilities located across the United States. As a result, instead of using a common structure, Allen-Bradley allows separate databases that are linked by a common data bus.

Before leaving the discussion of this point, it should be noted that there is a constant need to review and revise information systems. Recently, Steelcase began to revise the structure of its corporate database. The reason was that management had discovered that this database was no longer supportive of how Steelcase now did business. The database was consistent with how the firm did business 20 years ago. It was basically an MRP-oriented database. It was not adequate for the needs of a company that was trying to coordinate the activities of manufacturing, suppliers, and customers. It did not support the current strategic aims of the firm. Steelcase is now undertaking a complete restructuring of the corporate database.

13. Use automation as a means of breaking weaknesses and bottlenecks.

A bottleneck can be viewed as any activity or resource that limits the capacity of the firm or the ability of the company to compete most effectively on lead time, cost, or quality. A firm is often no better than its bottlenecks. By eliminating bottlenecks management can improve the competitive position of the firm.

In studying the experiences of the participating firms, it was noticed that bottlenecks fell into one of two categories: *indirect* and *direct* bottlenecks. Indirect bottlenecks were those that while not directly associated with the problem under study had to be solved first. In contrast, direct bottlenecks were those that most managers saw as being the major limiting factors. For example, if the intent of management was to improve the productivity of its engineers (the direct bottleneck) by introducing a CAD system, management had to first ensure that the manufacturing and design databases were properly structured. The lack of appropriate structure in these databases was the indirect bottleneck.

In the firms studied, bottlenecks occurred at different positions. In some firms the bottlenecks were found on the shop floor in operations. In other cases bottlenecks were observed in the planning system or in the order system. For example, at Cone Drive, order processing and the generation of bids were major bottlenecks. Bottlenecks often became ideal candidates for computerization.

The process of identifying and eliminating bottlenecks is an iterative process. Solving the first set of bottlenecks does not eliminate all of the bottlenecks. Often, the original bottlenecks are replaced by a new set of previously hidden bottlenecks. One conference participant described this process as being similar to shaving ice off the top of an iceberg. As the top layer is removed, the iceberg becomes lighter and more ice rises to the surface. The process is then repeated.

14. Always target and attack first those bottlenecks that are most critical from a corporate or strategic perspective.

The first bottlenecks that should be attacked using CIM are those that have the greatest impact on the ability of the firm to carry out its corporate and strategic objectives. This principle was best illustrated by the experiences of Ingersoll Milling. Ingersoll Milling initially began the implementation of CIM as a means of improving the productivity of its engineering group. Ingersoll competed on its ability to take a design from the concept stage and convert it into a real product delivered to the customer in a timely fashion. Top management recognized long ago that the supply of engineers was not keeping up with demand. For Ingersoll Milling to survive, the productivity of its existing engineering staff had to be improved. As a result, the CAD/CAM system was first introduced as a means of breaking this strategic bottleneck.

OPERATING PRINCIPLES

Operating principles are those guidelines and procedures that govern the implementation and subsequent use of a computer integrated manufacturing system. The major difference between the prerequisite and operating principles is that the prerequisite principles help prepare the corporate and manufacturing environment for the advent of CIM. In contrast, the operating principles assume that the prerequisite conditions have been satisfied. The operating principles help management succeed in making a successful and viable manufacturing system.

15. Select standard hardware, technology, and software whenever possible.

Whenever management begins the process of acquiring and implementing new technology, there are usually two basic options open. The first is to invest in standard technology. The major drawback of this approach is that standard technology is by definition generic. It is not designed to be used by any one firm but by most firms. Often the use of such technology requires that management change its practices so that they are consistent with the requirements of the new technology. In contrast, management can choose to draw on customized technology. This is technology (hardware, software) that has been configured to the specific requirements of the user. With customized technology, the technology is changed, while the underlying management practices are not altered. Often, customized technology is the product of in-house development.

It was the consensus of the participants that the implementation and use of CIM is greatly simplified whenever standard components are used. Several reasons were cited for this principle. First, such components are frequently less expensive to acquire and implement. They are usually better than anything that could be developed in-house. There is often third-party support for training, maintenance, and enhancements. Standard components also reduce the level of in-house support needed for maintenance. Finally, standard components add flexibility. As new features are added, the process of upgrading the system is simplified. It becomes a task of removing the old component (e.g., a piece of software) and replacing it with the new component.

Management must always be aware of the critical trade-off being made when using standard components. Current practices may have to be changed to accommodate the requirements of the new standard system. Cone Drive provided the best example of this principle in practice.

Cone Drive consciously made the decision to go with standard hardware and software whenever possible. Many of its current practices did have to be changed as a result of this decision. In one, these changes took the form of the revision of the parts numbering scheme. Management had found that the current parts numbering structure was too long for the software package that the firm had purchased. Engineering wanted the package modified so that it could accommodate the longer part numbers. Management took a different position. The existing part numbers had to fit within the field length offered by the software package. Part numbering was revised so that it could fit into the part numbering fields of the software.

16. CIM should be planned from the top down and implemented from the bottom up.

Top management is responsible for ensuring that everyone in the firm has a clear understanding of the needs of the system vis-à-vis corporate objectives and of the role of CIM in satisfying these needs. If CIM is to be the firm's response to a significant market threat, top management should ensure that everyone in the firm is aware of the threat and its potential impact on the firm. The responsibility for planning the CIM implementation (i.e., setting milestones, determining objectives, and setting the general sequence of activities) must come from the top and move to the bottom. The guidelines for integrating the various components of CIM must also be done by top management.

Actual implementation, in contrast, must be done from the bottom up. It must be carried through by the ultimate users of the new system. It is the users who must be responsible not only for implementing the system but also for the ultimate success or failure of the system. The CIM system must become their system, not top management's system.

During the course of the conference, this notion of making the users responsible for the ultimate success or failure of the system by making them feel that the system was theirs was described as *user ownership*. Each of the participants stressed the importance of user ownership in developing and implementing CIM systems that were acceptable to the users. Once the system was acceptable to the user, the probability of success was greatly increased.

At Steelcase, for example, users took over the major roles in implementing new systems. When these systems succeeded, the users were given credit for making the system. They were also given visibility. There were numerous instances, for example, of shop floor workers briefing other employees and top management about the features of the new systems and the requirements needed for them to operate.

Islands of automation or stand-alone automated products (e.g., robots, CAD, CAE systems) that were not integrated into a cohesive whole were often viewed by the participants as being indicative of management's failure to do adequate top-down planning.

17. The successful implementation of CIM is dependent on top management and the role that it takes.

Top management plays a critical and special role in the implementation process. Top management allocates the resources needed to implement CIM. Top management should also be seen as the champion of the new system. More important, however, top management should be prepared to play the role of system enforcer if necessary.

CIM can introduce corporate-wide changes in how every task is carried out. It can affect everyone's life. As a result, people may resist the move to CIM because they are not comfortable with the changes taking place. Sometimes, this resistance takes the form of people trying to change CIM practices to ones that are more consistent with what the users have done in the past. Such resistance is often evident in how people try to implement CIM.

Implementation from the bottom up is often done through committees utilizing participative management. Decision-making by committee, however, is not always effective. First, these committees may not reach a consensus position within the relevant time horizon. In other cases the position that they have taken may be inconsistent with the goals and objectives of the overall CIM implementation plan. Finally, the people involved may decide to resist certain changes demanded by CIM.

When such actions take place, the CIM project is exposed to the threat of *incremental deviations*. Very small changes or deviations are introduced in the implementation. If enough of these small changes take place, the resulting CIM is no longer the same as the system for which management had planned. The changes are not the result of one or two large modifications in certain areas. Rather, these changes are the result of continuous minor alterations being done everywhere.

In such cases top management must be prepared to step in and *authoritatively* restore conformance to the agreed-on CIM implementation plan. For example, in one company the engineering group decided to oppose any changes to the existing part numbering scheme. The group was convinced that the new part numbering scheme would not work. The reasons proposed were not really adequate. Top management reviewed the position taken and decided that it was unacceptable. The head of engineering was called in and told that either he would change the part numbering scheme or his successor would. The scheme was changed. It also worked.

18. Keep the number of "fires" that have to be put out down to a manageable limit.

When introducing change in the form of new technology or modifications to current production practices, management should try to control the number of problems that must be resolved. Not all changes resulting from new technology or practices create the same number of problems or problems of equal intensity. For example, the problems of trying to bring in a new assmbly line based on a new technology are often greater than those encountered when installing a simple pick-and-place robot. Management should be aware of the potential problems that accompany each of the changes being considered.

When there is a change with a high potential for problems, the number of other modifications simultaneously taking place should be reduced. This principle allows those responsible for the implementation to devote the appropriate amount of attention to problems as they occur.

Creating more problems than can be successfully controlled affects the morale of the people implementing CIM. It can also significantly hinder the overall implementation of the CIM system.

19. The implementation of CIM should be done incrementally.

Computer integrated manufacturing, like Rome, cannot be built in a day. Rather, CIM should be built up module by module. The development of a CIM system should begin with the prerequisites. Next, it should involve those elements that represent the most critical bottlenecks. When building CIM, each subsequent module should be added only when the previous module has been successfully implemented and integrated.

This approach offers several important advantages. First, the investment requirements of this approach at any point in time are not as high as those imposed when management tries to do it "all at once." Second, it gives both employees and management the opportunity and, more important, the time in which to learn about the new modules, their operation, and their capabilities and limitations. Finally, the number of potential problems present at any point in time is greatly reduced. As a result, the risk exposure of the firm is greatly reduced.

This incremental approach was evident in all of the participating firms. For example, Ingersoll Milling began with CAD/CAM. From that base it passed through the stages of CIM, CIFM (computer integrated flexible manufacturing), and is now in the process of moving toward CIOM (computer integrated optimized manufacturing). Similarly, Steelcase has moved from being a Class A MRP to a Class A MRP II firm and is now becoming a Class A CIM system.

20. System integration cannot be purchased from a vendor or supplier.

In any successful CIM implementation, there is a clear understanding of the role of the vendor and the role of corporate management. In general, there is no such thing as a "turnkey" CIM system that can be purchased from a vendor. In no case can a vendor provide all the elements of integration needed by a firm that is currently lacking this requirement.

Instead, vendors support the activities of the firm's management in the implementation of a CIM system. They provide goods and services needed by management. The obvious function of vendors is to provide components in the form of hardware (e.g., automated guided vehicle systems, flexible manufacturing systems, robots, pallets, and automated storage and retrieval systems) and software (e.g., for CAD, CAE, inventory control, and material planning). Vendors also provide training and education in the use of these components. They can also act as consultants to the firm.

The responsibility for designing, implementing, and using the firm's new CIM system, however, lies with the firm's management alone. As a prerequisite to implementing CIM, corporate management must ensure that the existing manual system is integrated. Finally, management must decide what CIM components to obtain and how these components are to be subsequently used within the firm.

SYSTEM PRINCIPLES

System principles are those guidelines that are presented to help the firm *as a whole* become a more effective user of CIM. These are also guidelines that address those CIM effects that have corporate-wide rather than manufacturing-wide implications.

21. A successful CIM system will affect all functions in the firm.

Computer integrated manufacturing is a corporate-wide system. CIM is not confined to manufacturing and engineering alone. As a result, once implemented CIM will influence every function in the firm. It will affect how marketing, sales, quality assurance, purchasing, and accounting do their tasks. It will also influence how top management executes its responsibilities.

It is important that the implementation of CIM be made a corporate-wide responsibility. By involving the various functions early in the implementation process, management can ensure that the functional groups become familiar with what CIM is and what it is not. More important, these groups can begin to anticipate and plan for the effects generated by CIM. They begin the process of taking advantage of the new or enhanced capabilities offered by CIM. They also begin to change current practices so that there is a better fit between their area's activities and the operation of the new CIM system.

22. CIM will affect the labor requirements of the firm.

Computer integrated manufacturing, if successfully implemented, affects the firm in two major ways. First, it will change the specific tasks of the people on the shop floor. Operators, instead of being responsible for one machine, may be responsible for running all the machines in a given area (i.e., work cell). The presence of CIM may also require that the firm upgrade the skill level of its work force or, in some cases, change the types of skills that its employees have. Employees may have to become familiar with procedures such as *statistical process control* (SPC). Employees may also have to learn to program the machines or to do their own general preventive maintenance on their work areas. Second, CIM may influence the firms demand for labor.

If successful, CIM will make each worker more productive. If the demand for the products manufactured by the firm does not keep up with the increases in worker productivity, management is left with a difficult problem to resolve. The firm's demand for labor has now fallen. The labor pool must be reduced. Management must recognize that labor reduction through attrition may not be enough. If the firm is successful with CIM, workers must be eliminated through layoffs or early retirements.

The possibility of a reduction in the firm's labor pool must be communicated to the employees before the firm begins the process of implementing CIM. The employees must recognize that such a possibility does exist. They must be given the chance to weigh the alternatives available. They must decide the path that they and the firm will ultimately follow.

In several of the participating firms, the situational choice presented to the employees was clear-cut. If the firm introduced CIM, fewer people would be employed. The jobs of these remaining people, however, would be assured because the firm would be more competitive. They and the firm could survive. However, if the firm did not introduce CIM, the firm would not be competitive. It would keep losing business. While the employees would have kept their jobs in the short run, they might lose them in the long term. Ultimately, if the firm failed there would be no jobs left for anyone. In each instance the employees decided that it was better for some to lose their jobs than for everyone to have work only temporarily.

23. CIM is dynamic.

A successful CIM implementation should give the firm a significant competitive advantage. However, management should recognize that this advantage is only temporary. The competitive advantage generated by CIM is influenced by such developments as

- Competitive reactions (actions taken by competing firms).
- Changing customer expectations.
- The development and introduction of new technology.
- Governmental actions.
- Internal changes (i.e., actions taken within the firm).

These changes will shape the future form(s) taken by CIM. The firm must be prepared to change its corporate and strategic objectives. The resulting new requirements will further influence the form, composition, and structure taken by CIM. The new system should be able to accommodate these recent developments. For CIM to be successful, management should become forward-looking, trying to anticipate the next set of changes and trying to shape CIM to accommodate these changes.

This principle was best illustrated by the experiences of Steelcase and Ingersoll Milling. In 1987, Steelcase saw CIM as being primarily an internal endeavor—something that was intended to make manufacturing more efficient. However, as a result of competitive actions, Steelcase had to reassess its position on CIM. To compete effectively management decided that it had to simplify the process by which customers placed an order. In essence, Steelcase had to offer *one-stop shopping*. That is, by placing an order through Steelcase, customers did not receive numerous invoices from the various suppliers. In addition, customers did not have to work with different groups. Instead, they would place an order with Steelcase for all the items needed to outfit their office. Steelcase, in turn, would manage all of the activities needed to satisfy the customers' needs. As a result, Steelcase had to change its view of integration to include closer ties with the customer and various suppliers.

A similar experience took place in the case of Ingersoll Milling. Ingersoll Milling has had to redefine its CIM system to include closer ties with its customers. Because of the highly customized and unique nature of its products, Ingersoll has had to involve its customers more heavily in its design process. This demand has in part resulted in the move of Ingersoll toward the implementation of *interorganizational* CIM.

There is an important downside to this principle that should not be ignored: a firm that is considered to be a leading edge CIM user in 1987 may not necessarily be considered even a CIM system in 1991. The implementation of CIM is often a very difficult, time-consuming process. However, the same cannot be said about the ease with which a firm can make the transition from being a leading edge CIM system to a nonleading edge system. One reason for this slide is found in a transition in corporate ownership.

This phenomenon was observed in two of the firms. One of these firms was Continental Can. When this firm was first studied in 1987, it could be best described as being a system in process. It was developing into a good CIM system. However, between 1987 and 1989 the firm was subsequently sold. The new owners decided to stop further investments in the CIM project. As a result, many of the people who played a key role in this CIM effort decided to leave the firm. Many of the gains made initially have now disappeared.

Other possible reasons include the loss of critical management personnel owing to retirement or resignations and the failure of management to maintain the pace of change introduced by CIM and expected by the employees. This latter factor occurs when employees expect change to take place. They expect management to introduce new and better ways of doing things. When such change repeatedly fails to occur, employees begin to lose their excitement for CIM. A sense of inertia now develops. Once this occurs, the CIM system stops growing. At this point it often begins to decay.

IDENTIFYING THE KEYS TO A SUCCESSFUL CIM SYSTEM

At first glance, the various principles and practices identified in this chapter as necessary for effective computer integrated manufacturing may seem to be too numerous to remember. These principles present management with a CIM jigsaw, with so many pieces that management must carefully fit together to construct an effective CIM system.

However, these principles are not as overwhelming as they may seem at first glance. They can be summarized into four major keys:

Corporate Policy. CIM should never be viewed as being the objective or ultimate goal. The purpose of undertaking the CIM journey is not to have a CIM system. Rather, the purpose of implementing CIM is to help the firm survive by developing a distinctive corporate advantage through its manufacturing capabilities. To this end, CIM requires, as a prerequisite, a well-developed, well-understood, and widely known corporate strategy. The specific type of CIM system that is implemented should be consistent with and supportive of corporate objectives and strategic directions.

CIM should be driven by corporate objectives; CIM should not drive corporate objectives.

Integration. While the heart of CIM is corporate strategy, its soul lies in system integration. Before any firm can seriously consider CIM, it must ensure that its operations and databases are integrated. Integration in the case of manufacturing requires that the activities of the various functional groups be formally coordinated. These groups must have developed the *practice* of integration (i.e., the practice of working together). The necessary interfaces supporting this coordination must precede the movement to CIM. Production and inventory control (PIC) and engineering, for example, must be able to work together effectively before either group can take full advantage of the integrating capabilities of CIM. Integration, in the case of the databases, requires that there be one common corporate database that is used by all groups in the firm or that there be a common data bus and a common data format form. CIM is no better than the database that supports it.

Manual Systems. The computerization of manufacturing operations should be considered only after ensuring that the manual system that it replaces (1) is efficient and effective and (2) has been simplified. Computerizing a system that is not currently operating well results in nothing more than computerizing the mistakes of the old system. Second, it is easier and less expensive to computerize a simple system than it is to computerize a complex system. Always remember: *CIM should be viewed as an enhancement rather than as a replacement for the current system.*

Sufficient Support. When implementing a new software package or introducing new technology, only acquire exactly what is needed to solve the problem(s). Never buy more capability than is needed. Second, standardize whenever possible. This approach avoids the problem of reinventing the wheel. It also ensures that there will be outside (vendor) support for the resulting CIM system. It offers the firm an effective defense against technolust: buying technology for the sake of buying the most advanced features. Finally, standardization is often the most effective tactic for any firm to adopt.

These four keys are summarized in Figure 4-1. They point to a process in which automation and technology, while important, are not the critical elements to success.

FIGURE 4-1
The Foundations of Successful CIM

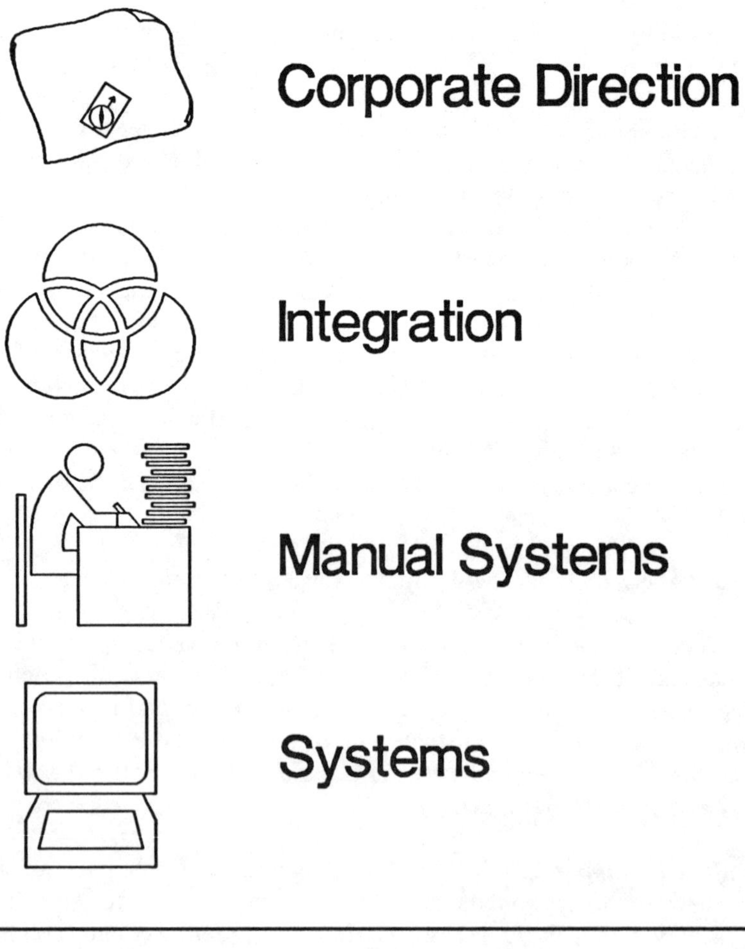

SUCCESS AND FAILURE IN CIM: THE CRITICAL DIFFERENCES

The discussion contained in this chapter has identified many of the principles associated with successful CIM systems. It has not, however, addressed a critical question: that of why certain systems are successful and others are not. This question was addressed to the participants. An interesting aspect of the resulting discussion was that the difference between a success and failure (however those terms are defined) does not lie in the activities undertaken but rather in the sequence in which these activities are carried out.

The Major Steps in Implementing a CIM System

There are four groups of activities that are common to all CIM users. These are

• *Automation.* Simply defined, automation involves nothing more than the application of well-defined computer-aided procedures to physical or information flows.

• *Simplification.* Simplification is the notion of eliminating or modifying any activity or organizational structure that in its current form either adds only cost without increasing value or that acts as a bottleneck.

• *Integration.* Integration deals with the introduction and management of coordination and cooperation between activities and groups.[5]

• *Focus.* Focus is in many ways the most critical and complex group of activities. It describes how management addresses the problem of how to compete in the marketplace. It is here that the issues of strategic consistency (the fit between corporate objectives and manufacturing strengths and weaknesses) are addressed. It is here that top management answers such critical questions as:

Who are our customers?

What is it that we sell or provide our customers?

How do we best meet the needs of our customers now?

How should we meet the needs of our customers in the future?

Does our system adequately support our customers and satisfy their needs?

What are the major strengths and limitations of the current system?

What is our system in its current form best able to do?

---

[5]This concept has been discussed in greater detail in the preceding chapter.

When dealing with the issue of focus, management is setting and communicating corporate and strategic objectives. It is also at this stage that top management determines how they will combine the elements of cost, lead time, and quality in such a way as to develop a distinctive competitive advantage for the firm.

Paths of Failure and Paths of Success

The Path of Failure. In unsuccessful firms the first step is automation. Frequently, small functional problems or bottlenecks are identified and automated. Questions involving the fit between the technology selected and corporate strategy and the effect of this technology on the other areas are seldom, if ever, addressed. Decisions are made using very myopic perspectives. A problem on the floor is identified. A CNC work center is put in. Whether or not this work center is compatible with the CAD system in place is frequently not considered.

The next step is simplification. Simplification is done to facilitate the use of the new technologies. This is followed by integration. With integration, the focus is often on *structural integration.* That is, there are two systems that cannot communicate with each other. Integration in this case involves using a communication protocol (e.g., MAP, TOP) to allow these two systems to share information.

Finally, there is focus. For management, focus involves answering the question of how the firm can best use the new technology it has acquired. At this point, the task of identifying is primarily reactive. The firm has invested heavily in certain types of technology. These past investments have effectively tied up most of the firm's capital. As a result, the firm's ability to invest in new developments is limited.

In defining focus under these conditions, management often responds by changing corporate strategy so that it is consistent with (i.e., driven by) the technology and automation now in place. Management is now in a position of damage control: trying to make the best of a bad situation. It must bring order to what has been up to this point a series of uncoordinated, myopic moves.

The Path of Success. A very different sequence of events is followed by successful firms. The first step is that of identifying focus. Top management begins by identifying what it must do well in its selected marketplace if the firm is to survive and prosper. These *key success factors* are then translated into those manufacturing capabilities needed to attain these objectives.

Next, integration and simplification take place simultaneously. At the end of this stage, management is left with a series of tasks or processes that are critical to the attainment of corporate objectives but that cannot be sufficiently improved through either simplification or integration. Only at this stage is automation considered.

In contrast to the automation decision observed under the path-of-failure scenario, automation is now driven by strategic considerations. The firm knows what features it is looking for when it goes looking for appropriate automation. Integration has been considered in advance (not as an afterthought). Any new automation must fit with the integration scheme developed by the firm. These four steps are summarized in Figure 4-2.

FIGURE 4-2
The "Correct" CIM Path

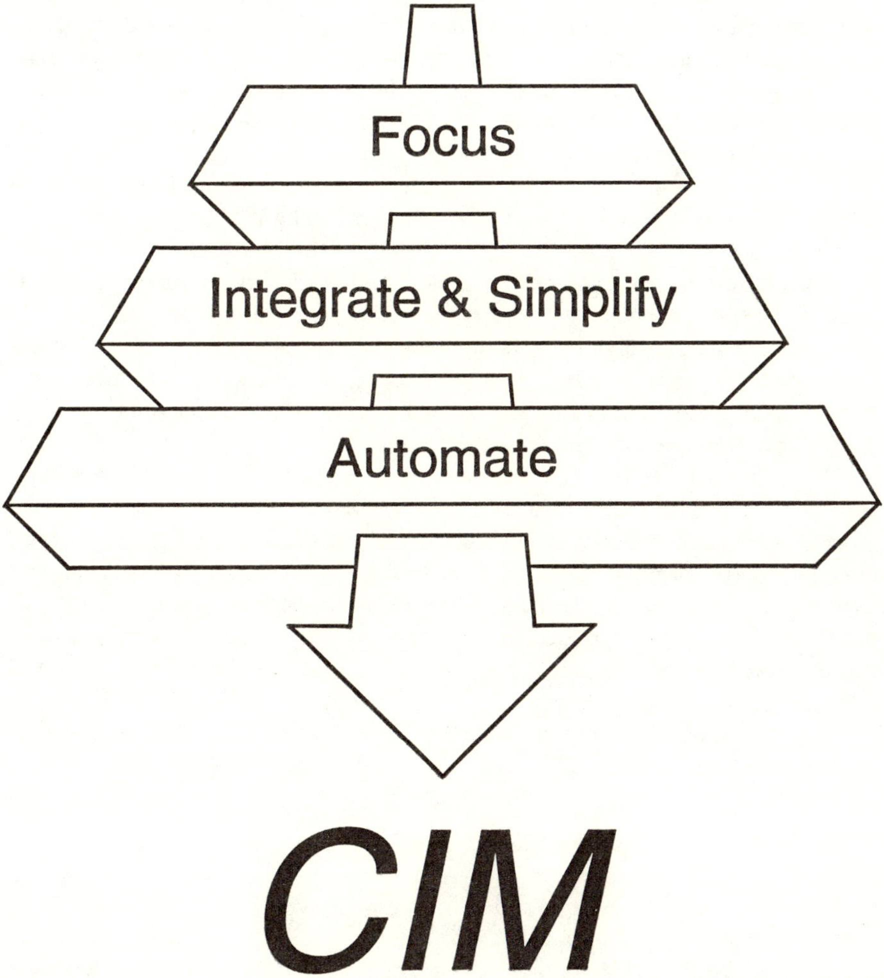

In this scenario, technology is an enhancement; it is not the focal point. This perspective was best presented by Ingersoll Engineers:

Technology *is* good, but it should come *last*.[6]

The CIM Journey: Another Perspective

There is another way of understanding the process of implementing CIM. The implementation of CIM is a process of change. Organizations tend to resist change unless they have to change. Such resistance can be overcome in one of two ways. Either there must be a crisis sufficiently critical so that the costs of not changing are now greater than the costs of change, or the organization must be conditioned to a tradition of change (as in the case of Steelcase). Once the process of change has begun, there must be a goal or set of milestones directing the process of change. Corporate strategy becomes the element responsible for guiding the transition. It also determines the capabilities and requirements to be satisfied by the new technology and the new organizational structure.

Problems inherited from the old organization (problems with data or planning) become obstacles and bottlenecks. They hinder the implementation of CIM. They force management to slow down or delay the move to CIM while these problems are resolved. As the implementation process slows down, enthusiasm for CIM starts to falter. People begin to blame CIM for the problems. Why? CIM has forced management to face a problem (or problems) previously hidden or buffered within the old organization.

The responsibility for directing the CIM implementation lies with top management; the responsibility for making CIM work lies with the users. Once the process of change begins, it is continuous. The pressure for change is generated internally and externally. Employees have become conditioned to change and expect it. As they identify new and better ways of doing things, they expect to see these changes incorporated into the new system. It also responds to such external forces as new customer requirements and demands, changes in corporate strategy, movements in technology, and competitive actions.

Whether or not the firm maintains the momentum of CIM is the responsibility of top management.

---

[6]Ingersoll Engineers, "'The Factory of the Future' Is a Myth." Rockford, Ill.: Ingersoll Engineers, Publication No. IEI 713-687-5M-TR, 1988.

CONCLUDING OBSERVATIONS

This chapter has identified a set of general principles associated with successful computer integrated manufacturing systems. In reviewing these principles, it is interesting to note that the heart of CIM is not technology. Instead, at the center of CIM lie three major elements: (1) a well thought out strategy, (2) integration, and (3) effective manual systems. Bad management practices are bad management practices, regardless of whether they are computerized or not. The burden for success always lies with management.

CHAPTER 5

THE PARTICIPATING FIRMS

The major objective of this book was to identify, describe, and summarize the principles and practices of effective computer integrated manufacturing systems by studying, through first-hand observation, systems in actual use and considered to be exceptional or noteworthy. It became evident early in the project that a rigorous and well laid out procedure had to be developed for identifying such firms. Since it was not possible to bring together and to study all of the firms considered to be leading edge, this procedure had to enable the authors to draw an appropriate sample for study. As a result, the authors settled on a three-step procedure consisting of (l) identification of the leading-edge firms, (2) a conference involving the selected firms, and (3) development of the case studies. The last two steps were critical features of the procedure.

There were several reasons for having a conference. The first is that this approach has been used previously by researchers interested in studying theory and practice in selected areas of production and inventory control. The master production scheduling study conducted by Berry, Vollmann, and Whybark[1] and the shop floor control study carried out by Melnyk and Carter[2] are two examples of the use of this approach. Second, the conference would provide a useful forum in which representatives from the invited firms could discuss their systems and share their experiences with the other participants. Third, the conference would enable the researchers to select specific firms for more detailed case study. Five of the seven firms were selected and their detailed case studies form the basis for this book. Finally, the conference would also provide a vehicle by which the participants, along with the authors, would develop an initial set of principles and practices associated with success in the implementation and use of computer integrated manufacturing.

---

[1]William L. Berry, Thomas E. Vollmann, and D. Clay Whybark, *Master Production Scheduling: Principles and Practices* (Washington, D.C.: American Production and Inventory Control Society, 1979).

[2]Steven A. Melnyk and Phillip L. Carter, *Shop Floor Control: Principles, Practices and Case Studies* (Falls Church, Va.: American Production and Inventory Control Society, 1987).

In the third stage the case studies presented by the selected participating firms would be written down and the resulting principles, practices, and frameworks for understanding CIM would be set down in a book. This book is the ultimate product of this third and final stage.

THE SELECTION PROCEDURE DESCRIBED

From the outset a major concern of the authors was to bring together for the purpose of this study those companies that had made CIM into a *corporate system* and exclude those firms that had simply implemented advanced technology or automation. The implementation of a bar coding system or an on-line data collection and reporting system might be an important achievement and it might generate significant benefits for the firm. However, consistent with the CIM paradigm presented in the previous chapter, if this denotes the full extent of the firm's activities with technology, it should be viewed as an implementation of only automation. There are numerous examples to be found in the literature that involve the implementation of automation; there are relatively few examples of firms that have made CIM a company-wide system and a strategic weapon. To help in choosing the appropriate sample of firms to be included in this study, the authors developed a three-stage procedure:

- Identification of the population
- Information gathering
- Screening and selection

Identification of the Population

The first task for the researchers was to identify all the firms that could be considered to be at the leading edge of the development, implementation, and use of CIM systems. To do this, three sources were used extensively:

- The CIM-related literature
- Expert opinion
- The steering committee

Each source yielded a set of potential candidate firms. In addition, these three sources were used to validate the correctness of the nominated population. That is, firms nominated or identified by more than one source were viewed as having a higher probability of being true leading edge CIM systems.

Computer Integrated Manufacturing-Related Literature

The authors began by undertaking an extensive review of the literature pertaining to the theory and practice of computer integrated manufacturing.[3] A partial listing of the various references pertaining to CIM can be found in Appendix III of this book.

There is a large body of literature pertaining to the theory and practice of CIM. CIM has been discussed by people interested in computer hardware and software, corporate strategy, operations management, production and inventory control, organizational behavior, organizational theory, industrial psychology, operations research, industrial engineering, artificial intelligence, international business, and computer science. In addition, these discussions of CIM have a dynamic quality in that the views of CIM have changed greatly in the last 10 years.

During this review a list of companies was compiled for the period 1980 to 1987. Frequently, these companies were the subjects of articles describing their experience with a specific aspect of CIM. Whenever a company was identified, it was added to a list of potential participating firms. In addition, its experiences with CIM were summarized and recorded.

After the list had been compiled, it was carefully reviewed. Thcre were several instances of companies that were once viewed as being CIM users which subsequently encountered problems with continuing the CIM effort. Often, these companies stopped their CIM efforts and scrapped much of what had been done in the past. These companies were deleted from the list. In addition, the reasons for designating the companies as CIM users varied depending on the journal, the writer, the audience addressed, or the time period. The review attempted to control for these variations by applying a common set of criteria:

Long-term experience with CIM. There had to be evidence that the companies had been working with CIM for at least two years.

Top management commitment to CIM. Top management had to have played a significant and highly visible role in the implementation and use of CIM.

---

[3]Many of the articles reviewed during this stage have been summarized in the form of a CIM source book. This CIM source book provides detailed abstracts of the various articles and references. Information about the source book can be obtained either from the *Education and Research Foundation* of APICS or by contacting the authors, c/o Department of Management, Graduate School of Business Administration, Michigan State University, East Lansing, MI 48824-1121.

Strategic impact of CIM. In describing the implementation and use of CIM, the major focus was on how the CIM-related developments affected the ability of the firm to compete and better serve its customers.

Technology and automation were not the critical elements. Companies in which the primary focus was on the introduction of a specific technology (e.g., bar coding, a robot, an automated guided vehicle system) were excluded since these firms represented the restricted view of CIM as automation.

Evidence of CIM effectiveness. There had to be some concrete evidence that (1) CIM was important; (2) that CIM had changed how the firm operated; and (3) that CIM had definite benefits (both tangible and intangible).

Using these criteria, the review resulted in the elimination of a large number of potential candidates. The firms not excluded became the initial pool from which a sample could be selected.

Expert Opinion

A second pool of candidates was identified by soliciting the opinion of CIM "experts." The experts fell into one of two categories: individual experts and association experts.

Individual Experts. The individual experts were industry consultants, practitioners, and researchers familiar with the theory and industrial application of computer integrated manufacturing. This list of experts was developed from a review of the literature.

Each expert was asked to nominate any company that he considered to be an example of a CIM success. When soliciting these nominations, the researchers did not provide a definition or description of CIM. Rather, they described their research task to these experts. Each expert was first asked to define what they meant by CIM. Next, each expert was asked to identify firms that they felt best represented the industrial application of CIM. For each company nominated, the experts were asked to explain their reason(s) for selecting that firm and to name a contact person within that firm.

In all, 25 researchers, consultants, and practitioners were contacted. Their responses were interesting for several reasons. First, while there were many examples of the application of specific CIM-related technologies (e.g., bar coding), there were very few firms that had company-wide CIM systems. Second, there was considerable disagreement among these experts as to what CIM was. To some, CIM and MRP II were synonymous; to others, these two concepts were distinct. To still others, a firm that had made extensive use of flexible manufacturing systems was a CIM user; to others, it was not. Third, and more important, there was a great deal of agreement over a small number of firms that were identified repeatedly as being "leading edge" CIM users.

Association Experts. This category was developed to describe one specific professional association—the Computer and Automated Systems Association (CASA) of the Society of Manufacturing Engineers (SME). Since 1981, this association has recognized excellence in the application of CIM by companies through the presentation of its LEAD award. LEAD stands for Leadership and Excellence in the Application and Development of computer integrated manufacturing. This award is presented annually to a single company that CASA/SME deems to be the best example of how to incorporate an "innovative, leading" CIM program into its operation. The researchers compiled a list of all of the firms that had been given LEAD awards.

The Steering Committee

A steering committee was formed early in the project. The committee consisted of eight people, including the authors (see Appendix I for a listing of the steering committee's membership). Of these eight, six were drawn from industry and were considered to be CIM experts by their peers. Each member of the steering committee was extremely knowledgeable in the theory and practice of CIM.

The steering committee played a prominent role in the CIM project. During the identification and selection stages, it identified specific attributes that it felt should be considered in identifying a true "leading edge" CIM user. These traits were incorporated in the list of traits used by the researchers. The members of the steering committee also identified companies that they felt were good examples of CIM in practice. In several instances, members of this committee were instrumental in securing access to these firms and even their participation.

After consulting these three sources, a list of 20 firms was compiled. Each firm was contacted to determine their interest in participating in the study. It was at this stage that the researchers encountered an unexpected obstacle: The level of interest in CIM is very high. There has been a great deal written about CIM, and several professional societies such as SME have taken a very active role in the study and dissemination of CIM-related information. As a result, it became evident to the researchers that the APICS-funded study was one of several CIM-related studies. Since the pool of firms is relatively limited, most of the firms in this pool had been contacted repeatedly. Because the involvement of these firms in similar studies had consumed a great deal of time and effort, the management of a number of these companies were no longer interested in participating in one more CIM study. Consequently, nine of the firms in our list declined to participate.

Information Gathering

The second stage dealt only with those firms that had initially expressed an interest in participating in the study. Each firm was subjected to a telephone interview. The purpose of this interview was to obtain more specific information about the companies, their products, and their CIM systems. The areas covered in this telephone interview included

- Demographic information (company size as measured in number of employees and annual sales).
- Product information (e.g., the type of product manufactured, how the firm competes in the market).
- A description of the CIM system (i.e., the structure and organization of the CIM, the operation of the system).
- A history of the CIM system.
- A summary of benefits and costs attributed to CIM.

Screening and Selection

In the third and final stage, the list of potential CIM candidate firms was reviewed and the final sample selected. This sample of firms was to apply only to participation in a conference on CIM at the leading edge. The selection of the firms for the casebook was to be made at the completion of the conference.

Of the eleven CIM candidates contacted, two were eliminated because their CIM systems were primarily technology oriented. In one case the plant manager informed one of the researchers that the company's CIM system consisted only of a set of flexible manufacturing cells. These cells were described as *islands of automation* by the manager. Their presence and use had not affected planning and the other areas of the firm. He felt uncomfortable with having the CIM label applied to the firm. One other firm was eliminated because the company had made a decision not to pursue further implementation and development of a CIM system. No one from the firm was interested in discussing the company's past experiences with CIM. One more firm dropped out of the sample just before the CIM conference (its management had changed its mind regarding participating in the study).

As a result, the process identified seven firms that had agreed to participate in the CIM conference, which was subsequently held at Michigan State University. These firms are shown in Table 5-1.

CONFERENCE OVERVIEW

The conference began on Monday, August 31, 1987, with the arrival of the participants and it ended in the afternoon of Thursday, September 3, 1987. The presentations were completed between September 1 and September 3. Each day

TABLE 5-1
List of Conference Participants

Firm	*Location*
Allen-Bradley	Cleveland Heights, Ohio
Cone Drive	Traverse City, Michigan
Continental Can	Chicago, Illinois
Digital Equipment Corporation (DEC)	Phoenix, Arizona
Frost, Inc.	Grand Rapids, Michigan
Ingersoll Milling	Rockford, Illinois
Steelcase, Inc.	Grand Rapids, Michigan

saw three company presentations and two breakout sessions. These breakout sessions played an important role in the CIM study by providing an opportunity for group discussion of the following areas:

- The obstacles to the successful implementation of CIM in American industry.
- Issues to be considered when justifying the implementation of a CIM system.
- Factors compelling firms to consider CIM as a strategic weapon.
- Defining and understanding the CIM concept.
- The principles governing the successful implementation, development, and use of CIM.

The CIM framework and the principles and practices presented in this book represent the culmination of work initially begun in these breakout sessions.

Thc conference and breakout sessions were *closed.* Participation was limited to the representatives from the seven firms, the two authors, and several Ph.D. students from Michigan State University. The students acted as transcribers. They took down comments and observations made during the presentations and breakout sessions.

To encourage free and open discussion, the participants were assured of confidentiality. That is, any comments made during the conference would not be attributed to any of the participants. Furthermore, every participant was informed that if selected for a case study, their company would have the right to review and edit the case study describing the company and its operation.

CASE STUDY OVERVIEW

At the start of the CIM project, it was decided that one of the hallmarks of this study would be the use of thorough case studies. Each company's experiences with CIM would be described in sufficient detail to enable the reader to understand the events that led to the decision to implement CIM and the changes that had to take place within the firm to make the implementation a success. Furthermore, the case studies would be uniform; each would follow the same format. As a result, the reader could easily compare and contrast the experiences of these firms with CIM.

As a direct consequence of this decision, it became obvious that writing case studies would be time-consuming and require a great deal of effort. This was an activity that could not be delegated to the participating firms. As a result, we decided to include only five firms in the case study book (see Table 5-2).

TABLE 5-2
List of Case Studies

Firm	*Location*
Cone Drive	Traverse City, Michigan
Continental Can	Chicago, Illinois
Frost, Inc.	Grand Rapids, Michigan
Ingersoll Milling	Rockford, Illinois
Steelcase, Inc.	Grand Rapids, Michigan

The five firms selected for the case study segment of the project represent a great deal of diversity in terms of such traits as company size, targeted markets, nature of ownership, and market position. More important, the CIM studies selected are very different. These differences are most apparent when studied along the following dimensions:

Initial reasons for CIM. In some cases CIM represented the last chance for corporate survival, whereas for other companies it was a method of maintaining the firm's position in the marketplace.

Objectives for CIM. The management objectives for CIM vary greatly across companies. In one case (Frost), CIM was introduced to help the company change its market from one that was mature and about to die to a new and different market segment that was beginning to enter the growth stage of the product life cycle. In two other cases (Continental Can and Cone Drive), management had implemented CIM in the hopes that it would enable them to maintain their position in an increasingly competitive mature market. For still another company (Steelcase), CIM was seen as a method of further enhancing value for the customer.

The Extent of the CIM system. The five companies illustrate one important aspect of CIM: CIM is not homogeneous. In one instance (Continental Can), CIM has consisted primarily of improvements to the manufacturing side of the firm. In other cases (e.g., Steelcase and Cone Drive), CIM has affected nearly every aspect of the firm's operation (including the office systems).

CIM as an internal or external system. In many cases CIM can be viewed as an internal system. The customer or the supplier is unaware of the existence of CIM or, more important, has not been affected by its presence. In other cases CIM has been used as the vehicle for bringing about better control of the entire supply chain (beginning with the supplier and ending with the customer). In these instances CIM has affected the activities of both the supplier and the customer. This latter aspect is best illustrated in the experiences of Steelcase.

The current status of CIM. In two of the firms the future of CIM is very uncertain. Despite its success, there is a real question of whether the firm will continue to make those investments necessary to maintain the evolution of CIM. In the other firms CIM is very much "alive and well."

The combination of diversity and similarity (each of the firms is an acknowledged leader in the implementation and use of CIM) gives the reader a unique opportunity to compare and contrast experiences.

As a final observation, it should be noted that the principles and the framework reported in this book are drawn from the experiences of all seven firms. It is only the experiences of five of the seven firms that are reported in detail in the case studies found in this book.

THE STRUCTURE OF THE CASE PRESENTATIONS

Ultimately, the usefulness of the five case studies presented in this study is dependent on the extent to which the reader can compare and contrast the reported experiences. One method of simplifying this process of comparing and contrasting is by having each of the case studies follow a common format. Each company's experiences with CIM are described using the following structure:

Executive Summary. The executive summary is a brief summary of the company, its CIM system, and those aspects of its experiences that are either unique or noteworthy.

The Company. Before discussing the CIM system, each study first introduces the reader to the company. This discussion covers such topics as

Products and services sold

Market segment served (along with some of the customers)

Competitors (both domestic and international)

Changes in the market (if relevant)

Market position of the company (position occupied by the firm in the marketplace, e.g., market leader, innovator)

Identification of the key success factors (i.e., those things that the company must do well if it is to succeed and grow in the marketplace)

Company history

Overview of the firm's manufacturing process, and the manufacturing planning and control system used to control it

Product breakdown (i.e., the extent to which the products sold are make-to-order, make-to-stock, engineer-to-order, or assemble-to-order)

Unique company features (i.e., any aspect of the company, its organizational structure, or its culture that is either unusual or exceptional)

This section describes the strategic view of the firm. It deals with how it competes in the marketplace. How a firm competes is described in terms of three sets of traits (using the framework developed by Hill[4]): *order winning, order qualifying,* and *order losing*. Order winners are those traits that the firm excels in and that are used to win orders in the market. Order qualifiers are those that the firm must do well to compete. These traits do not necessarily win orders for the firm; they only result in the firm being considered for the order. Finally, order-losing traits are those that can result in the firm either losing the current contract or not being considered for any subsequent contracts.

This strategic perspective provides the background against which any CIM-related activities must be presented. In this perspective the rationale for the move to CIM is often found.

The Events Triggering the Move to CIM. Next, the case study identifies and examines those factors that caused management to consider changing the prevailing way of doing things. Included in this section are such issues as when the decision to change was made, the major players involved, the inadequacies of the old system, alternatives evaluated, and the factors used to justify the move to CIM.

---

[4]Terry Hill, *Manufacturing Strategy: Text and Cases* (Homewood, Ill.: Richard D. Irwin, 1989).

CIM Overview. No two firms will view CIM in exactly the same way. As a result, it is important to understand how each company defines CIM. This section covers issues such as the elements that each firm includes in its view of CIM and the fit between how the firm competes in the marketplace and the capabilities offered by the CIM system.

A Description of the CIM Implementation. This section describes the sequence of events covered by the CIM implementation. It also identifies any factors that affected the implementation (either positively or adversely). It also indicates the time span covered by the implementation process (i.e., when the project began and when it was completed).

CIM Type. In the previous chapter, the reader was introduced to the four basic types of CIM system. Using this framework, the firm's CIM system is categorized and the extent to which the CIM type implemented is appropriate (given corporate objectives) is evaluated.

CIM Benefits. Next, the benefits directly attributable to the CIM system are summarized. Included in this list of gains are both quantitative (measurable) and qualitative benefits.

Lessons Learned. The management of each company was asked to review their experiences with CIM and to identify anything that they had learned from this undertaking. These lessons are summarized in this section.

The Next Stage. Each company was asked to identify the next step in the CIM evolution. The responses found in the case studies fall into one of two categories. For some companies, the future of CIM is very much in doubt. CIM has worked in the past, but because of recent changes (e.g., ownership) there is a real question in the minds of the users whether the current rate of change will be maintained. For the other companies, CIM is only a temporary resting spot. It is only one stage in an ongoing process of evolving and enhancing the manufacturing and strategic capabilities of the company.

AN OVERVIEW OF THE PARTICIPATING FIRMS

The five firms whose experiences are summarized in the case studies found in this book are described using two sets of characteristics. The first set of traits reflects important company and manufacturing system characteristics, while the second set focuses on distinctive or unique features of each firm's CIM system. These descriptions are brief and are intended to provide the reader with only a broad overview of the participating firms. For more information, the reader is referred to the detailed case presentations.

The general company and manufacturing traits are described using a classification scheme similar to that used by William Berry, Thomas Vollmann, and D. Clay Whybark in their master production scheduling study[5] and by Steven Melnyk and Phillip Carter in their study of shop floor control systems.[6] The following are some of the major characteristics used in describing significant company and manufacturing traits:

The orientation of the organization. This trait describes the breakdown of production according to whether it is MTS, ATO, MTO, or ETO.

Major product lines. This is a generic description of the central product lines made and sold by the firm.

Major markets. This trait identifies the major customers and the channels by which they are served.

Firm size. This attribute is described either in terms of the number of employees or annual dollar sales.

Other. Under this trait, we include such information as whether a union is present, the nature of corporate ownership (private or public), and any other factors thought to be relevant.

Thc second set of attributes focuses specifically on the features or aspects of the firm's CIM system that make it unique or of interest. Included under this category is the type of CIM in place (see Table 5-3).

SOME CAUTIONS ABOUT THE STUDY

The reader, in reviewing the findings of the study, should be aware of the presence of certain limitations. These limitations are best viewed as cautions that should be applied when realizing the results.

The first is that the study consists primarily of firms involved in *discrete production.* CIM can take place in other production settings such as continuous flow or line. Firms representing CIM experiences in these other settings were

---

[5]W.L. Berry, T.E. Vollmann, and D. Clay Whybark, *Master Production Scheduling: Principles and Practices* (Washington, D.C.: American Production and Inventory Control Society, 1979).

[6]S.A. Melnyk and P.L Carter, *Shop Floor Control: Principles, Practices and Case Studies* (Falls Church, Va.: American Production and Inventory Control Society, 1987).

not included in the study. Second, nearly every firm studied has developed a CIM system that involves MRP in one form or another. However, this fact should not be interpreted as indicating that if a firm wants to implement CIM it has to first have an MRP system present. A more appropriate assessment begins by first viewing MRP as being simply one example of a *formal* material and priority planning and scheduling system. Given this perspective, the position of the researchers can be stated as follows:

> Any firm wanting to implement CIM must have an integrated and formal manufacturing planning system. An important element of such a system is a formal, effective material and priority planning and scheduling system. Such a system can take several different forms, of which MRP is simply one.

TABLE 5-3
Firms Selected for the Case Studies: Key Company Attributes and Features

Company	*Company Attributes*	*Key CIM Traits*
Cone Drive Traverse City, Mich.	Small firm Unionized employee (UAW) Mixture of ATO, MTO, and ETO Market is in the mature/decline stage Publicly owned Wholly owned subsidiary Presence of foreign competition LEAD award winner (1985)	Type III CIM system Good example of application of CIM principles to office systems Good illustration of importance of top management in implementation process CIM needed to ensure corporate survival CIM as an evolutionary development CIM implementaton and corporate downsizing Future status of CIM uncertain
Continental Can Chicago, Ill.	Large firm Unionized Market is in the mature/decline stage Two major markets: can-making machines and replacements Publicly owned Very competitive marketplace Market very cost sensitive Demand primarily MTO	Type I CIM system CIM needed to ensure corporate survival Close fit between CIM structure and the basis on which the company competes Importance of doing the manufacturing basics right before automating System simplification and CIM Importance of a system champion Future status of CIM uncertain (change in ownership of company

TABLE 5-3 (continued)

Company	*Company Attributes*	*Key CIM Traits*
Frost, Inc. Grand Rapids, Mich.	Small firm Privately held Market at time of decision to implement CIM in the mature/decline stage Very competitive marketplace Primarily ATO, MTO Primarily supplier to large customers	Type III system (evolving into Type IV) CIM introduced not simply to enable company to survive but to change its market (from one in decline to one in growth) CIM used to change focus of company from selling a product to selling a corporate system to its customers Organization restructured to better support CIM Importance of information system to CIM Role of top management in supporting the CIM implementation CIM strong and improving
Ingersoll Milling Machine Co. Rockford, Ill.	Classic CIM system Large firm (over 1,500 employees) Nonunionized Privately held Professionally managed Introductory to growth stages of the productive life cycle Extensive use of flexible manufacturing system (FMS) Extremely small production patches (1–2 units) Little repetition of parts production (every part is unique) Almost 100% ETO Dominance of engineering Company sells not only product but more important its familiarity with new and advanced technology LEAD award winner (1982)	Type IV system Good illustration of the use of CIM in integrating the supply channel (from supplier to Ingersoll to the customer) Continual evolution of CIM system Role of top management in implementing CIM Extensive computerization of engineering system and the interface between engineer and manufacturing Need to revise corporate database Developing a CIM system flexible enough to accommodate new technologies as they become available.

TABLE 5-3 (continued)

Company	*Company Attributes*	*Key CIM Traits*
Steelcase, Inc. Grand Rapids, Mich.	Extensive fabrication and assembly facility Traditional job shop structure Well known Class A MRP II system Privately held Professionally managed Market in transition from growth to mature stages Mixture of MTO (85%) and MTS (15%) Very competitive market Change in the purchasing process (through the addition of architecture and design houses)	Type IV system CIM as a means of maintaining and enhancing leadership position in the market Importance of continuous improvement Importance of people as sources of improvement and change Need to reexamine and change, if inappropriate, the information paradigm that underlies the corporate information system Seamless transition from MRP II to CIM Integration of CIM with just-in-time manufacturing An example of a firm that sells the process to its customer, not simply the product (and the role of CIM in this sales process) CIM system still evolving

Finally, all of the systems studied have been built around IBM hardware and software systems. There are many software and hardware systems provided by non-IBM suppliers that are currently in use and yielding good results. The focus on IBM systems is nothing more than an unanticipated side effect of the sample that was identified and studied by the authors.

PART 2

CASE STUDIES

Ingersoll Milling
Cone Drive
Steelcase, Inc.
Continental Can
Frost, Inc.

CHAPTER 6

THE INGERSOLL MILLING MACHINE COMPANY

OVERVIEW

Ingersoll Milling is a pioneer in the area of computer integrated manufacturing. One of the first companies to implement CIM on a company-wide basis, Ingersoll has enjoyed a great deal of success.[1] In Ingersoll Milling we see a company that is continuously pushing the evolution of CIM forward. Currently, Ingersoll is pursuing the implementation of computer integrated optimized manufacturing, the third and latest development of the CIM concept. In this presentation we examine the experiences of Ingersoll Milling with the CIM concept. CIM is the major factor explaining Ingersoll Milling's success in a unique and very complex manufacturing environment.

INTRODUCTION

The Ingersoll Milling Machine Company (Ingersoll Milling) of Rockford, Illinois, is recognized as one of the foremost and most progressive implementers of computer integrated manufacturing in the United States. In 1982, Ingersoll Milling received the LEAD award from the Computer and Automated Systems Association of the Society of Manufacturing Engineers (CASA/SME), for its accomplishments in making CIM a corporate way of life.

At Ingersoll Milling, CIM is the glue that binds all aspects of the firm together. CIM links the various departments of Ingersoll Milling, and ties corporate strategy to the activities of the shop floor. It is a dynamic entity that is continually undergoing change, refinement, and enhancement.

CIM at Ingersoll Milling has evolved from a simple manufacturing and engineering system to computer integrated flexible manufacturing, which describes the integration of the CIM business and corporate structure with the

---

[1]This presentation was originally done by Richard Smith, Managing Director, Ingersoll Engineers, Southfield, Mich.

capabilities of the flexible manufacturing system. The next step in the CIM evolution is that of computer integrated optimized manufacturing, which seeks to eliminate all activities that do not directly contribute value or help in the production of an end item for the customer.

CORPORATE OVERVIEW

Ingersoll Milling Machine Company is a subsidiary of Ingersoll International, which is a privately owned holding company. In addition, Ingersoll International includes (see Figure 6-1)

Ingersoll Cutting Tool Company. Located in the United States, Ingersoll Cutting Tool is a manufacturer of specialized cutting tools.

Ingersoll Maschinen und Werkeuge. A West German–based manufacturing company building machinery and cutting tools with its subsidiaries, Waldrich-Siegen and Bohle.

Together, Ingersoll International employs over 4,500 people and ships some $400 million a year.

Ingersoll Milling is primarily an ETO firm that competes by designing and building heavy special machinery systems. The overall objective of the company is to help its customers reduce costs and improve competitiveness. For Ingersoll Milling, this objective has meant a heavy emphasis on engineering and manufacturing.

Ingersoll Milling Machine manufactures large-scale, custom-made metal-cutting machines and machining systems for special applications in its 959,000 square feet plant in Rockford. The systems are built for a wide range of manufacturing settings ranging from job shops to flow shops. Its customers include General Motors, General Dynamics, Caterpillar, Ford, Lockheed, Boeing, and McDonald Douglas.

Production falls into one of two categories. The first category consists of four types of machines or machining systems:

Portal-type milling machines. These are large numerically controlled machines that mill large parts. Historically, milling machines account for the largest and most constant percentage of sales.

Transfer lines. This type consists of groups of metal-cutting machines, each designed for a specific range of operations, connected together by an automated material handling system and controlled by a central control system. Transfer lines are designed to produce a given machined part (e.g., an automobile engine cylinder block) at very high volumes. They are generally considered to be "hard automation." Increasingly, however, computer control and flexible workstations are being incorporated into transfer lines. The result is the creation of *flexible transfer lines.*

Flexible machining systems (FMS). These systems consist of collections of numerically controlled machines, each designed for a broad range of functions, that are connected by a material handling system and controlled by a central computer system. Unlike transfer lines, which are best suited to a small range of high-volume parts, the flexible machining system can efficiently machine a wide variety of small-volume parts.

Composite manufacturing machines. These are numerically controlled machines that are designed to manufacture continuous-length structural shapes such as airframes out of composite materials such as graphite tape.

Of these, the last two are of special interest because they represent the latest in manufacturing technology. They represent the commitment of Ingersoll Milling to remaining at the forefront of production technology.

The four lines that make up the first category represent proven technology. That is, it is certain that products drawn from these four lines can and will oper-

FIGURE 6-1
Ingersoll International Major Operating Units

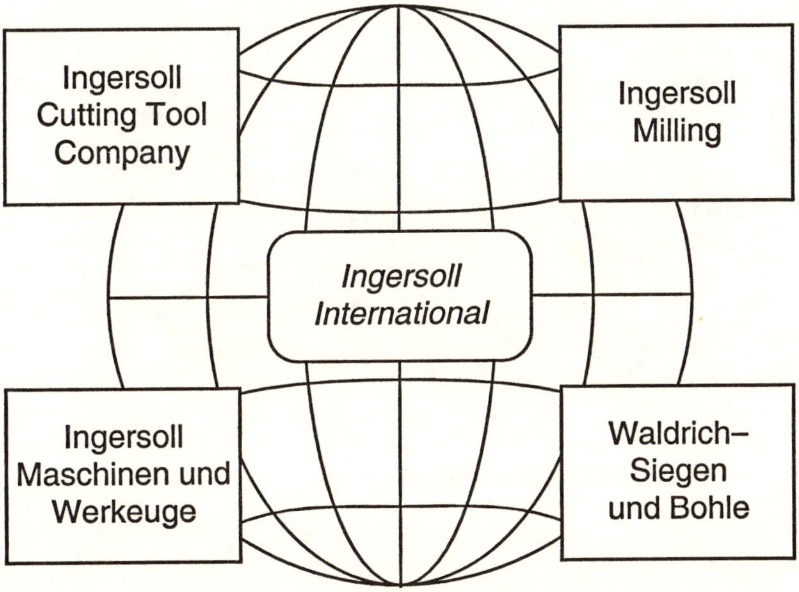

ate. In contrast, the second category represents products that have never been built before. They represent the leading edge of manufacturing design and technology. They draw heavily on all of the manufacturing and engineering expertise that Ingersoll Milling can offer. Often, these systems are so technologically advanced that there is a real question of whether they can be built.

These two categories are evaluated using different criteria. For products taken from the first category, the emphasis is on quality and on-time delivery. This focus is consistent with Ingersoll Milling's position in the marketplace. Ingersoll Milling products are considered by most industry experts to be the "Cadillac" of the industry. Customers who buy from Ingersoll Milling are not looking for the cheapest systems, but are buying a package that draws heavily on the engineering and manufacturing capabilities of the company. They pay a premium to gain this access. In return, Ingersoll Milling provides them with specialized equipment that is extensively engineered, of extremely high quality, and delivered on time. Quality to the management of Ingersoll Milling is taken to mean a product that meets the expectations and needs of the customer and helps to make them the low-cost producer of their products.

For products belonging to the second category, there is only one measure of performance that is emphasized: quality. Because these products are at the leading edge of technology, the due dates established are at best tentative. The due dates can be moved if any problems (typically engineering) are encountered. Failure to meet the due dates on these products is acceptable *provided* that the final product satisfies the needs of the customer.

For both categories, Ingersoll Milling has developed a reputation within the machine tool industry as a technological leader, one of the foremost innovators and a producer of systems that are continually extending the technology frontier.

SUMMARIZING THE KEY SUCCESS FACTORS AT INGERSOLL MILLING

For Ingersoll Milling to succeed in its market, it must offer its customers products that are of high quality and that are delivered on time. To Ingersoll Milling, quality is the extent to which the product designed and built for the customer is able to satisfy the customer's requirements. It is also a reflection of the success with which automation has been used to help the customers become the lowest-cost producers in their specific market segments.

As a key success factor, on-time delivery ranks as a close second. In many cases the large special equipment produced by Ingersoll Milling is part of a larger project (such as the renovation of a department or plant). Failure to deliver the product on time can delay the completion of the larger project (thus increasing costs). In other cases the customer depends on the delivery of the equipment so that it can be brought on stream as quickly as possible. Delays can adversely affect the ability of the customer to quickly bring the equipment on line and to take advantage of its capabilities immediately.

Acquisition costs, while important, are not the overriding consideration. Often the customers are not as interested in the initial acquisition costs of the equipment provided by Ingersoll as they are in the cost-reduction capabilities offered by it.

Ingersoll Milling must satisfy these requirements while remaining a technological leader. The products manufactured by Ingersoll must help the customers reduce costs by drawing on the capabilities offered by advances in manufacturing and computer technology. For Ingersoll Milling, these requirements demand that management develop in its own system certain critical capabilities and features. The following are the most important of these features:

Effective Engineering. Success for Ingersoll Milling depends to a large extent on the strength of the engineering function. Engineering is responsible for designing a product that exactly satisfies the needs of the customer and for ensuring that changes requested by the customer are quickly incorporated into the product. Engineering must also check the design of the product for problems that might occur during its operations (e.g., the collision of heads as they move back and forth on the rails). Ingersoll does not have the luxury of long lead times, which requires that the design be correct from the outset. Finally, the ability of Ingersoll to offer products that are technologically advanced is dependent on the skills and knowledge of its engineers. In short, the critical resource for Ingersoll Milling is engineering.

An Effective Manufacturing Function. The manufacturing function must be able to take the designs created by engineering and convert them into actual products. It must be able to accommodate changes in designs at any point during the manufacturing and deliver the product by the due dates. Manufacturing and engineering must work closely together. Problems encountered on the shop floor must be communicated quickly back to engineering. Similarly, design changes requested by the customer must be incorporated into the design and sent down to the shop as soon as possible. Engineering and manufacturing at Ingersoll are closely coupled so that the success of one depends on the other.

Experience with Advanced Technology. Ingersoll Milling has a unique challenge in that it must not only use advanced technology in its design but it must also demonstrate to its customers that the new technology does work. As a result, the manufacturing facilities of Ingersoll Milling must become a laboratory for the design and manufacture of new technologies. Before

Ingersoll can offer the technology to its customers, it must first implement the technology in its plant. It must gain experience with the management and control of the systems based on the new technology. It must also experiment with the structure of manufacturing systems incorporating this new technology. When customers buy a product from Ingersoll Milling, they are not only buying the product but also access to the implementation experience of the company. The customers are dealing with people who have made the technology work.

This requirement is best illustrated by Ingersoll Milling's approach to flexible machining systems. When it began selling flexible machining systems to its customers, Ingersoll built such a system in their Rockford plant. For a period of over a year, the management and engineers worked with this system and examined the problem of how to increase the amount of time spent by the system in actually cutting metal. They increased this percentage to over 50 percent. This was quite an achievement given that the average tool spends a very small percentage of its time cutting metal. The experiences gained from the building and running of this system were then incorporated into the designs of systems developed for its customers.

Given the unique production problems and system requirements faced by Ingersoll Milling, management has been forced to seek out new and different approaches to the problem of designing and building its products. One approach that has had a great impact on the firm is computer integrated manufacturing (CIM). The decision to implement and use CIM was based on the desire of management to maintain and enhance the firm's competitive position in the marketplace. It was not a technologically driven decision.

Ingersoll Milling: Selling the Process, Not the Product

Ingersoll Milling does not sell its customers simply a product. That is, its customers do not buy flexible manufacturing systems or transfer lines from Ingersoll. Instead, they buy access to the process that Ingersoll Milling has developed for taking abstract ideas and needs that the customers have and converting them quickly into products and processes that are able to satisfy these needs.

This process is far more important than the product. Ingersoll is primarily an ETO company. It must work with the customer in developing the product. The process developed is the major means by which Ingersoll offers its customers greatly reduced risk and uncertainty. For Ingersoll, CIM is another component in the process that management sells its customers.

PRODUCT DESIGN AND MANUFACTURING: AN OVERVIEW

Ingersoll Milling has between 1,800 and 1,900 employees. These employees, who are nonunion, work at one site in Rockford. The workers are evenly divided between the light shop, which produces items measuring up to 1 cubic meter in size, and the heavy shop, which produces all larger-sized orders. Both shops consist of metal-cutting machines. All employees are treated as salaried personnel. They are equally divided between touch labor and knowledge workers.

Because it is primarily an ETO firm, Ingersoll Milling continually faces certain design and production issues unique to it:

Very small lot sizes. In a typical year Ingersoll Milling produces approximately 25,000 different parts. Of these, over 70 percent are built in unit lots. Few of the production lots go beyond two.

Little repetition of production. Of the 25,000 parts produced in a typical year, about 50 percent will never be made again.

Extensive engineering. The average engineering cycle for a typical product is five months or greater and engineering costs account for about 5 to 15 percent of the selling price. Engineering is critical because it is responsible for ensuring that all parts are made right the first time. This requirement is central to Ingersoll's strategic position in the marketplace.

The importance of design learning as compared with production learning. In many traditional production systems, managers make extensive use of the detailed insights of their shop floor personnel. They rely on shop floor personnel to improve system efficiency by taking advantage of similarities in setups or processing. The management at Ingersoll Milling cannot draw on such insights because there is no learning on the shop floor. Due to the unique and changing nature of the product mix, shop floor personnel are always at the first stages of the learning curve. The only real learning that does take place is in the design stages in engineering. There is a strong tendency to reuse engineering extensively. Designs completed for past jobs are frequently called up and used as inputs to new jobs. It is this extensive reuse of past design experiences that is a major source of efficiency for Ingersoll Milling.

Long cumulative lead times. The total lead time for an order often exceeds 12 to 18 months. This lead time typically breaks down as follows: 5 months in engineering; 6 weeks in process routing; 16 weeks in build; 16 weeks in assembly and test; and 10 weeks for customer changes and revision and engineering cleanup.

The total lead time depends on the type of order received. For example, in the case of orders belonging to the second category, lead times often exceed this 18-month time frame.

Frequent design changes. The design of any given product can be altered at any point in the production process. It can change even after manufacturing on the product has begun. These changes originate from either engineering or the customer. In spite of these changes, manufacturing is still expected to complete the order on time.

No manufacturing time standards. Ingersoll Milling uses no time standards because of the unique nature of the products made and the processes used. Instead, estimates provided by engineering are used. Because of the lack of time standards, capacity planning is difficult and has a very strong subjective character.

COMPUTER INTEGRATED MANUFACTURING AT INGERSOLL MILLING

Before discussing the development of CIM at Ingersoll Milling, it is necessary to understand how management views the CIM concept. Computer integrated manufacturing is a corporate and strategic system at Ingersoll Milling. It is a system that management feels best addresses the unique problems and constraints faced by Ingersoll Milling. CIM is viewed by management as the glue that binds all aspects of the firm. It links together the various departments and coordinates their activities. It ties corporate strategy directly to the actions taken in engineering and on the shop floor.

The goal of CIM for Ingersoll Milling is not CIM. Instead, it is to enhance management's ability to serve effectively the needs of its customers. By achieving this objective, CIM helps the firm develop and sustain a unique competitive advantage—an advantage that cannot be easily duplicated by its competitors. Indeed, the ultimate purpose of CIM is viewed as the continued survival of Ingersoll Milling.

Ingersoll is now implementing computer integrated optimized manufacturing. CIOM marks the third major stage in the CIM journey. The first stage was computer integrated manufacturing. The second stage was that of computer integrated flexible manufacturing. Ingersoll Milling entered this stage in 1983. This is the stage at which it currently is operating. Understanding the practice of CIM at Ingersoll Milling requires an appreciation of the evolution from CIM (the past) to CIFM (the present) to CIOM (the future).

The Move to CIM: The Background

The move to CIM began in earnest in the early 1980s, when the president of the company, Edson Gaylord, with help from the corporate staff, decided that what was needed was a complete revision of the system present at Ingersoll Milling. This decision was a reaction to the problems facing the company at that time: excessive shortages, manual priority lists (hot sheets), and too much expediting on the shop floor.

Mr. Gaylord also recognized that a change was taking place in the machine tool industry. He recognized that those firms that were to survive in this industry would have to offer more than simply the ability to build tools. They would have to offer their customers a unique competitive advantage. CIM was to Ingersoll management a vehicle for enabling Ingersoll Milling to survive and to prosper.

The implementation of CIM between 1980 and 1981 was notable because it occurred at a point when the industry was in a downturn. At a time when other firms were trying to save money and cut costs, Ingersoll Milling was investing. In hindsight these investments enabled Ingersoll Milling to meet the demands of the market when the upturn occurred.

CIM at Ingersoll Milling: The First Stage

The winning of the LEAD award from SME in 1982 marked the culmination of the first step in the CIM process: the building of the CIM system. This process began in 1957 with the installation of the first two numerical control machines (used for drilling). Some of the key events in the development of CIM are summarized in Table 6-l.

The CIM system that was recognized by SME as being outstanding in 1982 was the result of four strategic decisions taken by management. These four decisions were

- The use of numerically controlled machines from the moment they were practically available. This first decision marked the start of computer-aided manufacturing at Ingersoll.
- The application of computer-aided design from the moment it was practical. This decision marked the beginning of CAD at Ingersoll Milling.
- The use of computer graphics as a way to accelerate CAD and build a solid foundation for CAM. This marked the start of the integration of CAD/CAM.
- The rewriting of all information system application software programs so that a company-wide database could be installed. For management, this fourth and final decision marked the true start of CIM at Ingersoll Milling. It also marked one of the major problems encountered.

Each of these decisions was aimed at eliminating or reducing a major corporate bottleneck. In introducing NC equipment and CAM, for example, management

TABLE 6-1
Key Events in the Development of the CIM System

Year	*Event*
1957	First two NC machines purchased for use in drilling
1960	First wire-guided vehicles in machine shop
1961	First storage and retrieval stacker crane
1966	"707 System" combined six machines into a manufacturing center with power carrier dispatch and return (expanded to 28 machines in 1968 and 42 in 1971)
1975	First NC machining center
1975	First installation of CAD computer graphics system
1978	Decision to go NC regardless of lot size
1979	First of four computer-directed CMMs
1979	Integrated database as heart of CIM system
1980	First full year of graphic CAM programming
1981	Decision to do all machining under CNC
1982	First DNC installation
1982	New CNC turning shop

was trying to reduce lead time and costs while improving quality through the use of advanced technology. CAD was intended to address the engineering bottleneck. As was mentioned previously, engineering is a critical resource at Ingersoll Milling. It was also a resource that was becoming increasingly scarce as fewer engineers graduated from North American universities. The challenge was (and still is), therefore, that of making every engineer highly productive. For an engineer to be productive, more time must be spent on designing. In many systems, the current practices discourage the effective use of engineering time. In these systems, all designs are done on paper with its inherent limitations. Since past designs cannot be easily retrieved, engineers often find themselves reinventing the wheel. That is, they redesign parts for which there are identical or nearly identical designs available as a result of past design work. Working on paper also discourages revision. To redesign a product (even for the most minor of changes) may require that all of the drawings required for the product be redone. This is a very time-consuming process. Even during the initial design of the product, engineers are often hesitant to commit themselves to paper just in case they are wrong. Also, there is no capability for testing the operation of the design. In many instances the only time that anyone knows whether the design will work is when it is out on the shop floor. This is the wrong place to be doing such testing.

A CAD system eliminates many of these shortcomings. It encourages engineers to commit ideas to the CAD system, where they become part of a database. They can now be retrieved quickly and scanned by an engineer to see if they can be used as a basis for generating new products. A CAD system, when combined with computer graphics, allows the engineer to use computer simulation and solid models to test the operation of proposed designs. In other words, engineers now spend more of their time designing new products that work rather than designing products that don't work or reinventing the wheel. The result is improved productivity and an engineering function that is truly supportive of corporate and strategic objectives.

Computer graphics, the third decision, were a part of nearly every aspect of the manufacturing and design process. Graphics are used extensively in predesign stage negotiations to ensure that the product and its requirements, as envisioned by the customer, correspond to the product being designed at Ingersoll Milling. It is also used during the design stage to develop and evaluate the product design. A computer simulation model of the proposed design is built on the computer screen and its operation dynamically simulated. During this simulation, any problems (e.g., two parts or cutting surfaces colliding owing to inadequate tolerances) are identified and corrected before the product is released to process routing and ultimately to the shop floor.

Computer graphics link engineering and numerical control programming. The design created by the engineer and captured in the computer graphics model is passed on to the numerical control programmer. During programming, the programmer can rotate the part on the computer screen and evaluate the effects of the proposed routing *before* the order is released. Any problems with routings due to product design can be referred back to engineering by the programmer. As a result, computer graphics have become part of the cement on which simultaneous engineering is built. Finally, computer graphics have helped management attain its objective of building the product right the first time.

The final decision, the revision of the corporate database, was one that was forced on management. That is, management found that the information used by Ingersoll Milling would not support the continued growth and development of the company. The organization of the data was not conducive to the activities of the company in the areas of CAD and CAM.

For example, it was noted that before this revision Ingersoll Milling was faced with some 1,300 different application programs and systems, operating from 225 different computer master files full of redundant and uncoordinated data. Many of these 1,300 application programs were nothing more than bridge programs taking data out of one file and feeding it into another file in an attempt to improve the integrity of the files, and therefore the usability of the management reports that came from them. In short, the database was "out of control." Yet, management realized that the database was one of the major foundations to the successful implementation and development of CIM at Ingersoll.

As a result, two years were spent by the design staff on rewriting the entire corporate database. It was a major undertaking but one that had to be done. The result was a completely integrated corporate database system that was accurate and without any redundancy. In building a single corporate database, many of the obstacles to cooperation were also eliminated. If Ingersoll Milling had not committed the time and effort to rewrite the database, it is doubtful that the CIM movement would have been successful.

These four decisions had also enabled Ingersoll Milling to build an integrated management and business information system. This has ensured that the activities of manufacturing and engineering are always coordinated with those of other functions in the firm, thus supporting the achievement of corporate strategic objectives.

There are two other features pertaining to computer integrated manufacturing at Ingersoll Milling that should be discussed before moving to the second stage in the CIM evolution. These involve the planning and control system and the implementation and use of simultaneous engineering.

Planning and Control at Ingersoll Milling

Management has done more than simply invest its resources in the implementation and development of very sophisticated computer-based systems for use by engineering and manufacturing. It has also developed an effective and integrated production planning and control system. This has been designed to support all of the planning and control activities necessary to process a customer order. In structure, the system is very similar to MRP II.

The major modules of the system are

- Master scheduling
- Engineering design
    - Assembly and piece drawing
    - Bills of materials (BOMs)
- Production planning
- Inventory control
- Purchasing and accounts payable
- Numerical control programming and postprocessing
- Assembly
- Job costing and management reports

- Shop floor control

These modules are important because they provide the structure required by computer integrated manufacturing. They are also a major reason why CIM is a corporate system rather than a manufacturing or engineering system.[2]

The Planning and Control System in Operation

The steps in the planning and control system follow the order-processing activity very closely. Each step in the process supports a major step in order processing.

The heart of the system is the bill of materials system. It is in this system that the product designed by engineering is stated in manufacturing terms. The bill of materials system generates the gross requirements needed to build a particular assembly. This is combined with the master schedule to derive a time-phased schedule of gross requirements. These are then compared against available inventory to develop the time-phased net requirements. These dated net requirements are split into "make" and "buy" items. These items in turn make demands on the manufacturing and purchasing organizations, respectively. When these two supply systems complete their work and the necessary material is drawn from inventory, the planning and control system then has everything needed to build the assembly.

Using this information, the system checks against the bill of materials to produce a shortage list of any additional requirements that may have come up since the original bill of materials was issued. This shortage list is available both in the form of a printed report and as an on-line terminal inquiry.

Simultaneous Engineering

A bottleneck that had to be resolved early in the CIM journey involved the organization of the design process. Initially, the process was sequential in nature with designing and delivering a product to the customer envisioned as a set of distinct steps. Each step began with the completion of the preceding step. For example, design began with the customer generating the product specifications. At the end of design, the product was turned over to process routing where the NC programs would be generated. There were two major difficulties with this procedure. First it lengthened the total lead time. The lead time for the overall design was the sum of the lead times for the individual activities. The larger

[2]More detailed information about the operation and structure of planning and control systems can be found in Steven A. Melnyk and Phillip L. Carter, *Shop Floor Control: Principles, Practices and Case Studies* (Falls Church, Va.: American Production and Inventory Control Society, 1987), pp. 429–470.

the number of activities, the longer the lead time. Second and more important, the process was inefficient, costly, and forced the design of products in which compromises were present.

In a sequential process, people involved in each step of the process focus primarily on issues of greatest concern to them. For example, design engineers are most interested in trying to develop a product that best meets the customer's requirements. Issues outside of this focus are often ignored. The design created by the design engineer may be very expensive to build or process on the shop floor because it does not lend itself to efficient NC programming or because it requires special tools. When these problems are identified later on in the process, changing the design to overcome these problems is time-consuming, difficult, and expensive. The sequential approach to designing products also created a great deal of frustration. Engineers who had provided a design that satisfied the customer's needs were now being asked to redo the design because it could not be built efficiently on the shop floor. Similarly, NC programmers and shop floor personnel were frustrated by the continuous arrival of designs that could not easily be built.

The real problem did not lie with either the design engineers or the NC programmers, but with the process. Given the sequential process, there was no real opportunity for other functions to provide input early. For example, the NC programming could only evaluate the design *after* it arrived in process routing. At this time it was too late. To eliminate these difficulties the process was changed. A new method of organizing the design and manufacturing activities was introduced, a concept now referred to as *simultaneous engineering*.

Simultaneous engineering is based on the view that the design of a product or process is not a sequential process but rather a set of coordinated parallel processes. The design of a product affects such functions as quality control, purchasing, process routing, NC programming, manufacturing, engineering, finance and accounting, production and inventory control, and marketing. As a result, these areas should be involved in the design stage of the product to provide input during the early stages of the product design. Modifying the product at this stage is far less expensive than introducing modifications at a later stage. It is also quicker to accomplish. Often, the result of simultaneous engineering is a cheaper, better-quality product delivered on time.

Technology in the form of computer graphics has enhanced the effectiveness of simultaneous engineering. The input of such people as the NC programmer is obtained in real time through the on-line systems linking the programmer with the engineer. Problems can be identified and communicated immediately as the programmer becomes aware of them. The response to the problem is just as immediate. The result is a better product delivered in considerably less time.

COMPUTER INTEGRATED FLEXIBLE MANUFACTURING

Computer integrated flexible manufacturing is an extension of the CIM concept in place at Ingersoll Milling. It represents the confluence of CIM and the flexible manufacturing system. It is a result of a strategic decision made by management to replace the "light" machine shop[3] by a $20 million FMS. The replacement of the light shop by the FMS is intended to satisfy several objectives.

The first objective is to enable management to increase overall manufacturing capacity while reducing the amount of floor space needed for the production facilities. Ingersoll Milling is a growing company that is located in an area where the physical expansion of facilities is no longer possible. The only economical option for accommodating the growth is to make more efficient use of current production floor space. By implementing FMS, production capacity can be increased without increasing the amount of floor space needed.

Second, in converting to the use of flexible manufacturing systems, management has attempted to improve in the shop overall product quality by gaining a new level of control over some of the most critical processes in the manufacturing process. The use of FMS has not only improved quality but also reduced manufacturing costs and increased the levels of production capacity available.

Finally, the flexible manufacturing systems not only offer increased production capacity but also provide management with a working laboratory. As the personnel operate the FMS, they learn how to take advantage of the capabilities offered. They also learn the limitations present and the problems that may affect operations. These lessons are then passed on to those customers who purchase a flexible manufacturing system from Ingersoll Milling.

The Flexible Manufacturing System

The FMS is now operational and consists of five Bohle-Ingersoll CNC machining centers and four Lablond-Makino CNC machining centers. The Bohle-Ingersoll centers have eliminated 17 older machines. Owing to a management policy that forces management to justify the continued use of any equipment that is at least five years old, the machines replaced were still relatively new. In some companies these replaced machines would have been considered "state of the art." However, these machines cannot match the productivity gains gained from the use of the new flexible manufacturing systems.

---

[3]The light shop is responsible for manufacturing all prismatic parts, 1 cubic meter and smaller.

The major difference between the Bohle-Ingersoll and Lablond-Makino centers involves the size of the parts produced. The Bohle-Ingersoll centers process parts that cannot be easily lifted by a worker and require some form of mechanized material handling system. In contrast, the Lablond-Makino centers treat parts that are easily transported by the worker. These machines are suitable for automatic, around-the-clock operation with provision for unmanned operation. In practice, these machines are never operated unmanned because of the low production lot sizes. The capabilities of these machine centers include

- Fail-safe seating accuracy of pallets through the sensing of foreign matter on pallet changer rest pads.
- Automatic release of the spindle drive in the event of overload.
- Automatic release of axis drives in the event of a collision.
- Automatic changing of a servo-driving boring and facing head with slide feed for taper turning, profiling, and other complex operations.
- Air-jet cleaning of the spindle and tool tapers during the tool change cycle.
- Diagnostic and maintenance alerts with the pinpointing of problem locations.

The system is still being expanded.

In addition to the CNC machining centers, the system includes wire-guided vehicles and pallet setup stations. When the implementation is completed, the system will also include coordinate index measuring machines (CMM) for inspection, wash and debur stations, and a complete tool and fixture management system. The CMMs are installed now, but they are in a separate temperature-controlled room. The FMS is controlled by a Digital Equipment Corporation (DEC) VAX-3400 cell management computer in direct communication with the corporate host computer, which is an HDS model XL-90 (operating at 42 MIPS).[4] The VAX cell management computer is in constant contact with the Allen-Bradley 8200 CNC controllers via the ABCAM two-way communications interface.

All of the components of the system (i.e., the machines, the carts, the tool setup and transportation system) communicate with each other through the cell computer. When a warning or emergency situation occurs, the system will try to correct itself. If it is unable to do so, it will signal a fault condition through the monitoring system to alert the cell manager. The system is designed to be as nearly automatic as possible.

---

[4]MIPS stands for millions of instructions per second.

The FMS software is on a database. The team that built the database management system underlying the CIM system has expanded it to include the full CIFM system. Currently, Ingersoll Milling's FMS shop is operating at level 1 software with the goal being to move to the advance level 2 software. Ingersoll's FMS level 2 software is organized around the seven basic resources that must be allocated and controlled in any flexible manufacturing system. These are

Parts
Pallets
Fixtures
Tools
Storage (AS/RS)
Transportation (AGVs, Robots, etc.)
Machines (cutting, deburring, washing, measuring)

Production in the Flexible Manufacturing System

Production in the FMS area is started as soon as all three of the following events have occurred:

NC programming is complete.
The bar stock is cut or the weldment is completed.
The appropriate tooling and fixtures are available.

Tool setters collect the tools identified by the NC program. After gauging and offsetting, the tool setters put the tools on a special trolley. The trolley keeps together all the tools required to machine a given part. This allows the tools to be easily moved around the department, while guaranteeing that they are available when the part is being machined.

The monitoring system tracks the use of each tool, as well as the current system and location of each tool. The toolroom operator is notified at the time a part is being set up on the setup standards regarding what tools will be required. The automatic guided vehicles move the pallet containing the fixtured part to the incoming table of each machine. Similarly, they retrieve a completed pallet from the outgoing table at each machine. This is all done automatically under computer control.

The machine, when ready, moves the part from the input station to the workstation. A set of machining tasks that can be completed without refixturing is called an operation. When an operation is complete, the machine moves the part to the output station. The AGV (automated guided vehicle) collects the part from the output station and returns it to the setup area, where it is removed from its fixture and, if finished, sent to assembly. Otherwise, the part is set up for the next operation.

The flexible manufacturing systems are used to machine a range of parts (i.e., the problem set) that is essentially open ended. In 1985, for example, Ingersoll Milling produced in the light shop some 7,000 different prismatic parts. Each part averaged 2.5 operations (i.e., new setups), so that the system saw a problem set of approximately 17,500 unique setups. Of these parts, 30 percent were machined in lots of one and 50 percent will not be manufactured again. Management soon realized that its FMS had to process thousands of different parts in very small lots. Planning and implementing an FMS in this setting meant planning for the most general possible application of flexible manufacturing systems. The large size and extensive variability of problem set has helped the management learn a great deal about the planning, implementation, operation, and management of a flexible manufacturing system—information that can be transferred to any of Ingersoll Milling's customers. As the CEO of Ingersoll Milling, Mr. Gaylord, put it;

If you automate this shop, you could automate *any* shop.

Before leaving the discussion of the use of flexible manufacturing systems at Ingersoll Milling, it is of interest to examine an unexpected result. The flexible manufacturing systems have been integrated into the overall operation of the shop floor. Currently, available uptime is averaging 80 percent (as compared with the 20 to 30 percent found in most shops). However, management cannot run the FMS at its highest level, since the assembly floor cannot use the parts as fast as the FMS can produce them. As a result, the FMS is being paced by the assembly floor needs.

Finally, as machining (and parts) get more complicated, this complexity will now appear in the software and test systems required to support production.

Classifying the CIM System

Using the classification scheme presented in Chapter 3, it can be seen that Ingersoll Milling has implemented a well-developed Type IV CIM system. It is very tightly integrated. Channel integration, the last integration component now being implemented at Ingersoll, is central to *interorganizational CIM*, the next stage in the CIM journey.[5] Channel integration is intended to bind not only the customer to Ingersoll Milling but also the various suppliers.

---

[5]This development is discussed in greater detail later in this chapter.

CIM AT INGERSOLL MILLING: THE BENEFITS

At Ingersoll Milling CIM and CIFM have been extremely successful in practice. Much of the success and reputation currently enjoyed by Ingersoll could not have been attained without CIM and CIFM. The direct benefits gained by the firm as a result of implementing these systems is presented below under overall benefits and specific examples. By presenting the benefits in this manner, a better appreciation of the effects of CIM and CIFM can be gained.

General Benefits

Overall, from CIM and, more recently (since 1983) CIFM, Ingersoll Milling is able to offer its customers better value for their money. The company is achieving its strategic objectives in large part owing to CIM and CIFM. In a highly competitive marketplace, Ingersoll Milling is surviving and prospering. Customers are buying from Ingersoll because they see it as offering the best value for the money. The ultimate proof of this consumer acceptance can be found by examining the backlog of orders. Since 1988, the current manufacturing capacity has been sold out for the next 12 months.

Specific Examples

The impact of CIM and CIFM on the operations of Ingersoll Milling is best illustrated by examining several examples of how these systems have affected operations at the plant.

Improvements on the Shop Floor

CIM and CIFM are the latest steps in a continuing program to improve operations in the machine shop. Between 1968 and 1986, ninety-five manual machines were replaced with 28 numerically controlled machines. This resulted in 285 operators being replaced by 84 operators. However, to achieve this reduction, management had to add 61 NC programmers. The net result was a reduction of 140 people. While the number of shop floor personnel fell, the production output rose dramatically.

Management is continuing to make improvements in the NC programming for multiaxis machining centers. For example, in the last four years the average programming time for a complex part has been reduced by some 40 percent from 26.6 hours per part to 15.5 hours per part.

Improvements in the Productivity of the NC Torch

An important feature that has been added to the system is the nesting system for the NC torch. The main advantage of the nesting system is that it is integrated with the entire corporate business system. The result is that the bill of materials, routing, and master schedule systems, working together, present to the nesting manager all of the requirements that are unfilled and waiting to be nested. He simply selects the problem set and sends it to the computer for nesting. With the problem set defined, the computer automatically withdraws from the CADAM database all of the geometric shapes for the parts placed in the problem set. It then prepares them for nesting. Next, the parts are placed in an efficient nest for the nesting manager to accept, reject, or modify. After the manager accepts the nesting pattern, the computer prepares the NC program for the actual nest with the most efficient remnant possible. This pattern is cataloged and stored in the computer for future use. The NC program, the labels for each of the parts to be cut out, and the manuscript are then available for the torch. The NC program can be sent down on the DNC system.

The result of this effort has been a 90 percent reduction in the manpower involved with this aspect of operations. Other benefits have included increased throughput by 125 plates per year and a reduction in turnaround time from 10 days to 3 days. Finally, the system provides for a direct and automatic updating of the routing, cost, and payroll systems.

Other Improvements

Other areas have benefited from the CIFM system. The routing department has been reduced by more than 50 percent—a direct result of the Ingersoll computer-aided process planning system (I-CAPP). The new on-line purchasing system, which is directly linked to the bill of materials system, has enabled management to reduce the personnel in the purchasing department by 43 percent. There has been a 34 percent reduction in the quality control organization. Finally, in 1988, the work-in-process in the machine shop was reduced by 50 percent as a result of the introduction of a new, on-line, just-in-time dispatching system.

CIOM: THE NEXT STAGE

In 1987 management began the transition to computer integrated optimized manufacturing. CIOM involves three major initiatives:

- Extending the CIM system to the suppliers (interorganizational CIM).
- Improving the productivity of such support functions as accounting.
- Investigating the suitability of new technologies such as artificial intelligence and expert systems.

Underlying these three initiatives is the fundamental desire of Ingersoll Milling to offer its customers better value for their dollars.

Interorganizational CIM

Almost everything that has been done to date has focused on internal implementation of CIM. The next step in the evolution of CIM is to expand it to encompass both suppliers and customers. The term Ingersoll is applying to this broadening of the CIM concept is "interorganizational CIM" (or alternatively, "external CIM"). There are strong indications that this concept is becoming reality. For example, companies are becoming linked as a result of such developments as the local area network (LAN), electronic data interchange (EDI), and standardized communication protocols such as MAP (manufacturing automation protocol) and TOP (technical and office protocol).

Further indications of this movement toward interorganizational CIM can be found in the efforts to standardize electronic product data definitions. Examples of such definitions are the Air Force programs called PDDI (product definition data interface) and GMAP (geometric modeling applications interface) as well as the standardization effort under way by the National Bureau of Standards called PDES (production data exchange standards).

Ingersoll Milling is now preparing for an environment in which the activities of suppliers, customers, and Ingersoll Milling are electronically linked and coordinated.

Improving the Productivity of Support Functions

In the past, the primary focus of CIM has been the engineering and manufacturing activities. Now that the productivity of these activities has been enhanced, attention is shifting to such support functions as accounting. In some ways accounting is a nonproductive use of resources. It occurs outside the manufacturing activity. Its results are often reported back to the manufacturing and design process after a significant time delay. If the productivity of these support functions is to be improved, the functions must be incorporated directly into the manufacturing and design activities. Cost accounting, for example, must move out of the cost accounting department and onto the shop floor. As the product is being built, the people on the floor should know how they are doing on cost. This information should be made available immediately.

Management is now exploring the question of how best to incorporate these support activities directly into the production and design process.

The Exploration of New Technologies

Manufacturing technology is continuously developing. Ingersoll Milling is exploring many of these new technological advances to assess their fit. In most cases these investigations are in the educational (learning) stages. So far, management has identified three technologies that are potentially useful: (1) artificial intelligence and expert systems, (2) solids modeling, and (3) training simulators.

Artificial intelligence and expert systems offer Ingersoll Milling the opportunity to capture and make use of human expertise in designing and building machines.

Solids modeling offers the possibility of incorporating more information into the products designed. Currently, most CAD/CAM systems are just advanced drafting systems. Many have progressed from two-dimensional drawings to either three-dimensional wire frame representations or three-dimensional surface models. However, these are often really nothing more than very attractive pictures. They do not adequately represent the complete product definition. True product definition is attractive because it not only conveys the geometry of the part but also the topology of connecting the elements of geometry together as well as the tolerances required for each element of the geometry. Product definition in the future will be represented in solid form and include features.

Ingersoll Milling is now working with Cimplex Corporation of Campbell, California, for the implementation of such a system. This system will allow personnel to define both construction and machined features (e.g., chamfer, countersink, threaded holders, and slots and grooves). This system, when it becomes operational, should result in a very advanced routing, generative process planning, and NC programming system.

Training Simulators

The final technology being investigated is a new graphics training simulator. As new advanced systems come on stream, training becomes increasingly more important. Based on the experiences of the military, which has demonstrated a 50 percent learning time savings with training simulations, Ingersoll Milling is now investing heavily in developing training simulation technology for large manufacturing systems.

Such systems are intended to train operators, programmers, and maintenance personnel. The systems substitute color graphics terminals for the actual machines. All machine motions are depicted visually in real time in response to the NC tape part programs or MDI (manual data input) commands entered by the operator. The graphics animation is an accurate representation of actual machine motion (based on 40-millisecond data updates) so that the operator can see the effect of his or her actions. These systems allow people to make mistakes and to learn from these errors. Using a real system, such mistakes could be devastating.

With traditional trial and error learning, it can take operators 2.5 years to develop full proficiency on a new machining center. By substituting simulation training for trial and error, the learning process is speeded up, the costs of learning errors are greatly reduced, and the operator attains a higher overall level of proficiency. The result is that the operator is better able to take advantage of all the machine's features. The operator learns how to anticipate problems and how to avoid them.

CIM AT INGERSOLL MILLING: LESSONS LEARNED

In reviewing the experiences of Ingersoll Milling, management has identified 10 principles (referred to by Ingersoll management as the "10 commandments" of CIM planning and implementation) that it believes are critical to the success of any CIM system. These are as follows:

1. *Always try to anticipate new or emerging technologies.* The implementation of any CIM system should be regarded as a long-term undertaking. If the system is based on mature or, in some cases, state-of-the-art technologies, such technologies may be obsolete and noncompetitive by the time the system is installed. Therefore, management must always be ready to look for emerging technologies. Management must identify those technologies that will be useful when the CIM system is ready to incorporate the technologies.

2. *There must be corporate-wide agreement on strategic directions and objectives.* Before the decision to implement CIM is made, there must be a clear and widely held vision of where the company is going and what capabilities the company needs to achieve its objectives. This vision of the future direction for the company is critical, for it forms the basis on which technology is evaluated and purchased and the necessary investments authorized. No automated system can ever be flexible enough to compensate for the lack of clear corporate objectives.

This overall corporate vision must be supported by compatible departmental objectives. Every department must be prepared to identify a set of objectives that its personnel are willing to commit to. In drawing up these objectives within the department, a sense of ownership is developed. That is, every department begins to feel that it has a part in the overall planning process for CIM and ultimately in its success or failure.

Such planning is iterative in nature. It starts in a top-down fashion with the corporate objectives. Each department must translate these overall objectives into a set of department-specific objectives that support the broad objectives.

3. *A single system-coordinating organization must be established.* Success in integrating the firm is greatly dependent on a strong coordinating organization. This organization can be the systems department or it can be the systems department operating under the umbrella of a corporate steering committee. The acid test of this organization is that it must be an organization bent on effectively serving the line functions.

4. *CIM must be a corporate system.* A successful CIM system can never be simply a system intended to be used primarily by the engineering and manufacturing functions. These functions, while important, are not the only ones that affect how a product is delivered to the customer. There are other functions involved—functions such as purchasing, logistics, personnel, accounting, marketing, and top management. These functions must be brought under the CIM umbrella. The CIM system must coordinate the activities of these other functions. Without their input the firm cannot successfully produce and deliver the goods demanded by the customer or the market.

5. *Operations management must determine the what of CIM.* The what of CIM pertains to the functions that the CIM system should perform. These define what the CIM system can and will do and what it will not do. These decisions must be made by those people and functions that will ultimately use the system. If they define the capabilities of the system and if the system accurately embodies these requirements, the system will be used. This is an important consideration because the CIM system will be successful only to the extent that it is used. When implementing CIM, management must ensure that the implementation process is very much "user driven."

6. *Systems management must decide on the how of CIM.* The how of CIM determines the manner in which the capabilities identified by the users as important are implemented in the resulting computer system. These decisions must be left up to that group that is most familiar with computer systems: the systems management group. They must become the coordinating function to ensure that the plans and capabilities fit together using the most advanced hardware and software and databases available to implement them. The systems management group must also maintain and safeguard the overall integrity of the computer system supporting the CIM implementation effort. Finally, the systems management group must ensure that the computer systems and interfaces implemented are transparent to the user. That is, the computer systems must not inhibit the user's access to information contained in the database. Rather, they must simplify the accessing of data.

7. T*he CIM implementation effort must be managed through the development and use of a comprehensive plan.* The implementation of a CIM system is a major undertaking that involves numerous people and functions. It is also a long-term undertaking. If not managed carefully, management can lose control of the process. To manage successfully such a project requires that an implementation plan be developed.

This plan must break the implementation process into a series of steps. At each step management must identify the tasks to be accomplished, the time frame in which these tasks are to be accomplished, the goals (objectives) to be satisfied, and the risks and uncertainties present.

The resulting plan should be in a series of linked, attainable, and controllable-sized projects. Each project should be assigned specific goals and target dates. Management must then track these projects from both a cost and a schedule perspective.

However, in developing this plan management should make allowances for changes in technologies. As a result, planning should be flexible. As new technologies emerge or as changes in the business environment take place, management must be prepared to evaluate these changes and be ready to exploit new ideas and modify the plans accordingly.

8. *Under the successful CIM system there is only one corporate database.* Any CIM system is built on a foundation consisting, in part, of an integrated corporate database. The corporate database is the repository of all of the information that the company needs to produce, design, cost, and sell the product. Because of the importance of good information to the success of CIM, management must centralize this information in one location—a corporate database.

Developing an effective corporate database is more than simply storing information in one location. The information found in a well-structured database must be accessible to all those who need access to it. It must have a high level of accuracy. It must be presented in a format immediately useful to the person requesting the information. It must be amenable to rapid revision and updating. Finally, it must have a high degree of data integrity.

A common database is a *must* for building an effective CIM system. It is also a major undertaking which is best started early in CIM implementation.

9. *The gains made in implementing CIM must be held. Holding these gains requires an extensive investment in training and participation.* The implementation of CIM must change the manner in which people manage the system. However, management must realize that people will use a new system when first installed, but if not encouraged constantly, people will tend to revert to the old system. Why? The old system is familiar to them. They feel more comfortable with the old system.

The most effective method of countering this tendency is through participation and training. If the people who will use the system are allowed to participate in its planning, design, testing, and implementation, they will *want* to make it a success. If they are given the proper training, they will *be able* to make it a success. Without participation and training, it is almost impossible for management to hold its gains with CIM and to make the system grow.

10. *CIM systems require and demand the full involvement and support of top management.* Like the corporate database, this is a must. There is no way to avoid it. Without top management support, the CIM system will not succeed. It will not become a *corporate* system. At best, it will simply become a better way or a glorified CAD or CAM system.

Top management involvement and support does not mean that management be willing to write blank checks or develop a trust in CIM that is based on blind faith. Rather, support means understanding and knowledgeable guidance. It means real leadership and involvement by *all* levels of management. It means courage to learn from the failures (and there will be some). Finally, it means the willingness to opportunistically exploit any success. In short, top management must become the supporters and champions of CIM. They must be willing to defend the CIM movement against any detractors.They must be willing to force the compliance of anyone in the firm with the requirements of the CIM system (where those requirements are reasonable or necessary).

Ingersoll Milling has been fortunate in that the architects for its CIM system have come from the highest levels of top management. The president, Mr. Gaylord, and the vice president of systems management, George Hess, have played a critical role in the development and growth of CIM. They have introduced many of the concepts now present in the CIM system. They have also been a major force in the continued evolution, extension, and refinement of CIM. In large part, they are the reason why CIM has been so successful at Ingersoll Milling.

CONCLUDING COMMENTS

It is difficult to overstate the importance of computer integrated manufacturing to the management of Ingersoll Milling. This system helps management fulfill its promise to the customer: the delivery of a well-engineered and well-built product to the customer by the promised due date. It also helps Ingersoll Milling survive and prosper in a very difficult and demanding manufacturing environment.

The best way to summarize management's view of CIM (and CIFM, the current state of CIM at Ingersoll Milling) is to quote the comments of the chairman and chief executive officer about the value of CIM to Ingersoll Milling:

> The gains our company is making through computer integrated flexible manufacturing represent our most important immediately available competitive edge.... Computer integrated flexible manufacturing, more so than with any other large investment in a manufacturing company, is based on people's brains and imagina-

tion. There is nothing at all automatic about the benefits. It simply depends on people and the directions they get.

In order to build the skills and gain the experience, one must get started. This is not a technique where you wait for somebody else to perfect the programs or the twenty-first-century version.

There is no way one can effectively put in computer integrated manufacturing with an eye on immediate returns on investment. People say American industry suffers because its managers pay too much attention to the short-term ROI. Well, if you go at computer integrated flexible manufacturing incrementally, looking to get every penny invested back within a short and definite time scale, this competitive advantage will be lost to you.

CHAPTER 7

THE CONE DRIVE OPERATION OF TEXTRON CORPORATION

EXECUTIVE SUMMARY

The Cone Drive Operation of Textron Corporation (Cone Drive) of Traverse City, Michigan, is a small batch manufacturing operation that produces double enveloping worm gears and speed reducers.[1] It is also a successful user of computer integrated manufacturing. Its experiences with CIM are unusual for several reasons. First, it is a relatively small company (less than 250 people). Second, for the firm, the use of CIM was a matter of survival. If management did not implement CIM and make this system work, the alternative for Cone Drive was a sure, slow death in the marketplace—due to aggressive foreign competition. Third, the introduction of CIM led to a reduction in the overall size of the firm (from some 550 employees at the start to some 250 employees now). Cone Drive manufactures a mature product with limited opportunity for growth. The market is very price sensitive. To survive in such an environment, Cone Drive had to emphasize productivity. However, with increases in productivity came the realization that fewer employees were needed. In Cone Drive we have a good illustration of the importance of using standard, off-the-shelf hardware and software. The use of such standard inputs forces the firm to change how it manages its business (to conform to the requirements of the hardware). Often, this is not a change for the worse.

Finally, in Cone Drive we see the importance of top management. At Cone Drive top management played a far more critical and extensive role. They not only provided financial support but were also instrumental in overcoming organizational resistance to CIM. They attacked bottlenecks in the firm. They also preserved the *integrity* of the CIM concept. They developed a vision of what CIM would be like at Cone Drive. Once Cone Drive began the implementation

[1]This case study is based on a presentation done by Paul Brauninger, CPIM. Mr Brauninger was a member of the management team that helped implement CIM at Cone Drive. At that time, his responsibilities included the overall coordination of the computerized systems as well as the installation of the flexible manufacturing cells. Since 1988, Paul Brauninger has been associated with GMD, Inc., of Grand Rapids, Michigan.

of CIM, they protected this vision against modifications introduced by others. Such changes, though small when considered separately, often result in a gradual distortion of the CIM concept, with the result that what is conceived at the start is often very different from what is implemented. These modifications are often introduced by people who are unwilling to conform to the demands that the new system imposes.

At present, the CIM system at Cone Drive has reached a plateau in its evolution due to a change in ownership. The new owners are not as committed to the CIM concept as the previous owners. One interesting effect of this transition has been an increase in turnover. People who had become accustomed to the changes and continuous improvement stemming from CIM have been frustrated by this slowdown. As a result, they decided to leave the firm rather than accept the situation as it currently exists.

INTRODUCTION

Cone Drive is a relatively small batch manufacturing firm that is engaged in the production of double enveloping worm gears. The engineering principles underlying these gears enable them to generate more power for a given size. Cone Drive has long been considered a pioneer in this line of products. The founder of the company developed the first cone drive in the 1930s.

From 1955 to 1987 Cone Drive was also a division of the Ex-Cell-O Corporation. It was during this time that most of the work in implementing the computer integrated manufacturing system took place (with the implementation beginning in 1980). In 1987 ownership was transferred to the Textron Corporation. Being a subsidiary of these two companies has helped Cone Drive to implement CIM. It has been able to draw on the financial resources of these two large firms, when needed.

MARKET DEMAND AND CONE DRIVE

Annual sales for Cone Drive are between $25 and $50 million. The sales are made primarily to the capital-goods segment (i.e., machine tools, conveyors, the steel industry, and mining equipment) located in the United States and Canada. Production takes one of three forms. About one-third of production is in the form of standards. These are primarily ATO in that they are assembled *only* after the receipt of the customer's order. The next one-third of production is MTO. The final one-third are specials. These are gears that must fulfill unique or very specific requirements, and as a result are ETO.

Irrespective of the type of production (ATO, MTO, or ETO), competition in this market has focused on four major attributes:

Developing the most favorable trade-off between power and efficiency. A cone drive is a product that offers more power in its size range than other gear systems. However, this power comes at a cost. Unlike other gearing systems, this increased power comes at the cost of increased power consumption. To be competitive, the trade-off between efficiency and power must be managed very carefully. Both elements must be improved. Ideally, the level of power should increase while the rate of energy consumption falls. Since the energy crisis of the early 1970s, this consideration has increased in importance. It has placed a premium on effective engineering.

Price. Because of the large number of near substitutes, the market is price sensitive. The customer's purchase decision is based not only on the initial purchase cost but also the cost of generating power. As a result, Cone Drive is under constant pressure to reduce all the costs associated with the purchase and use of a cone drive. With the increased presence of foreign competition, cost control has increased in importance.

On-time delivery. This attribute encompasses a whole spectrum of activity. It includes the physical delivery of the product, the design time, and the time to respond to a customer's request. It also deals with the issues of average lead time and variability. That is, reducing the average lead time for customer response is not enough by itself. Reductions in lead-time variability must also take place. Together, these two activities create both timeliness and consistency. They reduce the risk associated in doing business with a company such as Cone Drive.

Quality. Of these four attributes, quality is typically treated as an order-qualifying attribute. That is, the buyer expects good quality irrespective of the supplier. The other three factors tend to play a more significant role as order winners. However, the importance of these attributes is very dependent on the market segment being served. Typically, for the MTO and ATO items, price and lead time are more important, whereas for the ETO segment power and efficiency and lead time are more important.

Cone Drive has been very successful in competing in this marketplace due to its extensive use of computer integrated manufacturing (CIM) systems. Cone Drive is an example of what can be done in a small firm with CIM to improve the overall effectiveness of the firm. In 1985 the Computer and Automated Systems Association of the Society of Manufacturing Engineers recognized the achievements of the management of Cone Drive by awarding it the LEAD award. In recognizing its achievements with CIM, Thomas L. Moon, the chairman of the Industry LEAD Award Committee, noted that Cone Drive was "a showcase model of computer integration for small manufacturing operations."

However, Cone Drive's implementation of CIM has coincided with a general reduction in the company's employment base. At the start of the CIM journey in 1980, Cone Drive employed some 550 employees. These were divided between 150 office and 400 shop personnel. By 1987 the total number of employees at Cone Drive had fallen to some 250 people. These two developments are not unrelated. Rather, they reflect the effects of the market that Cone Drive serves.

Cone Drive competes primarily in a mature market. Its customers all compete in markets that are themselves in decline. With fewer sales and no change in the number of suppliers, overcapacity will tend to develop over time. This, in turn, will increase the level of market competitiveness. Under these circumstances Cone Drive's customers will become more sensitive to cost. Increasingly, the firms in this market will be forced to reduce costs and to improve corporate productivity. Without growth potential, this will mean that the firms must downsize. They cannot have excessive or unproductive employees. These influences are very evident in the experiences of Cone Drive with computer integrated manufacturing.

Given the overall importance of the market, it is very useful to understand the developments that took place in this market between 1970 and 1987. These developments have made CIM a necessity for Cone Drive's survival.

Changes in the Marketplace: 1970 to 1987

From a *product life cycle* perspective, the period between 1970 and 1987 was a time when Cone Drive's product moved from the late stages of growth into maturity and from maturity, into decline.

The period from 1970 to 1973 was a period of growth. During this time, Cone Drive's gears were in demand because of their advantage in the power-to-weight area. The higher energy consumption of the gears was not considered to be much of a problem. This view changed in 1973.

In 1973 the U.S. faced the first of its energy crises. Managers became more aware of the need for efficiency. For Cone Drive, this meant that the design of its cone drives had to be reexamined and changed to improve its energy consumption. Demand began to slow down; Cone Drive's share of the market began to fall.

In 1977 another significant change in the market took place. Prior to 1977 there were almost no foreign suppliers in the American market. American suppliers served American customers. In 1977 this situation changed when several European manufacturers entered the market. These firms came primarily from Germany, Great Britain, and Italy. They attempted to gain market share by identifying specific *niches* in which they had a significant competitive advantage.

They decided to compete on lead times and reliability. They promised customers delivery lead times that were shorter than those offered by American suppliers. Furthermore, they consistently met these deliveries—something that the American suppliers could not match. Finally, they offered products that were of good quality and competitively priced. Effectively, by focusing on these areas the European firms had targeted the market segment that was more standard.

The presence of foreign competition was further increased in 1983 by the entrance of the Japanese. More important, however, were the fundamental structural changes that took place between 1973 and 1983. This was a period during which business was very good. Demand was high. Suppliers were selling everything that they could produce. Since there was not enough capacity at American companies such as Cone Drive to satisfy all of this demand, customers began to buy from foreign suppliers. It is this development that provided the foreign suppliers with the needed opening into the American market.

In 1983 the demand for cone drives began to decline precipitously. As a result, there was a great deal of unused manufacturing capacity. Customers began to be far more selective about where they placed their orders. This was in part a reaction of the buyers to what had been seen as poor service in the past.

In general, customers were not pleased with the American suppliers of cone drives and other similar gearing systems. They had complained about the poor quality of service, the poor delivery reliability, and long delivery lead times. By 1983, however, Cone Drive was well on its way to implementing its CIM system. Management was aware of the same problems as far back as 1980. They had selected CIM as the most effective system for improving customer service and for attacking high costs created by the old production planning and control system. By 1983 the benefits of this system were beginning to be evident.

Cone Drive prior to CIM (1977 to 1980)

By 1980 management had become aware that the current system for controlling production was no longer adequate. This system incorporated a Min/Max system for inventory control. Future production was often based on past usage. Production priorities did not reflect true needs. The system in place was often like the "horror stories" told by manufacturing consultants. It was the root cause for many of the customer-delivery problems affecting Cone Drive. It was a system that created a large number of backlogs and delays. It required the frequent intervention of expediters. In 1980 there were six to eight full-time expediters whose job was to simply get the production out.

Since the system could not get the work out on time, people started to release work earlier to the shop floor. The only effect of this action was an increase in the inventory on the floor. By 1980 it was estimated that between 33 and 50 percent of the floor space was used to store inventory. At its worst, railroad box cars were rented and used for inventory storage. Any attempt to remedy these problems by increasing the planned lead times became a lesson in relearning *Wight's syndrome*: As planned lead times were increased, backlogs were increased. Inventories and shop queues rose. This increased actual lead times. In the end, nothing had changed except for the presence of more inventory.

Another indication that things had to be changed was the increasing dissatisfaction expressed by Cone Drive's customers. Customers complained about Cone Drive's poor service stemming from long lead times and poor delivery integrity. The people involved in scheduling and production were complaining about the time spent fighting fires.

Cone Drive's management recognized that Cone Drive was no different from its competitors. Its service to the customer was no worse than its competitors; it was also no better. Aside from its product, there was nothing really unique about the firm. Unless something was done quickly to give Cone Drive a distinctive competitive edge in the market, management realized that the company might not be in business in the long term.

The final factor that brought about the demise of this system and which began the move to CIM involved the actions taken by corporate headquarters. In 1980, with the cost of borrowing money running between 15 and 20 percent, holding inventory was prohibitively expensive. As a result, the decision to look for a system for managing production was initially based on the need to reduce inventory.

COMPUTER INTEGRATED MANUFACTURING AT CONE DRIVE

To the management of Cone Drive, computer integrated manufacturing is a corporate-wide system that draws together the four major elements of the firm (see Figure 7-1):

- Shop equipment automation
- Office automation
- CAD/CAM
- Production and business systems

The management of each of these elements by means of computerized systems is critical to the *long-term* success of the firm. Furthermore, CIM is a system that combines the management of information with the automation of work. This perspective is very evident in Cone Drive's definition of CIM: *the integration of information within a business in conjunction with the mechaniza-*

tion of work within the business, to provide the most efficient method of supplying a product to a customer and ensuring the long-term viability of the business. This view of CIM reflects over eight years' experience with CIM. It is a definition that treats CIM not simply as an information or an automation system; it treats CIM as a system that encompasses both concepts. It also treats CIM as a method of better serving the customers. To understand this perspective, it is first necessary to understand the development of CIM at Cone Drive.

1980: The Setting for CIM

In 1980 everyone at Cone Drive realized that something had to be done to ensure the long-term survival of the firm. At this time, management had not heard of computer integrated manufacturing. What it was looking for, rather, was a better *total* business system. The new system had to integrate the operation of the entire organization. It had to manage not only the activities of the manufacturing system but also the activities of the front office. The new system had to draw on the power of computers to create a more effective total production and business system.

FIGURE 7-1
Computer Integrated Manufacturing at Cone Drive

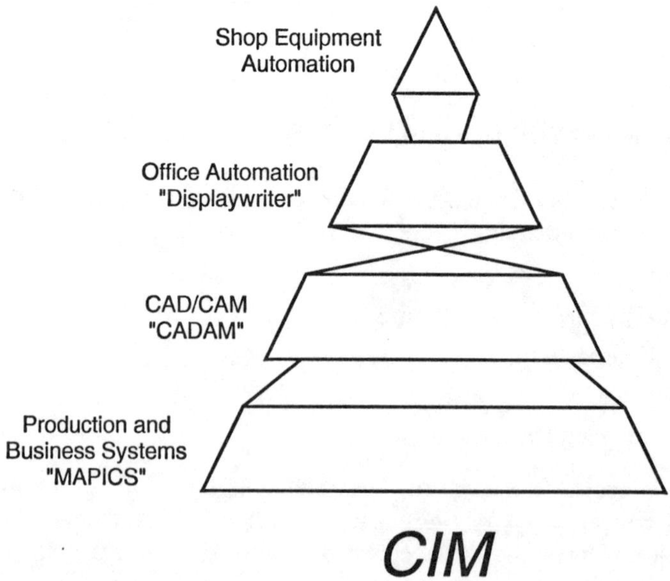

One factor that helped bring about the recognition that the current system was no longer adequate was the hiring of new employees in the late 1970s. Many of these new hires were young people. Being young, they were naturally inquisitive. Being new, they were unfamiliar with the Cone Drive system. They began to question the appropriateness of many practices at Cone Drive. The questions and concerns they raised began to affect the older employees. Their presence also helped to change the power structure at Cone Drive.

All of these factors (the recognition of the need for change, the pressure from corporate management for inventory reductions, the dissatisfaction of the market with the current level of customer service, the discontent of shop floor employees with the continuous fire-fighting, the introduction of many new employees) helped to create the right environment for change. They helped simplify the introduction and acceptance of a system such as computer integrated manufacturing.

Preparing for the Move to CIM

Prior to 1980 Cone Drive had very limited experience with computer systems. There had been some time-sharing with corporate headquarters (Ex-Cell-O). However, these activities had been primarily devoted to activities such as accounting, order entry, and inventory control. In 1980 the management at Cone Drive agreed that whatever solution they selected had to enable the company to discontinue the use of any computer systems located at corporate headquarters.

In January 1980 the first action taken was the formation of a task force consisting of the data processing manager, the controller, and the administrative manager. It was to select and implement a new system. The composition of this task force was notable because of the leadership role taken by top management. The members of this committee were taken from the top levels of the firm. It indicated to everyone in the company top management's commitment to changing operations at Cone Drive.

The system that this task force recommended had to satisfy certain minimum requirements:

- Improve delivery integrity.
- Shorten lead times.
- Reduce inventory (primarily work-in-process).
- Maintain and enhance the high-quality image of the company.
- Provide for better inventory management.

From January 1980 until early April 1980 this team investigated several approaches and eliminated alternatives. By June 1980 the team had formulated their strategy for improving operations at Cone Drive. This strategy would be built around certain decisions:

• *The new system had to be fully integrated from the outset.* The new system had to bring together under one computer-aided umbrella the areas of manufacturing, accounting, engineering, and order entry. Cone Drive was a small company; it did not have the resources needed for a piecemeal implementation approach.

• *The new system could not add any new staff people.* The new system had to work with the existing staff. Additions to staff were to be avoided, since they would do nothing more than add to the overhead.

• *The new system would be based on existing hardware and software.* Whenever possible, existing "canned" software and hardware were to be used without modifications. There were several reasons for this decision. In introducing the new system, one of the objectives was to improve current business practices at Cone Drive. By using existing hardware and software, the people at Cone Drive would be forced to change their practices to conform to the requirements of the technology. This might not be all bad, since they might learn from the experiences embodied in the software. After all, if the software had been changed to conform to the needs of Cone Drive, this would imply that the practices at the company were better than those found in the software. This was not the case. Furthermore, by requiring that existing hardware and software be used unmodified, the cost of implementing and maintaining the new system would be greatly reduced. Finally, the risks present in the use of such hardware and software were very low. The task force wanted Cone Drive to be at the leading, not the "bleeding" edge of technology.

• *The system had to be installed and in production currently.* This was one way of ensuring that the system did operate. By requiring that whatever system selected be in production, the task force hoped that Cone Drive could draw on and benefit from others' experiences with the system. Users tend to be more objective and critical than salespeople.

• *There would be single-vendor support.* This guideline was developed specifically for software. Whatever system was purchased had to be supported by one vendor. Cone Drive did not want to deal with different vendors and have to continuously coordinate their activities. In practice, this guideline resulted in the decision to purchase all of the software from one source. While this practice meant that Cone Drive did not get the very best system for each area of the firm the benefits were significant. First, purchases were greatly simplified. Second, Cone Drive did not have to worry about whether the various system components would be integrated and "talk" to each other. Finally, training was greatly simplified.

• *The vendor had to make available user training and education.* From the outset the task force recognized that if the new system was to be successful, the users at every level of the company had to be trained and educated in its uses. Cone Drive did not have the resources to train its personnel. That task was better left to the supplier. The vendor had to demonstrate to Cone Drive and its task force a plan for training its personnel.

• *The system had to have growth capabilities.* Irrespective of the system selected, it had to be evolvable. Growth was not limited simply to changes resulting from increases in sales. Instead, the task force viewed growth as embodying both changes in sizes and changes in requirements (the addition of new products, the deletion of old products, the modification of existing products, changes in customer requirements). By requiring that the system have growth capabilities, the task force wanted to avoid the turmoil that accompanied a change in system.

• *The system had to be easy to use.* The use of any system in practice is directly related to its ease of use. Systems that are easy to use are always used by company personnel. Conversely, systems that are hard to use are rarely used. The more frequently a system is used, the greater its chances of being successful. This was a difficult guideline to follow and evaluate in practice.

The first goal of the implementation plan drawn up by the task force was to improve the operation of the basic business areas (i.e., production, accounting, engineering). The first area with which they dealt was the production and manufacturing area. Here their intent was to find and implement a good manufacturing and materials requirements planning system. Before they could begin this task, the task force had to convince everyone in the manufacturing system to "buy into" the new system.

Creating an Atmosphere for Change

From the outset, the task force, and management in general at Cone Drive, recognized that the limiting factor to success was not the technology but rather the ability of the people to accept change. Being a small company, Cone Drive had a very important advantage over large companies: the number of people that needed to be influenced was relatively small, so that the task was relatively easy.

To get the people to accept the need for change, the task force began by listing the activities that had to be done better to satisfy the needs of customers both now and in the future. Next, the capabilities of the current system to satisfy these needs were evaluated, which showed that changes had to be made. In reexamining these capabilities, a list of changes that had to be made was also identified.

In carrying out this reexamination, the task force tried to ensure that everyone looked at problems from an overall business perspective, not from a department point of view. This change in perspective can be better understood by looking at the example of lead time. The objective for Cone Drive was to shorten the lead time between the time when a customer thought about ordering a product from Cone Drive and the time that the company collected the invoiced amount. Examining this area from a corporate perspective required the examination of the activities of many different departments within the company (e.g., order entry, sales, engineering, accounting, manufacturing, purchasing, and shipping).

Changing the Production System

Owing to the poor record on delivery integrity and the rising cost of holding inventory in the early 1980s, the first area that the task force targeted for improvement was the production and business system. This system was built around an MRP system. The transformation of the production and business system began in March 1980 and was completed by January 1982. The milestones in the implementation of this system are summarized in the accompanying table. Included in this implementation period was the rental of an IBM 34 computer for employee education and training and to test out the software.

The heart of the new production and business system was MAPICS from IBM. MAPICS was chosen for several reasons. First, Cone Drive had decided

TABLE 7-1
The Production and Business System at Cone Drive: Major Implementation Milestones

Date	*Milestones*
March 1980	Ordered system (MAPICS)
June 1980	Installed interim S/34
September 1980	Developed use of MAPICS methodology
January 1981	Began training office employees
April 1981	Established new databases
September 1981	Began training shop employees
	Installed S/38 and MAPICS
December 1981	Reentered files
January 1982	Began production

to standardize on IBM as the single vendor. Second, MAPICS was an integrated system that drew together the financial, data collection, engineering, sales, and production sides of the business (see Figure 7-2). More important, it satisfied all of the guidelines previously set down by the task force.

The MAPICS system implemented provided the following capabilities. For manufacturing, it provided capacity requirements planning (CRP), MRP, inventory control, production control, and costing and product data management. For marketing, it provided the capability for sales analysis, order entry, and invoicing. Accounting benefited from its abilities to manage payroll production control and costing, accounts payable, accounts receivable, the general ledger, and order entry and invoicing. Finally, engineering could use MAPICS for product data management.

From 1980 to 1981 the implementation of the new production and business system focused on training, education, system debugging, and the writing of some programs. Training and education were heavily emphasized. They provided a way by which the shop floor personnel could learn, question, and accept

FIGURE 7-2
MAPICS Overview

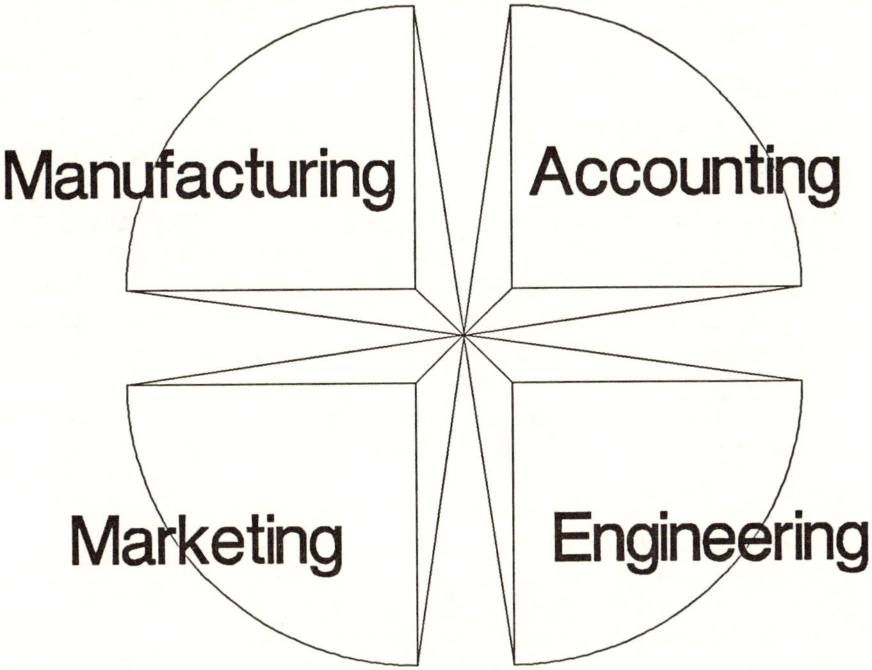

the new system. Everyone was encouraged to participate and to identify any improvements to the new system. This emphasis on education and training still continues at Cone Drive.

The implementation of this first building block of the new system had to be done right. If this first element did not work, the implementation of the rest of the system would be in jeopardy. Furthermore, the introduction of this first system would provide a testing of the resolve of management's position. How management reacted to challenges from the users would greatly influence how the company personnel would view the implementation of the rest of the system. One such challenge involved the part number system.

Prior to 1980 Cone Drive had used a parts numbering scheme based on 22 digits. When work on the implementation of the new production and business system began, it was decided to simplify the parts numbering scheme and use one that required only 15 digits. There were several good reasons for this move. First, with fewer digits, there was less chance for data entry errors. Second, it would be less expensive to maintain a 15-digit parts numbering scheme as compared with a 22-digit scheme. Finally, the software program that had been selected could not handle a 22-digit part number.

Many users still refused to accept this new parts numbering scheme. This resistance to the new numbering scheme was strongest in the engineering group. Management responded by having the users study the purpose of the parts numbering system. It was agreed that a part number was simply an identifier. The number of digits was not critical as long as the number assigned could satisfy this basic purpose. The general manager of Cone Drive had a meeting with the engineering group. He made it clear to the head of engineering that the parts numbering scheme would be changed. If the existing head of engineering could not do it, his successor would. The parts numbering scheme was successfully changed.

This incident communicated to everyone in the company top management's resolve to effect changes and to adhere to the guidelines that had been set down in 1980. Management had a concept of what they needed and of what they viewed as being acceptable. This incident proved to everyone that management would protect the integrity of the CIM concept.

In 1984 the economy weakened. By the spring of that year, Cone Drive laid off about 100 factory and 30 office workers. These layoffs accounted for a 25 percent reduction in the company's total work force. These layoffs were justified on the basis of the economy. Top management tried to be open with its employees. Information regarding incoming sales and the general nature of the business was made available. The reductions were accepted. The employees did not blame the transition to the new system for the lost jobs.

Identifying the Benefit of the New System. By early 1982 there were unmistakable indications that the new system was beneficial:

- Inventory had been reduced by 40 percent.
- Delivery integrity had been improved by 90 percent.
- Quoted lead times had been reduced by 50 percent.
- The materials control department had been reduced by 50 percent.
- Expediting had been eliminated.

More important, the new production and business system provided Cone Drive with the fundamental blocks on which the rest of the CIM system was built.

Changing the Office System

In early 1982 management decided that this would be the best time to expand the transition begun in production and business systems to the rest of the firm. At this stage, management began to investigate the office system.

The office area presented management with several critical challenges. The most important were

- The need to reduce equipment costs.
- The need to reduce communications time and costs.
- The desire to improve the quality of office output (e.g., reduce typing errors).
- The pressure to improve customer correspondence processing time.

The first step in addressing these challenges was to study the office system as it was.

It was found that the major departments (personnel, marketing, engineering, and general administration) each had its own office area, its own pool of secretaries, and its own equipment. Most of the typing was done on typewriters. This arrangement was not efficient. There were often situations in which secretaries were over- and underutilized simultaneously. That is, there would be an excess of secretarial staff in one area and not enough in another. Furthermore, typewriters did not represent the most efficient method of storing, retrieving, and managing textual information. A good deal of the material that was typed was essentially the same material (in bits and pieces). It was also very difficult to share information. Something done in one area, if needed by another, had to be retyped. Unless something was done to improve the efficiency of the office systems, these areas could become a corporate bottleneck.

To resolve these problems and to improve the overall efficiency of the office system, management introduced several important changes. First, all of the office areas were consolidated into one general area, which served all of the firm. This area prepared employee letters for personnel. It also did general typing for general administration and it prepared proposals for engineering. For Marketing, office systems prepared customer quotes, proposals, and product price lists. It also did customer mailings.

Next, the typewriters were replaced by computers running IBM's Display Writer word processing program. Everyone used the same program. The sharing of information now became very easy. The various activities assigned to the office systems area were evaluated and the tasks that were repetitive or routine were automated. Automation took the form of the use of standard "boiler plate" paragraphs that were stored and retrieved when needed. The secretaries were now freed to concentrate on the other tasks. Every piece of correspondence produced by this area was now stored. This information would be called up when needed. The need for enormous filing cabinets and endless paper files was now greatly reduced.

Identifying the Benefits of the New System. The implementation of this new system in the office area was done between 1982 and 1983. By 1983 many of the benefits of this system had become apparent:

- Customer correspondence processing time had dropped from four days to two days.
- Quality had improved. The documents now looked better. There were almost no typing and spelling errors.
- The office area was able to avoid the future costs of adding new equipment and personnel. The equipment and personnel now in the office area were used more effectively and efficiently.

Changing the Engineering System

In 1982 management became aware that the engineering system was becoming a corporate bottleneck. With the efficiency of the manufacturing system improved, management now found that in some instances it was taking longer for engineering to release an engineering drawing than it took for the shop floor to build the part. If the efficiency of the engineering system was to be on a par with that of the manufacturing system, this system had to be revamped and automated.

In revamping the engineering system, management had identified several objectives for the new system:

- Reduced internal paperwork lead time.
- Improved drawing and NC (numerical control) tape accuracy.
- Reduced cost

- Improved professional and technological image.

This latter objective was important in that the engineers had to see themselves as being better engineers and at the "leading edge" in terms of the use and development of computer-based engineering tools. They had to develop pride in both the system and in their work with the system.

Before changing the engineering system, the selection criteria for the new system were first identified and agreed to. The most important of these criteria included:

- *Ease of use.*
- *Integrated with other systems.* The new engineering system could not be confined to engineering alone. It had to be fully integrated with the production and business systems. Specifically, the new system had to link the engineering system and the NC machines. It had to provide a method by which the designs created in engineering could be quickly transformed into NC tapes that were accurate and complete. The vehicle for this transfer of information had to be the computer system.
- *Single-vendor support.*
- *The availability of training and education.*
- *The provision of computer-aided mechanical drafting.* The pencil and paper method of drafting had to be replaced by a terminal, keyboard, and drawing tablet.

These criteria were mirror images of the ones that had been developed for the production and business systems. The concern with engineering reflected certain changes that had taken place in the marketplace. By 1983 market competition had increased, owing in part to the presence of foreign competitors. The emphasis now was on preserving Cone Drive's existing market share.

Before any changes in the engineering system took place, it was decided that the MAPICS bill of materials and parts number file would be the master and that the engineering computer system would handle the drawing and NC programming areas only.

The implementation of the engineering system began in 1982. After testing several different systems, the management team selected the CADAM system from IBM. This system was selected for several reasons. It was a true CAD/CAM system. It offered a system for data management. It facilitated computer-aided design by its use of a geometric interface. Finally, it could produce acceptable hardcopy. The implementation of this system began in September 1982 and was completed by April 1983 (see Table 7-2).

One effect that resulted from the use of this CADAM system involved the change from single-shift engineering to multiple-shift engineering. The new

TABLE 7-2
The Engineering System at Cone Drive: Major Implementation Milestones

Date	*Milestones*
September 1982	Ordered system
December 1982	Began training of engineers
February 1983	Installed interim 4331 with CADAM
	Started internal on-site training
April 1983	4341 installed with CADAM
	Began production

CAD/CAM system brought with it high fixed costs. In this respect, it was similar to an NC machine tool. Just as the NC machine was operated three shifts a day, it was decided to apply the same approach to the CADAM system. To maximize the output of the system, engineers began to work multiple shifts. This change in staffing helped to reduce engineering lead time.

Identifying the Benefits of the New System. By 1983 the new engineering system was in operation. Cone Drive began to notice the benefits of this new system:

- Drawing time was reduced by 60 percent.
- Facility layout time was reduced by 50 percent.
- The accuracy of the resulting drawing was significantly improved.
- The engineering lead time was shortened from five days to one.
- The professional image of the engineering group was greatly improved.

Automating the Shop Floor

Automating Communication Flows. In late 1984 it was decided that communication flows on the shop floor had to be improved. The installation of the MAPICS system had helped to reduce scheduling problems and provide better order status information. However, the information system was still paper-based. The introduction of CADAM had greatly reduced the engineering throughput time. Again, all drawings were printed out on paper. The NC part programs were distributed using paper.

At this time one of the employees suggested using a personal computer (PC) to download NC part programs directly to the machine tool on the shop floor from the CAD/CAM system. A project was initiated to explore the suitability of personal computers for data collection and management.

This project was one of several undertakings started during this time period. Management was interested in identifying faster and more accurate data communications systems and in getting better utilization of manufacturing capacity. To achieve these two objectives, management approved the following shop automation projects:

- Replacing paper tape with personal computer input to direct numerical control machines (DNCs).
- Using personal computers to create a *paperless* factory.
- Using bar codes for part identification.
- Automating material handling and machining cells.
- The creation of NC programs automatically from drawings.

In general, most of these projects were successful. One of the most interesting aspects of these projects was their heavy reliance on the simple IBM PC. These computers were introduced on the shop floor without any significant modifications. Suppliers were surprised by the decision to use such machines. After all, these machines had not been modified so that they could better cope with the harsh operating conditions found on the shop floor. However, to the people at Cone Drive, these PCs were the appropriate technology. There was no need for anything more rugged or sophisticated.

This factor was not as well understood or appreciated by the vendors. One vendor tried to convince the management at Cone Drive that a more rugged computer system was needed. Such a computer could better resist problems due to the high levels of dust present in the shop. Its keyboard could still perform even if a cup of coffee was accidentally spilled on it. These arguments did not impress management. To the first, management replied that the problem of dust could be eliminated by putting a filter in front of the computer. To the second, management replied by mounting the keyboards at an angle.

Later on, another vendor tried to convince management of the need for color screens. To show the benefits of color, an RGB[2] monitor was introduced on the shop floor. Before long, everyone on the floor became convinced that they needed a similar screen to do their job effectively. Management studied the need for these screens. They came to the conclusion that a color monitor did not really have any impact. As a result, the RGB monitor was removed and the old green monochrome monitor returned. The demand for RGB monitors went away.

---

[2]RGB stands for red, green, blue. It is a simple three-color monitor.

The use of personal computers as communication vehicles was a success at Cone Drive. As a result of a program written by industrial engineering and run on the IBM PC, shop floor personnel were able to display CADAM drawings on the PCs along with routings, bills of materials, and other data from the MAPICS system. This allowed the system to display shop packet information on the shop floor via the personal computer. Cone Drive had taken a major step toward becoming a true *paperless* factory.

Other savings generated by the use of personal computers included:

- Lower paper tape costs.
- Lower engineering drawing duplication costs.
- The reduction of operator error in data input.
- The reduction of downtime due to program changes.
- The reduction of computer hardware costs.
- Lower costs to upgrade the memory of NC systems.
- The reduction of shop paper costs.
- Lower transmission costs of CADAM drawings.

Overall, the time and effort devoted to improving data communications on the shop floor resulted in improved data accuracy, reduced cost of data input, and faster feedback to and from the shop floor. Improvements in data accuracy came from several different areas:

- The elimination of manufacturing parts lists and drawings prior to the latest engineering revision level. This eliminated the many out-of-date drawings left in the system and helped to dispose of confusion regarding which drawing to use.
- The use of correct fixtures and tools.
- A reduction in errors caused by old prints in the shop.
- Lead time shortening due to *"no wait for paper."*
- Faster NC part program changes.

More important, this work helped to greatly improve employee morale. People on the floor now felt that they were an integral element of this new overall system.

The communications network that has evolved over time at Cone Drive (see Figure 7-3) is truly a company-wide system.

The Introduction of Flexible Manufacturing Cells. In 1987 a flexible manufacturing cell was installed to machine cast iron housings (see Figure 7-4). Since its introduction, this FMC has been integrated via a personal computer with the other systems.

The automation of the shop floor is an important next step at Cone Drive because it will enable the company to run parts economically in lot sizes of one. Ultimately, it will move the company to achieving its goal of machining today what is to be shipped tomorrow.

Continuous Improvement

The last step in the implementation of CIM at Cone Drive has come about as a result of the users of the system. As a result of the changes that have taken place since 1980, everyone has come to expect and, in many cases, welcome change. The users have begun to question current ways of doing things. They have begun to look for better ways of accomplishing their tasks. As they have become familiar with the new computer systems and their operation, they have

FIGURE 7-3
Cone Drive Computer Network

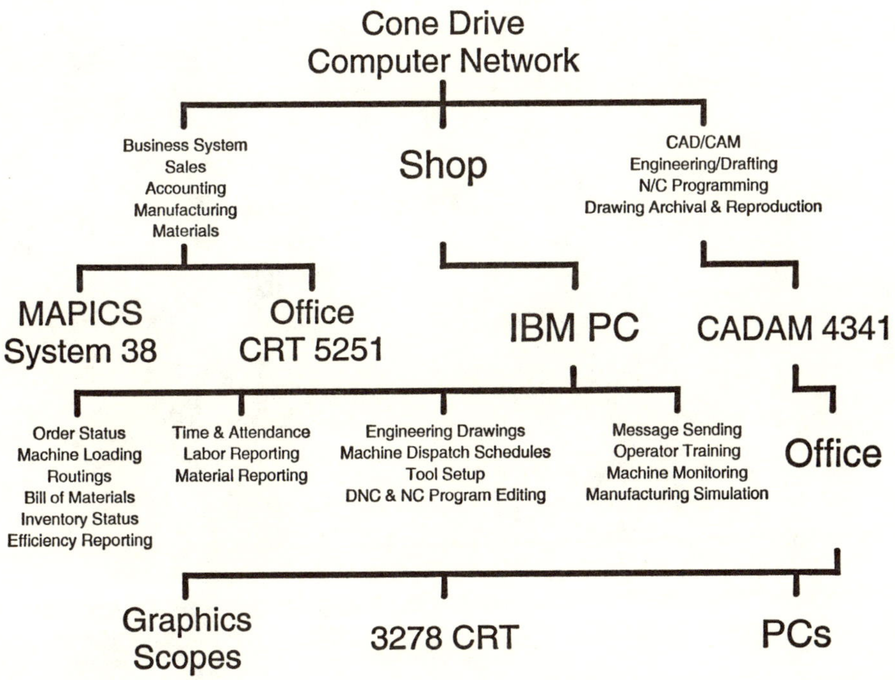

FIGURE 7-4
Flexible Machine Cell

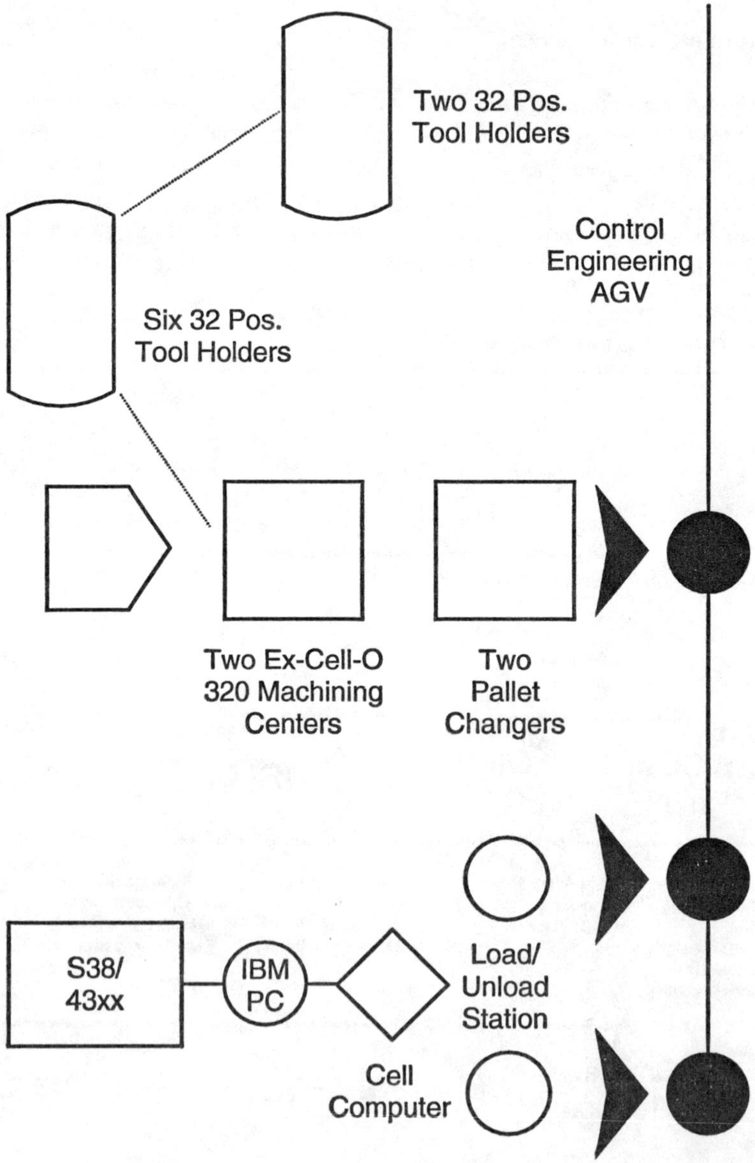

begun to look at how these systems can be used to improve operations in the company. In many cases the users have begun to identify uses not previously envisioned for these computer systems.

The resulting improvements in the operation of the firm are now coming from an "ethic" of *continuous improvement*. That is, every day, the people at Cone Drive try to identify and recommend at least one improvement that can be made in the system. Continuous improvement is important for several reasons. It is an indicator of the extent to which the users have developed a sense of *ownership* over the current system. Second, continuous improvement emphasizes the notion that change and improvement are processes that are always ongoing. These are also processes that are not necessarily linked to, or driven by, hard automation or technology. Third, continuous improvement is yielding benefits. These benefits are as important as those created by MAPICS or CADAM. Finally, continuous improvement is keeping alive the drive to manufacturing excellence that was first begun in 1980.

Becoming a CIM System

From 1980 to 1984 Cone Drive was simply trying to draw on the computer tools that were available to improve its overall effectiveness and competitiveness. These developments were not associated with the CIM label in the minds of the people at Cone Drive.

However, by 1983 the management at Cone Drive had begun to hear more and more about the CIM concept. More important, IBM, the major system supplier to Cone Drive, was beginning to point to Cone Drive as an example of the successful use of CIM in a small company. From 1983 to 1984 Cone Drive began to receive a lot of publicity from IBM. As a result, people began to accept the notion that Cone Drive was a CIM system.

This acceptance of Cone Drive as a CIM system reached a peak in 1985 when the work of Cone Drive in the implementation and use of CIM concepts and systems was recognized by CASA/SME. In that year CASA/SME awarded Cone Drive the LEAD award.

These developments had several important effects on Cone Drive and its people. First, it made many of Cone Drive's customers aware of the changes that had taken place within the company. CIM became an important marketing tool. Second, it boosted employee morale. People could see a payoff to the years of hard work and to the various changes that had taken place within Cone Drive. They could point to a system that was successful and a system that they had helped build and shape. As a result, this initial success with CIM helped breed success.

IDENTIFYING THE TYPE OF CIM SYSTEM IN PLACE (1988)

By the beginning of 1988 most of the work on CIM at Cone Drive had come to an end.[3] The resulting CIM system was now operating at two levels. At the first or formal level, Cone Drive had implemented a Type III CIM system. Internally, the organization was very tightly integrated. The activities of manufacturing and engineering were closely coordinated with strategic and long- range planning (*strategic integration*). Activities were also coordinated across the various functional areas (*organizational integration*). Finally, the manufacturing and engineering systems were closed loop (*functional integration*). A Type III CIM was consistent with those objectives that management had first set for it in 1980.

However, the CIM system was informally moving toward being a Type IV system. The people realized that if they were to effectively serve the demand of the engineer-to-order and make-to-order segments of their market, linkages had to be established between Cone Drive and its customers. To support these forward linkages, suppliers had to be more closely tied to the Cone Drive system. These links were established informally. That is, people within Cone Drive would regularly contact suppliers and provide them with information on schedules, sales trends, designs, and so forth. There was a general sharing of information. However, this sharing of information was not supported by such systems as electronic data interchange (EDI). There were also similar attempts to build closer and more regular ties between Cone Drive and its customers. These ties were being introduced because they made sense and they made the task of serving the needs of the customer easier.

This linking of Cone Drive with its major customers was to be achieved by means of three initiatives. The first was to allow major customers to access Cone Drive computers directly so that they could get information about such areas as order status. Second, management began to think about letting major customers enter their orders directly into Cone Drive's computer system. Finally, management started to examine the potential for transferring engineering data and drawings directly to customers and vendors electronically through personal computers.

IDENTIFYING THE BENEFITS OF CIM

By 1987 most of the implementation effort on CIM had been completed. More important, the CIM effort was beginning to yield benefits for Cone Drive.

[3]This slowing down of CIM-related activities at Cone Drive is discussed in greater detail in the section entitled "The Future of CIM at Cone Drive."

These benefits were evident in nearly every aspect of the firm. The most important were

• *Reductions in inventory.* Inventories have fallen in excess of 40 percent. These reductions in inventory have not affected any of the other aspects of customer service (i.e., lead times).

• *Reductions in lead times.* Lead times have been cut by 50 percent. This reduction in lead times is not specific to the shop floor. Rather, it is a system-wide reduction in lead times. It has come about as a result of improvements in the design stages, in the office, and on the shop floor.

• *Improvements in delivery integrity.* Delivery integrity has increased from about 50 percent in 1980 to over 90 percent. More important, it is still increasing. From the customer's perspective, these improvements in delivery integrity have meant that Cone Drive has become a more reliable supplier. The risk (in the form of missed ship dates or unpredictable deliveries) present in doing business with Cone Drive has fallen enormously. The resulting reduction in uncertainty has given Cone Drive a new means of differentiating itself in its market. Cone Drive does not simply sell gears; it sells certainty.

• *Improved morale.* CIM has helped turn the working environment around at Cone Drive. Employees have a sense of pride in not only the CIM system but the company as a whole. There is an awareness that they have survived adversity. They have gone through a period when many positions were lost at Cone Drive. They have seen their market fall. However, they have been able not only to survive this period of adversity but also to become the best in the industry. Everyone in the firm believes that Cone Drive is truly number one. As a result, by 1985 turnover had fallen to about 1 percent.

More important, people are now constantly looking for new ways to improve the current CIM system and to become even more effective. There is a belief that nothing is impossible. As a result, people at Cone Drive are talking about cutting lead times in half again, reducing inventories another 25 percent, and running orders in the factory with a batch size of one.

• *Improved planning and execution.* One of the biggest benefits gained from CIM has been the ability to plan and work together as a team. This has resulted in better plans being generated and effectively implemented on the shop floor.

• *Increased awareness of waste.* The process of implementing CIM has forced everyone at Cone Drive to reexamine how they carry out the various tasks of designing, planning, and building products. This process has forced them to study these activities to identify those actions that add little or no value. In doing so, people have come to realize that such tasks are nothing more than *waste.*

• *Survival.* This has been the single most important benefit created by CIM. CIM has enabled Cone Drive to survive and thrive in a declining market that is becoming increasingly competitive. Without CIM, Cone Drive would not be in business today.

These benefits are real. Everyone agrees that these benefits have been gained as a result of the hard work and commitment of everyone at Cone Drive. More important, everyone recognizes the connection between computer integrated manufacturing and the improved health of the company.

It should be emphasized that these benefits came about because Cone Drive made CIM a *company* system, not a *manufacturing and engineering* system. Problems in manufacturing are only symptoms of system problems. To improve manufacturing performance, the corporate system must be improved.

WHY HAS CIM BEEN SUCCESSFUL AT CONE DRIVE?

What makes the CIM experience at Cone Drive so unique is that a small company such as Cone Drive has been able to make this complex concept work. In reviewing the experiences of the company with CIM, there are several factors that have helped it succeed with CIM:

Being a Small Company. Being a small company has its advantages. Being small gives the company the ability to get things done quickly and to change very rapidly. There is not a large bureaucracy obstructing efforts to improve. However, being small does have its disadvantages. The risks of failure are higher and the competition more severe. What this means is that the preparation and planning needed for a change such as CIM have to be done more carefully. Once the decision is made and the path set out, the company can move very quickly.

Trust. In the period from 1980 to 1987, Cone Drive underwent a complete and radical transformation. Not only was there a new way of managing the company but employment dropped by almost 50 percent. Such changes taking place simultaneously could have destroyed Cone Drive. However, this did not happen. One reason was the trust that the employees had in management. This trust did not occur by accident. Instead, it was carefully nurtured by management.

Management was always open with its employees. It showed the employees the sales figures and the trends. It gave them a clear choice. As one manager described it, the employees were given the choice of having a system in which there were short-term jobs for 550 employees (with the prospect of no long-term jobs) or a system in which there were permanent jobs for 300 employees. The employees were allowed to "buy in." Top management never forced them to accept CIM.

As the implementation took place, the employees were always kept informed of what was happening and was about to happen. They were actively involved in the implementation process. Their suggestions and recommendations were actively solicited by management.

Top Management Support. The top management of Cone Drive played a critical role in the success of the CIM system. They were actively involved in the implementation. Their presence was very visible. Their support of the new system was also visible to everyone in the company. They also preserved the integrity of the CIM system. That is, they established certain rules that were to govern the implementation and use of the CIM system (such as the use of standard systems). Whenever anyone tried to challenge these rules by trying to introduce "minor" modifications, top management would step in and preserve the original implementation plans.

Corporate Support. During the implementation of CIM, Cone Drive benefited from a very supportive corporate headquarters. The Ex-Cell-O Corporation accepted the need for CIM at Cone Drive. It recognized that without change in the form of CIM, Cone Drive would not survive in the marketplace. It provided the necessary financial support. It also give the management at Cone Drive the time in which to make CIM work.

Without any one of these factors, Cone Drive would not have been successful in its attempts to make CIM work. When combined, these factors greatly simplified the implementation process.

THE CIM STRATEGY

Cone Drive has been working at CIM since 1980. In reviewing its experiences with this concept, management has identified a CIM strategy. That is, they have identified a sequence of steps that, if followed, can help enhance the probability of successfully implementing CIM:

1. *Implement reliable and effective business systems for the basic elements of the firm.* All manufacturing systems require five major functions: manufacturing, materials, marketing and sales, engineering, and accounting. These five are the foundations on which everything else is built. Their operation must first be improved. This is then the first focal point for all CIM implementation activities. Unless the firm can develop effective designs that it can effectively and efficiently produce and sell for a profit, it cannot hope to compete in the marketplace. No amount of automation can ever adequately compensate for ineffective and unreliable business systems.

2. *Integrate the basic systems*. It is not enough to simply put the pieces of the business system in place. These pieces must be made to work together as a unifed whole through *integration*. Integration must consider the coordination and linkage of information flows and activities. People in the various areas must develop the belief that they must work together. Accountants must work with engineers; marketing with manufacturing. Integration of activities is important because no one has access to all parts of the problems or can see all the implications of a proposed solution. Instead, each of the functions sees one important component.

Integration can never be bought from a hardware or software vendor. Integration is not a package. It is something that the firm must develop. It is a management task.

3. *Establish a manufacturing EOQ*[4] *of 1*. This is the ultimate goal that should drive every manufacturing system. An order or run quantity of one should provide the firm with the highest degree of flexibility, the shortest lead time, and the lowest costs. It also gives the firm one of the strongest competitive weapons. This goal should drive all of the activities of the firm. Investments should be selected and made that are consistent with this goal. Nearly every investment made at Cone Drive since 1980 has tried to help the firm achieve this objective.

4. *Make business and engineering information accessible to the shop floor*. While there are certain areas that should be kept off limits (e.g., payroll), most other types of business and engineering information should be available to users. There are advantages to having such an open system. First, it encourages trust. It also allows users to see what is happening to the company. More important, however, openness results in good data. An operator calls up a design and reviews it. He or she identifies a flaw in the design (something that the designer might not be aware of). The information is communicated back to the designer and the flaw corrected. The system is now working with good designs and good data. This approach enhances the operation of such an information-intensive system as CIM.

5. *Achieve the lowest manufacturing costs and shortest lead times by the use of appropriate technology and "wise" investments*. In reducing costs and lead times, management can draw on four major tactics. The first is the upgrading of existing machinery. Often, new technology is not required. What is needed is to make the existing equipment more efficient. The second is to acquire new technology if and only if it is needed. Third, all equipment should be integrated by a simple system. At Cone Drive the integrating device was the IBM PC. Fourth, strive to create a "paperless" factory. Paper should be avoided whenever possible. It adds to lead time; it increases clutter; it can be lost; it increases costs; it decouples flows by introducing a paper delay. The purpose of

[4]1 EOQ stands for Economic Order Quantity.

paper is to convey information. There are better ways of achieving this same purpose. Ultimately, paper should be regarded as *waste.*

6. *Do these tasks with internal manpower (retrain as required).* Adding people is not the solution when implementing CIM. Often, the people that are added know the technology but not the company. It is better (i.e., faster and cheaper) to keep the people who know your company and to retrain them. This approach draws on their insights and adds this aspect to the development of the new system. It also reduces the perceived threat present in the new system. Finally, it provides employees with a clear indication of their importance to the company.

7. *View each automation project as part of the overall plan.* There is no surer way of creating incompatible islands of automation than by treating each automation project as a separate and distinct element. Rather, the issues of fit (i.e., how does this project fit with and support corporate goals and objectives?) and integration (i.e., the sharing of information, the movement of material) should be explicitly addressed in advance. It is far cheaper and easier to address problems in the planning stage than it is after the fact, when the project is over and the machine operating. Finally, by relating the automation projects to an overall plan, the chances of making a wrong or inappropriate investment are greatly reduced.

8. *Be on the leading edge of technology, but don't wait for the ultimate.* Technology is a moving target. It is always changing; new advances are always occurring. One problem for management is that of timing or determining when to buy new technology. One strategy is to wait until the latest and best version is created. The problem with this strategy is that it inhibits action. You are always waiting. The more logical approach is to accept that the future will always bring with it better technology and to look for that technology that best meets the needs of the company. It is always better to solve a manufacturing problem now with less-than-perfect tools than to leave the problem unresolved while management waits for the arrival of these perfect tools.

9. *Use standard hardware and canned software whenever possible.* The costs of acquiring, implementing, and running a CIM system are greatly reduced by using standard hardware and software. Debugging is less of a problem; solutions are more readily available; upgrades are easier to implement. The purpose of any manufacturing firm is to build products for customers, not to write code or to debug programs. Finally, as was noted previously, in canned software the firm takes advantage of the experiences of other users. What others are doing (as represented in the software) may be better than what the firm is currently doing.

10. *Show customers how the firm is using CIM to benefit them.* There are several reasons for making customers aware of a firm's CIM efforts. First, it is a good marketing tool. It is also an indication of the firm's commitment to remain in the market and to be competitive. Finally, it serves as an excellent education tool. CIM does change the entire process of producing a product. These changes may affect the buying process. By introducing the customer to the CIM system and showing how it can benefit the customer, the firm has an opportunity to explore with the customer all of the implications of CIM for such activities as order placing, scheduling, design, and shipping. The result is a better fit between the customer and the firm.

LESSONS LEARNED

Cone Drive has been working with CIM since 1980. Over this time, management has learned a great deal about what in CIM works and what does not. This experience has resulted in the identification of a set of lessons that can be offered to others interested in CIM. The following is a discussion of the most critical of these lessons:

1. *Understand your business first.* The starting point for the implementation of any new system such as CIM should be the company and its strategy. Management must take the time to study and understand what it is that the company offers the customer. They must identify those factors that are instrumental in qualifying orders; those that are critical in winning orders; and those that can result in lost sales. They must identify the segment of the marketplace targeted by the company.

This evaluation must not simply look at the market today. It must look into the future. It must look at where the business is going. It must assess the future viability of the current market. It must identify any changes in the requirements of the market.

The information and insights gained from this process are then used to shape the CIM system. The technology acquired and the system changes introduced by the new CIM system should be consistent with the business requirements of the rlrm. They should also be supportive of the strategic objectives. The hand of corporate strategy must be seen in every step of the CIM implementation process.

2. *Use existing software and hardware.* When implementing CIM, the major focus of people should be on changing the operation, not on writing code and debugging. There is enough software on the market to support most aspects of CIM. As a result, by using existing software and hardware, the implementation process is greatly simplified. In addition, the firm gains from the experiences and practices embodied within the software systems.

3. *Encourage extensive user involvement.* The success of the CIM system is ultimately influenced by how the system is seen by the users. If the users are not interested in the system or do not have a vested interest in the success of the system, the system will fail. To succeed, there must be a sense of user ownership. The users must feel that the system is theirs. One very effective mean of developing such a sense of ownership is through extensive involvement. The users must participate in the development and implementation of the system. They must have the opportunity to evaluate the system and to understand its operation. They must learn how it operates. To achieve these objectives, there must be user involvement.

User involvement offers several major benefits. Not only does it enhance the chances for success, it also provides a medium for improvements. As users become familiar with the system, they can identify new opportunities for improving the operation of the CIM system. The system and the firm now benefit from the users' knowledge of the company, experience, and insights.

4. *Emphasize extensive training and education.* This cannot be emphasized enough. Extensive training and education is one of the keys to success with CIM. It encourages user involvement; it helps overcome the uncertainty surrounding the introduction of change; it reduces the errors of working with any new system. However, everyone must recognize that education and training is not simply limited to the introduction of a new system. Rather, it is ongoing. It is required whenever someone new is hired; it is also needed to make sure that users are employing the system most effectively.

5. *Provide employee exposure to the system prior to installation.* One of the most interesting lessons learned came about as a result of the decision to rent an IBM System 34 prior to the changeover. The people at Cone Drive used this computer to test out the new software and system. In effect, they "stress" tested the system. What happened was that as the users worked with the system and saw how it responded to different inputs and conditions, they began to accept the system. When they saw that the system could work in their company, the users developed a sense of confidence in the system. They knew that the system would work as planned. Finally, because the new system was not yet fully operational (i.e., it was not yet managing operations), users were more willing to try different thmgs out on the system. They were more willing to make mistakes. As a result, they learned a great deal about the new system that was to help manage Cone Drive.

6. *Good planning leads to success.* To be successful, a plan must be in place. The plan fulfills a number of important functions. It identifies the ultimate goal, the intermediate steps needed to reach the goal, the exact sequence of steps, and the time horizon needed. Furthermore, good planning helps to identify the resources (time, manpower, money) required. It identifies potential pitfalls and it sets the milestones. Finally, good planning provides the structure within which the implementation effort takes place.

7. *Have regular progress meetings.* This is the corollary to the preceding principle. Once a plan for the implementation of CIM has been identified, agreed to, and put in place, the next step is to carry it out. At regular periods (quarterly or monthly), progress should be reviewed with the groups involved.

These meetings should deal not only with the successes but, more important, with any problems encountered. The reasons for these difficulties should be discussed. Contingencies should be developed and evaluated. In light of these discussions, the feasibility of the current plan should be reviewed. In short, regular progress meetings ensure that everyone involved is focused on meeting the objectives agreed to in the plan.

8. *Have management support.* Everyone in the firm takes their cues from top management. If top management is not supportive, CIM will fail. More important, there are other reasons for requiring top management support. First, it is ultimately top management that wlll provide the resources required to make CIM a corporate reality. Second, top management support is important in convincing others within the firm that CIM will become a corporate reality. Third, top management support and involvement is often required to overcome potential resistance within the firm to CIM. In spite of all the arguments made and logic advanced, there will be some people who will not support the move to CIM. There is often one last resort for these people and that is to have top management present them with a simple alternative: either change and support the new system or leave the company.

9. *Recognize that some people will have to be moved.* Ultimately, not everyone in the company will accept the changes that are required to make CIM a corporate reality. Such people will never use the new system. Whenever there is a problem, they will blame it on the system. They will tend to be continually looking for problems with the new system. Management must recognize that such people are going to have to be moved (and the earlier, the better).

10. *Recognize the need for a system champion.* Every system requires someone who is willing to protect it and to defend it against others within the company. CIM is no different. The most effective type of champion is one who is highly visible, has a high level of authority, has a driving personality, and is goal oriented. At Cone Drive, the controller was the system champion.

11. *Measure bottom line results.* It is critical to keep track of the benefits that the changes are making in the firm and its operations. Such benefits provide feedback that things are indeed changing. Making CIM work is not easy. Management has to show its people that the time and effort spent on CIM are paying off. Furthermore, these results provide incentive. Success breeds success. Finally, tracking and recording these benefits forces everyone in the firm to identify and measure all the possible effects created by the changes.

THE FUTURE OF CIM AT CONE DRIVE

By the start of 1987 the CIM system was in place. However, in 1987 the Ex-Cell-O Corporation (and its divisions) was acquired by the Textron Corporation.

This change in ownership has had a significant impact on the continued growth of CIM at Cone Drive. The new owners have not had the same opportunity to grow up with the CIM system. They have not yet come to accept the continued need for CIM at Cone Drive. As a result, the rate of investment in CIM has slowed down greatly. In addition, changes in the operation of the system (changes triggered by employee suggestions) are meeting more resistance from the new owners. Fewer changes are now taking place.

One interesting result of this change in ownership has been an increase in employee turnover. Between 1987 and 1988 turnover went from 1 percent to some 6 percent. One possible explanation for this increase lies in the expectations of the employees.

From 1980 to 1987 the employees were encouraged to become critical of existing procedures. They were asked to identify new and better ways of doing things. Furthermore, the employees became used to seeing these recommendations quickly implemented. With the new owners and the general slowdown of CIM-related changes, the employees now find themselves questioning systems and making recommendations but not having these suggestions accepted as quickly. As a result, some of the employees have become disillusioned and have decided to leave Cone Drive.

Whether or not CIM will continue to grow and develop at Cone Drive is now very much an open question. How this issue (and the ultimate fate of CIM) is resolved will depend on the position taken by the new owners. It remains to be seen whether they will accept the continued need for CIM at Cone Drive.

CONCLUDING COMMENTS

By any measure of success, the introduction and use of computer integrated manufacturing at Cone Drive has been a complete success. It has enabled a small company to survive and grow in an increasingly competitive market. It has provided Cone Drive with an important and decisive competitive advantage in its market. It has enabled the company to offer its customers the timely and reliable delivery of quality and low-cost products.

In examining the experiences of Cone Drive with CIM, there are several aspects that are significant. The first is that Cone Drive, a small company, has been able to implement such a system. CIM is not something that only large companies can afford to do. It is something that every company can do. Being small does have certain advantages. The number of layers of management required for a decision is relatively low. As a result, the company can react very quickly; there are not as many people to convince of the need for change; the barriers to change are relatively low. However, being small also means that there is very limited tolerance for error. The implementation has to go right the first time; the implementation also has to be system-wide.

Second, the success of CIM at Cone Drive is due in large part to the role played by top management. Top management played a very visible role in the implementation process. The top managers convinced people of the need for a new system. They provided the necessary resources. They also defined the CIM concept and guarded against any unwanted modifications to this concept.

Third, the success at CIM is not based on the size of the company or the amount of money expended. CIM is not technology. Buying the latest and best in manufacturing technology does not guarantee the presence of a CIM system. CIM cannot be purchased or achieved as a quick fix. Rather, CIM is part of a long-term plan for constant improvement.

Fourth, CIM at Cone Drive has truly been a system-wide development. Its presence has touched every aspect of operations at Cone Drive. Not only has it affected the areas typically associated with CIM (engineering and manufacturing); it has also influenced the operation of the office systems. The use of CIM has significantly improved the overall productivity of this often overlooked area of the firm.

Finally, CIM must be viewed as a strategically driven response to the need of American industry for a more effective and efficient business system. These options are summarized as:

- Automate
- Emigrate
- Evaporate

CIM has worked at Cone Drive. Whether it will continue to provide the same competitive advantages in the future at Cone Drive remains uncertain, however.

CHAPTER 8

STEELCASE, INCORPORATED

THE PARABLE TOLD

Steelcase, Incorporated, is a privately held corporation headquartered in Grand Rapids, Michigan.[1] Long recognized as a market leader in office furniture, Steelcase has developed a well-deserved reputation for being an innovator in the development and use of new manufacturing technologies. In 1971, for example, Steelcase implemented an MRP system. Between 1971 and 1987 Steelcase successfully made the transition from MRP to closed-loop MRP to MRP II. Since then, Steelcase has begun the movement to computer integrated manufacturing. This change in corporate direction reflects a move by management to better satisfy the needs of its customers.

In Steelcase we see a company different from the other companies described in this book. For many firms, CIM has become a way to maintain corporate survival; a long-term response to a set of basic underlying problems. In Steelcase we see CIM used as a means of maintaining and strengthening its position in the marketplace. It has become an instrument by which management can better serve the customer by anticipating and quickly responding to their needs. Management has decided to build a CIM system at Steelcase because CIM is the *right* solution for those problems currently facing the firm. As these needs change, the solutions that Steelcase will implement will also change.

Another feature unique to Steelcase is the company's ability to accommodate the demands that the CIM implementation places on the firm. This ability is a result of a corporate culture that is built on change. Change is expected; it is viewed as something that brings improvement rather than crisis. Finally, in Steelcase we see a firm that draws on a number of different technologies and systems (such as just-in-time manufacturing) as needed in its pursuit of giving its customers the best possible service.

---

[1]This case is based on a presentation initially done by Cecil Bradshaw of Steelcase. Additional information was provided by John Dubois of the CDC (Corporate Design Center) of Steelcase.

INTRODUCTION

Steelcase was established in 1912 to manufacture fireproof safes and has since grown to be the world's leading manufacturer and designer of office furniture. With current annual sales in excess of $1.5 billion, Steelcase, along with its two nearest competitors, accounts for approximately one-third of total industry sales.

Major product lines are panels, desks, files, chairs, task lights, tables, video display terminal (VTD) stands, and system furniture for the open office.[2] Worldwide, Steelcase and its joint-venture operations have a combined work force of over 16,000 employees; more than 700 independent dealers; sales offices or dealerships in 58 countries; and 16.5 million square feet of manufacturing and assembly facilities located at 25 sites. The manufacturing sites are located in the United States, France, Africa, Canada, West Germany, and Japan. The primary production location is still located in the Grand Rapids–Wyoming, Michigan, area. Over 8,000 of the firm's employees work in this area.

Steelcase is engaged primarily in the production of make-to-order office furniture. The MTO component accounts for approximately 85 percent of its sales. The remaining 15 percent consists of make-to-stock. The MTS component is centered on basic office furniture. The demand for its products is not seasonal but is closely linked to other economic trends. Steelcase sells to a variety of customers, ranging from small local insurance offices to large multinational corporations such as Citicorp.

THE INTERNAL ENVIRONMENT DESCRIBED

Steelcase's ability to adapt relatively quickly to developments and changes in its market is a direct result of its corporate culture. It is a culture characterized by the following:

Employees Are Treated as Highly Valued Assets. For Steelcase, when management says that its people are its most valued assets, this is not publicity "hype." It is a fact of corporate life. Management works hard at building a good working environment for its employees.

First, its employees are well paid as a result of several payment schemes. There is a very generous discretionary profit-sharing bonus that is distributed twice a year. There are various incentive pay systems. These systems reward

---

[2]Steelcase's product offerings are discussed in greater detail later in this chapter.

people for doing their jobs well. They also encourage identification with corporate goals. For example, there is a direct link between corporate performance and reward. As the firm does better, so do the employees. As a result of these schemes and a generous base pay system, the earnings of the typical employee are far above average for the Grand Rapids area.

Second, management treats its employees well—not simply as workers but, more important, as people. It provides all employees with access to a wellness center located nearby. It also acquired a recreational facility that provides exercise facilities for running, swimming, and other such activities. This facility is close enough to the main plant location so that most employees can make use of the facilities during their lunch breaks. For those employees with children and in need of a day care facility, Steelcase provides a very progressive and extensive day care center. This center has been recognized by other groups as being outstanding. Finally, there is also an extensive employee counseling program.

Third, management respects the opinions and views of its employees. Any employee is free to voice his or her concerns without worrying about retribution. An "open door" policy is in effect to reinforce this view: employees, if they are not satisfied with something at Steelcase, can go up as high as the president of the firm. The president makes available a regular time slot during which he is available to talk with any employee. It is understood, however, that employees should first try to work through their immediate supervisors before seeing the president.

This respect for the opinions and views of its employees even extends to hiring practices. An employee can nominate anyone for employment at Steelcase. When that person is hired, the nominating employee is expected to become a mentor.

Fourth, Steelcase has an informal policy of promoting from within. It is management's view that its people know the requirements of the firm better than do strangers who come in from other companies.

Finally, Steelcase invests heavily in its *human capital.* Employees are sent out at company expense to take courses in materials management, material requirements planning, CAD/CAM, robotics, and general management practices (e.g., supervisory training). They are also actively encouraged to take the APICS certification examinations (with the aim of gaining either a CPIM or CFPIM).

An Emphasis on Long-term Objectives and Goals. This is a direct result of the nature of ownership at Steelcase. Steelcase is privately held but professionally managed. It is understood that any investment that improves the competitiveness of the firm can be easily justified. There is no pressure on management to generate short-term results at the expense of the long-term competitive stance of Steelcase. Everyone knows that the owners of Steelcase are more interested in long-term performance.

Failure is Tolerated; Success Is Rewarded. The management at Steelcase encourages its people to do more than simply complain. If someone sees a problem or identifies an area for improvement, he or she is encouraged to do something about it. If that effort does not succeed and if the effort was well intended and well conceived and if the person learns from the failure, the failure is not punished. In management's view, something was being done to attack a problem—that feature is extremely important.

However, if the effort is successful, the person responsible is rewarded. More important, however, the reward process is visible to everyone in the firm. This approach to problem solving encourages employees to view the process of change as being a "win/win" proposition.

The Customer Is King. It is the goal of Steelcase to serve the needs of its customers (current and future) better than any of its competitors. In the past this commitment to the customer was best summarized in the sign shown in Figure 8-1—a sign that was promptly posted everywhere in the company. It is also recognized that customers and their expectations are dynamic. As a result of competitive actions, governmental actions, foreign actions, economic conditions, and the expectations of customers, it is expected that what customers regard as being acceptable or superior can and will change. As these changes become evident, the system at Steelcase must be able to accommodate them.

There are numerous indications that this corporate culture is not only well received by its employees but very successful in practice. The low annual employee turnover rate (less than 3 percent) and the absence of a union are indications of employee acceptance of this culture. Furthermore, Steelcase has developed a reputation for being a good place to work, as indicated by its inclusion in the following books and articles:

100 Best Companies to Work For

100 Best Companies to Sell For

Forbes magazine, "The Steel Behind Steelcase" (October 1985)

Management Review, "Nice Guys Finish First" (November 1985)

Finally, Steelcase is considered the market leader in the office furniture industry. As was previously noted, it is also considered to be a leading market and manufacturing innovator.

FIGURE 8-1

THE SCHEDULE MUST BE MET!!

It is our Boss! We must fight to
the Absolute Limit to ship

QUALITY PRODUCTS, ON TIME.

Any action is warranted to accomplish this end.

THE SCOPE OF PRODUCTS OFFERED

Steelcase can be best described as being a provider of office environment systems. Consistent with this perspective, Steelcase is organized into the following five major product groups:

Metal products. This is the primary focus of Steelcase. In the past it has been the major product line offered by Steelcase.

Wood products. This line is offered by Stow and Davis, an established supplier of wood furniture and a relatively recent acquisition of Steelcase.

Software. Systems software (programs and support) is available from either Steelcase or Hedberg Data Systems.

Speciality lines. This line describes a group of furniture products that are offered as a result of a design partnership with such companies as Vecta, Brayton, and Metropolitan.

Plastic injection moldings. This is the final general group falling under the Steelcase umbrella.

These five groups, taken as a whole, provide the following products and services:

Systems furniture

Computer support furniture

Seating

Lighting

Desks (these first five products form the bulk of Steelcase sales)

Media storage

Research (specifically into office environments and their problems)

Dealer and customer software. Subcategories: (1) acoustical; (2) space planning and facility design; (3) lighting visualization; and, (4) electronic product catalog.

Product Structure: An Overview

The Grand Rapids complex offers some 21,000 *standard* offerings. Complicating these numerous offerings is the wide range of multiple finishes. The variation in finishes is a product of differences in paint, hardware, upholstery (fabric type), laminates, plastic, and veneers. The result is that there are literally millions of allowable combinations. For Steelcase, options are an unavoidable fact of life. To succeed, the company must find a better way of managing these options—of converting them from a liability to a corporate strength.

The 21,000 standard offerings are built up from 32,000 assemblies, 17,600 piece parts, and 14,000 purchased parts.

STRATEGIC DIRECTIONS AND COMPUTER INTEGRATED MANUFACTURING

Strategically, Steelcase has implemented and followed two distinct approaches to building and maintaining its market position. These two approaches can be classified as follows:

1968 to 1987. Product oriented.

1987 to Today. Process oriented.

In the first phase, management was concerned with two issues. The first was managing the growth rate facing the company. The second was building a *better* product. By 1987 Steelcase had succeeded in addressing these two issues. However, it was recognized that the past strategic stance of the company was no longer adequate. The market had changed. As a result, a new strategic stance was introduced—one that focused on the corporate process used in delivering the product to the customer. These two approaches had a significant influence on how the firm introduced computers and automation.

Competing on Product (1968 to 1987)

In 1968 the management at Steelcase recognized that they were faced by an upcoming period of major growth in sales and production. This growth would be beyond the capabilities of their current manufacturing system. The system had to be changed to accommodate the anticipated growth.

Several independent consultants were brought in. As a result of their recommendations, management began the implementation of what would ultimately be an MRP II system. The new system involved primarily three functional areas: marketing, manufacturing, and distribution. It was primarily *product* oriented. Its ultimate goal was to enable Steelcase to deliver a better product.

What made for a better product? Such a product was defined in terms of three major attributes: quality, on-time delivery, and service. If Steelcase could build the best quality product, ensure that it was delivered on time and help its customers with the process of laying out their offices, it would have a distinctive competitive advantage in the marketplace. To achieve these three attributes would require the cooperation of marketing, manufacturing, and distribution.

MRP 11: The Initial CIM System

For Steelcase, MRP II was the most appropriate form of CIM to implement. It made sense given the market, the specific time frame, the product, and the expectations of the customer. This system consisted of the following major elements (see Figure 8-2):

FIGURE 8-2
Production Planning System Planning Levels

Forecast
Sales Operating Plan
Master Schedule
Material Requirements Planning
Capacity Requirements Planning
Production Activity Control

Material Requirements Planning. The bills of materials were computerized in 1969. MRP was implemented in manufacturing between 1970 and 1971. The MRP logic was subsequently extended to purchasing in 1972. Finally, the entire MRP system went on-line in 1976. By 1978, Steelcase was recognized as a Class A MRP user.

Capacity Requirements Planning. Effective management of capacity was critical if Steelcase was to achieve its objective. As a result, in the period between 1971 and 1976, Steelcase introduced CRP. The system consists of three elements. The first element is a 10-year forecast used by top management in planning capacity requirements. The second is a l-year detailed capacity forecast. The third is a quarterly quota. This quota fixes capacity needs and sets an upper and lower bound that determine how much above or below the quota actual demand is allowed to fluctuate.

To flag potential capacity problems, top management decided that the capacity utilization would never exceed 75 percent. Since Steelcase is a two-shift operation,[3] this decision translated into 100 percent use of the first shift and 50 percent use of the second. In general, if capacity utilization ever increased more than 75 percent, this would be a signal to add more capacity. For more expensive pieces of equipment, new capacity is added whenever the machine is used for at least one and three-quarters shifts. To management, such a decision made strategic sense, since the costs of insufficient capacity were far greater than the costs of excess capacity.

Sales Operating Plan and Master Production Scheduling. The entire manufacturing system is driven by two linked plans: the sales operating plan (SOP) and the master production schedule. These plans form a two-level MPS with the SOP driving the MPS.

The SOP, the first plan formulated, is based in part on the sales forecasts. These forecasts are created in a meeting that brings together the director of production and inventory control, representatives from Business Research, and representatives from marketing. At this meeting the forecasts are formed into the annual plan.

The SOP is a critical plan, since it is derived at the top management level. It is here that top management agrees on how capacity at Steelcase is to be used. The SOP is a monthly plan that breaks down production for a 13-week period (the quarter) into a detailed schedule expressed by product group and by plant. The SOP is established by a committee that consists of the vice president of

[3]A shift is 45 hours per week.

operations, the vice president of marketing administration, the vice president of manufacturing, the director of production and inventory control, and the director of marketing administration. This committee meets regularly every month. At these meetings, the committee not only deals with the formation of the current SOP but it also reviews changes to the previous SOP and any other additions that occurred in the intervening period.

The SOP links and coordinates corporate strategy. It also coordinates the operation of the other functional areas. It is also the rough-cut capacity plan for the entire firm. The SOP establishes the output quotas for each plant.

The master schedule is driven by the numbers taken from the SOP. Marketing Administration is responsible for the master schedule. It maintains the customer backlog and uses the backlog to smooth out production. The MPS is a more detailed restatement of the SOP. It breaks down the numbers present in the SOP and restates them by plant and individual lines (grouped by commodity or product group). It is stated in enough detail so that it can effectively drive the detailed material and capacity planning systems.

A critical feature present in the process of generating these two plans is the requirement for agreement. The plans are not released until everyone on the committees involved has agreed to abide by the objectives stated in these plans. To indicate agreement the members are required to *sign off* on the plans.

In addition, two other features have been added to this basic planning structure. The first is the introduction of automated scheduling of certain components and lines. The second is improved subsidiary interfacing. This latter feature is important, since there are certain critical components and activities such as chrome plating that are done by its subsidiaries. This interface provides for an explicit scheduling linkage between Steelcase and these subsidiaries to better coordinate the flow of material and information.

Inventory Management. Inventory management is used to establish and maintain tight control over inventory. This control focuses on accuracy (by part and location) and tracking. The following systems have been introduced to enhance inventory management:

- Locator systems (1971)
- Cycle counting
- Part and count verification systems
- An automated storage and retrieval system (AS/RS) (1973)
- An AS/RS interface to manufacturing applications (1980)
- Radio frequency
- Bar coding

Production Activity Control.[4] This system, more commonly known as *shop floor control*, is responsible for the execution side of operations. It is responsible for taking the orders generated by the planning system and, by drawing on the resources available to it (manpower, machine capacity, tooling, and materials), converting these orders into a set of end items completed in a timely and cost-effective manner. The production activity control system is responsible for (1) maintaining accurate inventory numbers, (2) the accurate reporting of schedule status, and (3) priority control.

The first systems developed in the 1970s were manual. By the early 1980s these had been replaced by on-line systems providing current information on such topics as

- Order status
- Input and output report
- Production status

In addition, the current system has been further enhanced by the introduction of an interface to the financial system and a mainframe interface to production control. This latter interface has focused on bar coding and two-hour production status. The two-hour production status system is a means whereby the status of all orders is regularly updated every two hours.

Interfaces. In building this system, management has attempted to build an integrated system that brings together the various areas of the firm. It has tried to facilitate this integration by using the MRP II system and its supporting computer systems as a facilitating vehicle. The following interfaces are indicative of this effort:

- Order scheduling to manufacturing planning.
- Traffic planning to manufacturing scheduling.
- Production status to final shipment planning.
- Label activity file to the personal computer.
- Manufacturing scheduling to AS/RS.
- The use of radio frequency.
- Production status to financial accounting.

---

[4]For a more detailed discussion of this aspect of operations at Steelcase, please see the chapter on Steelcase found in Stephen A. Melnyk and Phillip L. Carter, *Shop Floor Control: Principles, Practices and Case Studies* (Falls Church, Va.: American Production and Inventory Control Society, 1987).

- Artificial intelligence (a recent development).

Just-in-Time Manufacturing. The last major element of the system was just-in-time manufacturing. The management became aware of just-in-time in the early 1980s. They began the serious implementation of this system in 1983. It was not viewed simply as a system for inventory reduction. Rather, it was seen as a manufacturing-wide system for identifying and eliminating all non-value-adding activities.

In practice, just-in-time has become a system for simplification. It has been an impetus forcing people to reexamine areas such as work flows on the floor, machine groupings, and inventory control procedures. In some cases it has resulted in changes in layouts. In others, it has introduced simpler inventory control procedures.

Classifying the First CIM System

This first system described in the preceeding section was essentially a Type III. The manufacturing function was closely integrated. In addition, there were strong functional linkages (primarily between marketing and manufacturing) and a strong strategic link (again between manufacturing and top management). There were some formal linkages between Steelcase and its suppliers, but these were not as well developed or as strong as the internal linkages.

Evaluating the First CIM System

By 1985 the first CIM system, MRP II, was completely in place. By all accounts, Steelcase had been able to successfully implement and use this new system. There were actual benefits attributed to this new system. Inventory turns had increased greatly (in some cases between 25 to 50 times per year). On-time was also very high (see Figure 8-3). Quality was up to the point that the field return rate was less than 0.5 percent (the best in the industry). Steelcase was known throughout the industry for its record on on-time delivery. Finally, Steelcase was recognized as being a Class A MRP II system.

It had managed to achieve its goals of building a better product and of managing growth. However, as early as 1983 management was aware that the actions that had made Steelcase a success to this point would not be adequate if Steelcase wanted to maintain its growth and market share in the future. A better system was needed.

Competing on Process (1987 to the Present)

By 1985, the marketplace had changed greatly. These changes were the result of three factors. The first was the general slowdown and leveling off in demand in the office furniture market.

FIGURE 8-3

Letter of Commendation

JULY, 1983

Over the past six years, the Chair Plant has completed 313 weeks out of 321 at 100% schedule completions.

Six years of RELIABLE production is a tremendous accomplishment.

The Chair Plant Manufacturing and Staff groups have demonstrated a strong schedule - conscious posture, and are hereby commended for this outstanding record.

The period from the late 1960s to the mid 1980s was marked by a strong and growing market demand. Contributing to this growth were factors such as the introduction of computer equipment (requiring its own type of office furniture), the increased use of the *open office layout* scheme, and management awareness of the need for improved white collar productivity (and the impact of well-planned offices on this productivity). By the mid 1980s there were strong indications that the demand for the product was leveling off. The product had become mature. Competition had increased. Profits were now leveling off or, for some very competitive lines, falling.

Contributing to this increased competition in the office furniture market was the second factor: the actions taken by Steelcase's rivals. The competitors (e.g., Hermann Miller, Haworth) had not been passive during the time that Steelcase was building up its MRP II system. They took similar steps. They were actively involved in building up their manufacturing planning and control systems. They were implementing MRP and CRP. They were improving the performance of their shop floors. By the time 1985 to 1987 arrived, these competitors had so improved their performance in the quality and on-time delivery areas that the differences between themselves and Steelcase were no longer as distinct as they had been in the past. In many instances the customer might be unable to discern any differences. The products were becoming very much like commodities to the customer.

The third and final factor was the general change in the customer's perception of what an office was. Companies began to see office systems as being more than places to work. Rather, they were statements about the company. As a result, the aesthetics of design assumed a new and more prominent role in the design of office systems and in the selection of office furniture. A new element was now introduced into the design process. This was the architecture and design (A & D) house.

The introduction of the A & D house had three impacts on the process of supplying office systems. The first was that it changed the buying process and the linkages with the customer (see Figure 8-4). Now Steelcase had to build linkages with this new element. Steelcase recognized that it was selling as much to the A & D house as it was to the primary customer. Second, the A & D house increased the demand for more unique and individual products. This became most evident in requirements for new and often unique fabrics and more finishes and options. To get the orders, Steelcase now found itself required to deal with products that were becoming more complex (owing to these options). However, the customer did not expect to see any differences between a standard product and a special. Third, the A & D house, in laying out the new office, specified not only the type of office furniture but often carpeting and the mix of office furniture. No one office furniture company could meet all of these different requirements. Rather, they were drawn from a large number of different suppliers. This aspect greatly increased the general complexity of the buying process. One buyer would have to deal with more than one supplier. The buyer would not only receive numerous bills but would also have to worry about coordinating the actions of these various suppliers.

While these changes in the market were taking place, the customer still expected to get a well-built product delivered on time. In many cases the pressure on the supplier to improve quality and to reduce lead times (the major focus) had increased.

Steelcase had achieved its position by relying on the raw power of computers (in the MRP II system) and capacity to reduce lead times and to improve quality. These actions were now no longer as effective. As a result, Steelcase had to redefine its strategic stance. It had to develop a new strategy by which it could generate a unique competitive advantage for itself and improved value for its customers. After studying the market, management decided that the key to this new strategy lay in Steelcase's taking an active role in managing the *entire* purchasing process.

For Steelcase to survive, it had to simplify for its customer the entire process of purchasing an office system. It had to offer its customers a simple "one-

FIGURE 8-4
The New Customer Network Facing Steelcase

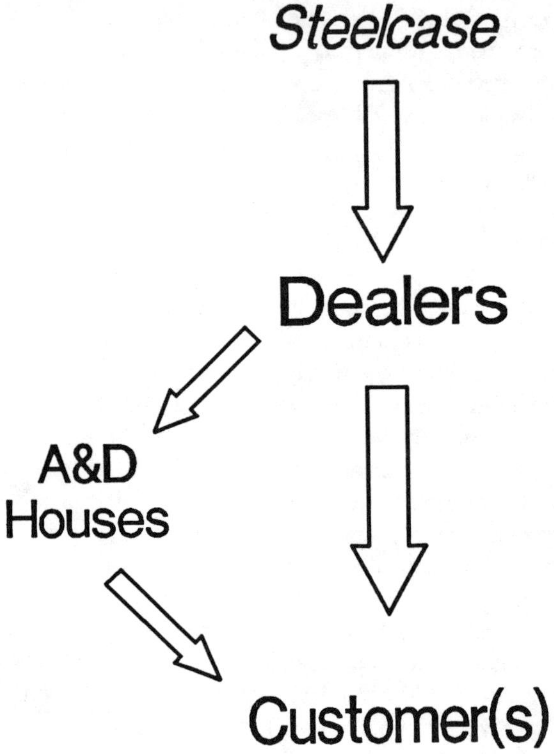

stop" shopping system. It had to enable the customer to specify the complete order and to know with *certainty* the completion date of the order. It also had to have in place the systems required to support the designer. A designer working on a plan had to have immediate feedback regarding the feasibility and availability of his or her plans.

To be effective, this system had to operate as follows: The Customer now deals with one firm, Steelcase, which then becomes responsible for coordinating suppliers. Steelcase also works closely with the A & D houses. The customer pays one bill (to Steelcase).

Furthermore, the new system had to be easy to work with. It had to be seamless. Customers did not have to know whether they were placing the order with Steelcase or with one of Steelcase's suppliers. The system had to emphasize service more than product. In the past, customers were responsible for the task of coordinating the activities of these various suppliers. This meant receiving 10 or 12 different bills and trying to ensure that everyone delivered products at the right time and at the right place. By simplifying this task for its customers, Steelcase could create a significant competitive advantage for itself.

This is not a minor undertaking, since outfitting an office requires the services of an A & D house, office furniture (wood and metal), carpeting, and other items. Many of the products required for these new offices are not offered by Steelcase. Rather, they are supplied by vendors. Steelcase now has to manage effectively a network consisting of suppliers, dealers (people who offer Steelcase products), and architecture and design houses. This factor has forced management to become aware of the need for strong channel integration.

However, simplification of the purchasing process is not enough. In addition, Steelcase must offer its customers help in designing the office system. Any resulting layout must make the office employee more productive. This has meant that Steelcase has had to invest in research and design capabilities. As a result, its corporate headquarters building has become an enormous laboratory for experimenting with different layouts. It also has required investing in computer software and hardware so that the resulting design capabilities can be computerized and these capabilities pushed out to the users (i.e., dealers and A & D Houses) to support their analysis.

Another implication of this change in strategic stance has been to make management more aware of the service implications of its business. Since customers expect a system that can accommodate options, the entire process of product design and order entry has also had to be examined and rethought.

In rethinking the order entry process, Steelcase became aware that what it was really selling its customers was not office furniture or office systems. Rather, it was access to an entire corporate system that was designed to deliver a set of products and services with the highest possible level of certainty. Dealing with Steelcase had to be seen as almost riskless. Steelcase had developed a system that brought together suppliers, marketing, engineering, manufacturing, and designers and tightly coordinated and controlled their activities in support of customers' needs.

SIFM: The Newest CIM System

This change in strategic posture has had a significant impact on how the management now defines CIM. Since there are many previous perceptions and misconceptions about what CIM is, the system now being implemented at Steelcase can be best described as *strategically integrated flexible manufacturing (SIFM)*. This term best describes the major focus and objective of this new system. SIFM is a system in which corporate strategy and the needs of the marketplace (and customers) drive the selection and use of technology and not the other way around. It is also a system in which all of the functions that affect manufacturing are actively involved in the process of designing, building, and delivering the product to the customer.

To Steelcase, SIFM is the appropriate solution for the problems now being faced. It is as appropriate today as MRP II was for solving the problems of the past.

At the heart of strategically integrated flexible manufacturing are two corporate concepts. The first is the *product concept*. To be successful in the market, Steelcase must be able to manage new product introductions. It must be able to reduce the lead time from market need to concept to design to product to market. It must be able to get its new product to the market fastest. The products that are created by this new and improved design process must also be built right the first time. Reductions in design lead time cannot come at the expense of reduced quality. Steelcase must be better at doing this than its competitors. In addition, this design process must be flexible enough to handle the increasing variations in fabric and finish.

Management has come to recognize that it is now dealing with two different markets: a standard market, which is satisfied with existing fabrics and finish, and a fashion market. The fashion market is both a major source of growth as well as the major area of difficulty. Increasingly, office designs are being driven by fashion considerations. These fashion considerations show up in requirements for nonstandard fabrics and finishes. For a company such as Steelcase to survive, it must be able to accommodate these requirements during the design phase. As one manager has put it: *"Those who cannot change are doomed to die."*

The second concept underlying SIFM is the customer concept. This objective does not simply require that Steelcase be willing to accommodate customer-requested options (such as unique finishes and fabrics). This objective is much broader. Steelcase must be perceived by everyone who works with it as being an easy company to work with. Order entry must be simplified. The time and effort required to go from the customer's desires to an order to scheduled production must be decreased. Critical to the success of this concept are improvements in the structure and operation of the corporate database. Changing the corporate database is not a minor undertaking, for the database is an embodiment of the paradigm that the firm has built up in guiding its business activities. To change the database often requires a rethinking and reformulation of this corporate paradigm.

These two concepts did not develop at the same time. The emphasis on the product concept preceded the stress on customer service. Management realized that a better product design system, by itself, was not enough. The advances in product design had to be supported by appropriate improvements in the level of total customer service offered.

The current system is the result of actions first taken in 1983. Its current form is the result of an evolutionary process aimed at trying to better understand the customers' needs and of trying to accommodate these needs. In part the evolutionary nature of this system reflects the dynamic nature of customer's expectations. What customers viewed as being acceptable in 1983 would not be acceptable in 1989. This evolutionary nature also reflects management's growing and changing understanding of technology and its roles.

Enhancing Product Development

In 1983 top management appointed a task force to audit the current system and identify problems with the current system. This task force, which was aided by an outside supplier, was charged with recommending special projects and identifying the associated costs and benefits. The focus of this group was on product design and manufacturing processes. It had recommended the introduction of CAD in product design and numerous process developments.

The implementation began in mid 1985. As part of this project, a technical group in Information Systems was established. In addition, a coordinator was appointed in Product Engineering. This person was supported by a staff. Management selected a coordinator for manufacturing. The task force also began the process of training the users. Common to all of these steps was an emphasis on the engineering side of product design.

The engineering activities associated with product design were seen as being the critical path for all product-design–related activities. As a result, in the period from 1985 to 1987 there were two major objectives for this area. The first was to improve and simplify new product design; the second was to reduce the design time. The major vehicle for achieving these activities was to be CAD.

The resulting CAD systems were used in the following three areas:

- New product design.
- Major engineering changes.
- Limited redraw of existing parts.

The CAD system was intended for the use of three groups: industrial design, research and development, and, product engineering. These three groups were selected because they were identified as playing a critical role in product design. Furthermore, by implementing CAD in these three groups, management could be assured that a mechanism (the CAD system) for integrating the process of product design from initial concept through product design and down to designing for manufacturability was in place.

Another action taken by management to integrate the process design of new products was the establishment of the Corporate Design Center (CDC). Located near the main facilities in the Grand Rapids area, the CDC occupies a new building. The CDC deals primarily with new products. Its purpose is to improve the design of new products and to reduce the associated lead time by bring together everyone who has an impact on the design of these products.

During this period, over 100 engineering workstations were put in place. There are approximately two engineers per workstation. Like operations in the various plants, the engineering workstations are used two shifts per day. By 1987 there were approximately 15,000 production CAD models in the system.

One of the concerns raised during these implementation efforts was that the engineers and their activities might become divorced from the realities of production. As a result, the new system might produce "nice" designs that were beautiful to look at but that did not make a great deal of sense when it came to building these designs on the shop floor. To overcome this concern, it was decided that the CAD system would be organized to permit decentralized product support. Furthermore, those engineers working with existing product designs would reside at the various plants. To ensure that their design activities were compatible, a central geometric database was implemented.

Enhancing the Manufacturing Process

In focusing on the product design process, management decided that design and manufacturing concerns had to be balanced and linked. Ultimately, whatever engineering designed had to be built quickly and cost effectively on the shop floor. In turn, the realities of manufacturing had to be reflected in whatever was designed by engineering.

To ensure that the product design and manufacturing activities were integrated on an ongoing basis, a design process was put in place that provided manufacturing with early communication of product specifications. This process was very similar in form and structure to simultaneous engineering. It was not called that, since the concept was yet to be developed. Steelcase was integrating the activities of these two areas because it made sense.

In addition, features were added to the CAD system that were of great interest to manufacturing. The CAD system had the capability to generate tool designs. The system also offered three-dimensional robot path simulation. The CAD system also had the capability to generate from the designs cost estimates and a process plan. It also allowed for the downloading of true geometry to the manufacturing system for NC programming. There was an interface between the coordinate measuring machine (CMM) and the CAD system.

One area emphasized when enhancing the manufacturing process was the bill of materials. The bill of materials is one device for integrating the activities of product design, research and development, industrial design, and manufacturing. These four groups must work with one common bill. As a result, the bill of materials became a method for communicating between these departments.

Consistent with this role, all bills of materials are on-line. This feature is important to manufacturing since it enables its personnel to preview preproduction drawings and to assess their manufacturing feasibility. The database system for managing the bills of materials also allows for viewing and printing of all production drawings. By putting on-line all of the bills of materials, management has achieved another objective—that of keeping the bills current and accurate. Since shop floor personnel can now review bills whenever needed, errors in the bills are flagged more frequently. This has resulted in an increase in the number of approved requests for engineering changes.

As a final step in enhancing manufacturing effectiveness, the layout of the shop floor has been changed. There has been an attempt to simplify the flow of work on the floor by moving work centers together and also by forming new work centers based on the part families produced. By moving work centers together, the distance traveled by parts has been reduced. Increasingly, the movement of work resembles that of a continuous flow rather than that expected of a job shop.

In addition to these investments in *soft* technology, management has invested heavily in *hard* technology. Since 1983 it has put in place:

Dedicated automation. This is not flexible and has been used primarily in machine and weld operations.

NC programmable equipment. This technology has taken the form of turret punch presses, fabric cutters, wood routers, and DNC machines.

Programmable robots. These are used for arc welding, material handling, spray painting, and machine loading.

Flexible automation. This type of technology is used for high-volume production involving families of like parts. The major attraction of flexible automation is quick changeover (and short setups). Flexible automation is used for metal fabrication, weld, and assembly tasks.

In selecting the types of technology to buy, management has followed a simple rule: *buy the technology that best fits the requirements of the task, not the one that is the newest or the flashiest.* There is a difference between being on the *leading edge* and the *bleeding edge.*

Enhancing Customer Service

Steelcase has tried to improve customer service by implementing a two-pronged attack. The first prong has focused on simplifying the buying process for this customer. The second prong has dealt with simplifying order entry. This section examines this second prong in detail.

By 1985 it was evident to most management personnel at Steelcase that order entry, or the process of moving the customer's order from the marketing system into the customer order database and from the database into the scheduling system and onto the shop floor, was unnecessarily complex. In part, this complexity was a side effect created by the projects aimed at enhancing manufacturing effectiveness.

The manufacturing database was a Tower of Babel. There were many manufacturing islands of automation. Each island was previously implemented as a way of resolving a specific manufacturing problem. Many of these islands, such as the automated Hi-Rises (the automated storage and retrieval systems), had their own associated manufacturing databases. The major focus was on whether the technology being considered for acquisition could solve the problem being faced. The information implications of introducing these islands was never adequately considered (or even recognized).

Over time, the manufacturing database had became a corporate bottleneck. There was no well-defined common interface between these islands. Maintaining the flow of information between these islands required a lot of manual intervention. This intervention resulted in a great deal of internal overhead. The problems encountered in circulating information in the manufacturing system helped create the perception both internally (i.e., within Steelcase) and externally that Steelcase was difficult to work or do business with. This perception had to be eliminated for Steelcase to maintain its position in the marketplace.

As people began to investigate these problems, they concluded that the problems with the manufacturing database were simply *symptoms* of a much deeper problem. The manufacturing database was based on a business paradigm that was over 20 years old. This paradigm was developed to support a company that simply supplied office equipment to its customers. It was organized around MRP and its needs. This paradigm was out-of-date. It did not satisfy the needs of a company that was trying to simplify the entire purchasing process faced by the customer. It did not support the activities of a firm trying to integrate and coordinate the activities of purchasing, manufacturing, environment, research and design, suppliers, architecture and design houses, and the dealer network into a tightly controlled system. Many people within Steelcase became aware of the need to change the method by which the company managed its information. This change, to be effective, had to be corporate-wide.

Few people get excited about changing a database. Most realize that it is a major undertaking, requiring a great deal of work. For others, it is an inconvenience. Line people view such activities as getting in the way of their jobs. Overcoming the resulting resistance and inertia requires a major effort. For Steelcase the pressure for change came from top management, which made it a corporate *mandate*. They realized that the current database was inadequate. If

Steelcase was to maintain its position in the marketplace, the information side had to be upgraded and upgraded immediately.

To improve management's control over its information resources, Steelcase has invested in several information technology (IT) techniques. The primary technique has been that of *entity-relationship modeling* (*ER*). In 1989 a cross-functional committee was formed and given the task of evaluating the suitability of ER for Steelcase. They laid out the various entity-relationship diagrams with the goal of soon having a working prototype.

In developing this prototype, care was taken to ensure that the use of the information technology was user driven, not database driven. To identify the needs of the company, a procedure was implemented (see Figure 8-5). This procedure began with an assessment of Steelcase, the market, and corporate strategy. Based on this assessment, the information requirements needed to support corporate activities were identified.

It is generally recognized that Steelcase is at the beginning of the journey to improve its control over information and to enhance its ability to offer better service to its customers.

Cost Justification

For many companies, cost justification has become a major obstacle to CIM. At Steelcase every expenditure in SIFM has undergone a structured *cost-benefit* evaluation. However, cost justification has been based on a mixture of "gut feeling," quantitative cost measures, and strategy. Where the technologies make sense and are needed to help the firm meet its strategic objectives, management is willing to make the necessary investments. Furthermore, because Steelcase is privately held, management is not under the same pressure to provide quick short-term results.

IDENTIFYING THE TYPE OF CIM SYSTEM PRESENT AT STEELCASE

From 1983 to 1989 Steelcase has moved from a Type III to a Type IV CIM system. In 1983 Steelcase was tightly integrated internally. There were well-defined, formal linkages between manufacturing and strategy; manufacturing and the other functions (notably marketing and engineering); and within the functions of manufacturing. With the addition of strong *formal* linkages to the A & D houses, the dealers, the customers, and the suppliers, Steelcase has made the transition to a fully integrated Type IV system. The activities now taking place within Steelcase (as described in the preceding sections) can be viewed as steps aimed at solidifying its position as a Type IV system.

FIGURE 8-5
The Process of Identifying User Information Needs

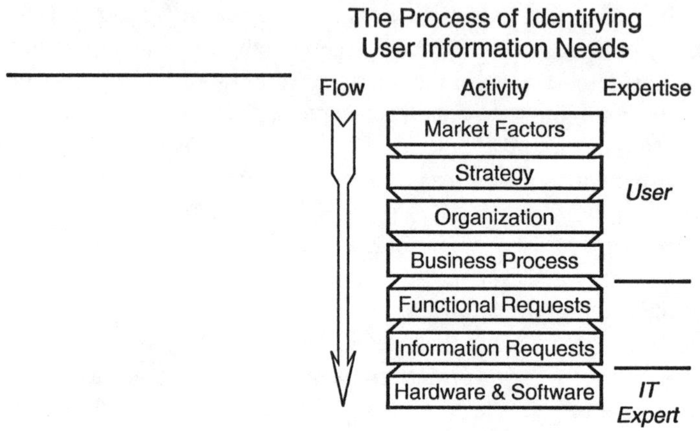

FUTURE DEVELOPMENTS

SIFM is a system under development. The people at Steelcase recognize that new technologies are constantly surfacing and these technologies must be explored *when* they appear to be appropriate solutions for problems facing the company. As of 1989 management was exploring the following major technologies:

- *Solids modeling.* This design technology is being evaluated to see if by developing and testing computer-based models of new designs the number of prototypes can be reduced. Building prototypes is a time-consuming activity.
- *Data communication to tooling vendors.* As new designs are proposed, one potential bottleneck that must be overcome to ensure the shortest design and manufacturing lead time is the supply of appropriate tooling. One means of reducing the tooling lead time is to communicate design requirements directly to the tooling vendors. The information is to be communicated by computer. Ideally, the tooling vendors should be able to work off the same computer-generated drawings being used internally.
- *Simulation of factory flows.* By building computer-based models of proposed floor layouts, management can study the effects of changes in product mix, work flow, and machine layouts. New layouts can quickly be built,

evaluated, and discarded if found to be unsuitable. Bottlenecks can be identified and eliminated. The result is an effective and efficient factory flow that works right and in which people have confidence because they have "stressed" on the computer before they have even placed one brick or inserted one bolt on the shop floor.

- *Factory networks and the use of MAP protocol.* The use of networks and a standard communications protocol is being evaluated as one means of linking the various islands of automation and overcoming the need for human intervention.

BENEFITS

Identifying the benefits directly attributable to CIM at Steelcase is very difficult. The CIM system is still evolving. The benefits are often not attributable to any one activity. The nature of these benefits is also long term. However, the following are some of the benefits that have resulted from the implementation of CIM:

- *Productivity increases.* These increases have been most evident in the product design areas, where significant decreases in design lead times have been observed.

- *System simplification.* There is now an effort aimed at simplifying the flow of work on the shop floor. This has been most evident in the reduction of distances traveled by work. By moving work centers closer together and defining work centers in terms of family membership, work does not have to travel as far. In one case, before the changes, work travelled over 5,000 feet without having any value added to it. After the change in layout, this has dropped to less than 600 feet. These changes have also had the effect of reducing shop floor lead times and the associated variability.

- *The acceptance of the dynamic nature of business.* By forcing people to study and understand business and strategic requirements and to compare these requirements with what the existing system can provide, people have become aware of several factors. The first is that existing information is not adequate. It no longer supports the business and strategic requirements. More important, people have come to recognize that the business in which Steelcase is located is dynamic. It will and must change. How Steelcase competes today is very different from how it competed 10 years ago; how it will in five years will be very different from how it competes today.

- *The identification and elimination of inappropriate projects.* In many cases projects have been stopped that were driven primarily by technology rather than by strategic or business needs. One such project involved automating NC programming. After this project was begun, management began to question its appropriateness. People could not identify the competitive advantage generated by "lights out" programming. As a result, work on the project was stopped. This saved Steelcase the need to make significant outlays in terms of time, money, and manpower.[5]

- *Better product designs.* The new process now creates designs that are easier to build and "right" the first time. The need for building prototypes has been decreased. Quality problems associated with product startup have also fallen.

- *Better information.* As a result of the review of the information system, the overall quality of the information system has improved. The system is becoming easier to use; the data that is needed can now be obtained much more quickly; the information is now starting to support business activities.

LESSONS LEARNED

Certain lessons can be drawn from a review of the experiences of Steelcase with CIM and SIFM. The following are the most important:

1. *Take advantage of outside help whenever possible.* The implementation of a CIM system is a major undertaking. It is a path strewn with many pitfalls and dangers. One way of avoiding these problems and of reducing the total implementation lead time is to take advantage of the expertise of people who have been involved in one or more prior CIM implementations. Often, this means hiring a consultant. These people have prior experience with CIM. They know what works and, more important, what does not work. They can identify the critical tasks. Their guidance and advice can help the firm avoid "reinventing the wheel." They also provide eduction for the firm and its people. Finally, they are also familiar with technologies or systems new to people within the firm. This was how the people at Steelcase became familiar with the operation and the concept of entity-relationship modeling.

2. *People skills are extremely important to the system's successful implementation.* The key to success with SIFM, CIM, or any other such broad system lies in how the human, not the technology, side of the implementation process is managed. As a result, it is important that management get the users to understand and accept the system and the proposed changes that will accompany its introduction. It is important that the users develop a sense of *ownership* over the new system.

---

[5]This example is discussed in greater deal as part of the section on lessons learned.

3. *There are no off-the-shelf solutions to the task of implementing CIM.* CIM cannot and should not be seen as a hardware solution. Buying more and newer technology does not often make sense. In the case of Steelcase one of the major keys to its implementation efforts has been understanding business and strategic requirements and translating these into the appropriate supporting hardware and software requirements. Because every firm competes on a different basis in the marketplace, its business and strategic needs vary. Given such a perspective, off-the-shelf solutions do not make sense. They may provide a starting point; they are not the end. Standard solutions are built for the general case. The purpose of strategy (the starting point for CIM or SIFM) is to differentiate.

4. *Begin CIM with a clear understanding of what you need to do well in your business.* The ultimate purpose of CIM (or SIFM) is to support how the firm competes in the marketplace. Unless management knows the basis on which they compete (i.e., how they win orders, lose orders, and become considered for orders), they cannot build an effective CIM system.

This feature is very evident in the shape taken by CIM at Steelcase. Steelcase competes on reduced product lead time, product flexibility, and good service. Steelcase also sells *process*. That is, Steelcase sells its customer the assurance that once an order is accepted, Steelcase has in place a coordinated and integrated *system* that will produce that product when required, at the place required, with the required product features, and at the level of quality required. These features have significantly affected the focus of CIM activities at Steelcase. Management has invested in only those developments consistent with these objectives. As a result, everyone can see the relationship between how Steelcase competes and the changes taking place in the company.

5. *The critical element (and a major bottleneck) to CIM is the information database.* It has only been recently that people at Steelcase have begun to appreciate the importance of information and the appropriate data structure. The data structure is important because it describes the firm and how it operates. It is also an integrating device for the firm. The problem for Steelcase was that while people were aware of the need for good information, they were unable to articulate it well. They were also faced by a manufacturing database that was essentially fragmented. To succeed with SIFM, they must overcome these and other information-related problems. This has been and continues to be a major task.

6. *Recognize that errors will occur and be prepared to admit and correct them.* In spite of all the planning that takes place, errors will take place. There are many reasons for such errors to occur. However, once you become aware of their presence, the most effective course of action is to admit their presence, stop working on them, and move on. Punishing people for making errors is not effective. The key issue is to get back on track, not to assign fault.

An example of this lesson in action can be drawn from the decision to stop work on a major engineering project. This project, if successful, would have resulted in a system that generated an NC program from a CADAM basis. This generation would have been totally automatic and would have been done without any human intervention.

Initially, the project seemed to make economic sense. However, after the project had begun, management began to rethink their decision. They saw that this project was essentially technologically driven. There was no real customer for the product. The product was also too complex. Finally, there was strong evidence that the product was not being accepted by product development. As a result, the project was stopped and the personnel involved reassigned. Everyone agreed that it was better to stop instead of pushing the project on to completion.

7. *Have top management support and commitment.* This point cannot be emphasized too many times. To succeed, CIM must have the support and commitment of top management. Such support and commitment is required for several reasons. CIM is a long-term undertaking; the returns from CIM take a long time to develop. CIM is also a major and expensive undertaking. Top management is responsible for the investments made in CIM; it is also responsible for getting people involved. Finally, it is top management that often ultimately breaks major bottlenecks and overcomes resistance within the firm. This last aspect was most evident in the recent efforts being undertaken to restructure the information system present at Steelcase.

8. *Take advantage of cross-functional groups when possible.* CIM is not a manufacturing and engineering system. It is a corporate undertaking. It requires the active involvement of all the functional groups (e.g., accounting, top management, engineering, quality control, marketing, human resources). By involving these groups early in the implementation process, corporate awareness of CIM is increased. Furthermore, these groups have a chance to evaluate CIM and to suggest improvements in it. They have the opportunity to become part of the process.

The structure of the CIM committee best illustrates this lesson in action. The CIM committee is charged with the task of overseeing the definition of CIM and its subsequent implementation. It consists of 12 members. These members are drawn from accounting (l), engineering (1), manufacturing (1), production and inventory control (1), information systems (1), marketing (1), sales and administration (order entry) (l), and human resources (1). In addition to these 8 representatives, there are 4 others who are "ad hoc"; they are representatives from other subsidiaries and other Steelcase product areas.

9. *Don't oversell CIM and its promise.* Finally, disappointment always adversely affects any implementation project. One way of avoiding disappointment is to not build up expectations by promising too much. When too much is promised, whatever is delivered is never quite enough. Instead, people must develop realistic expectations. They must understand the potential advantages and disadvantages present in a system such as CIM. They must understand that CIM is a relatively new development and that, as a result, there is a great deal that is not known about it. They must also recognize the possibility of failure. There is a delicate balance here. On the one hand, realistic expectations must be developed. On the other hand, the people should not become so "realistic" as to expect the system to fail.

CONCLUDING COMMENTS

This case study has described the CIM journey of Steelcase. It has provided a glimpse into a system that is still evolving and developing. It has tried to show how at Steelcase technology has become the servant, not the master, of strategy.

Before we conclude this discussion of Steelcase and CIM, it is important to recognize certain of the major traits found in CIM at Steelcase:

• *Evolutionary development.* Our current place in the CIM journey has been very much influenced by what we have done in the past. Steelcase began in the early 1970s as a major user of MRP. MRP was introduced as a means of improving the manufacturing system. Because of limitations in MRP (MRP is after all nothing more than a material planning and scheduling system), Steelcase subsequently introduced a closed loop MRP system and an MRP II system. Each system, as it was introduced, built on a foundation inherited from the prior systems. Each new system extended the current system by overcoming critical limitations in that system. Old systems were not thrown out because of limitations. Rather, they were added to. This evolutionary development is very evident in the current CIM system.

What Steelcase is doing now with CIM cannot be described as being a radical departure from past developments. Rather, it is a logical and consistent next step. By following such an evolutionary path, management has been able to avoid the turmoil that frequently accompanies the implementation of new systems. It has also built on its strengths. MRP, the closed-loop MRP, and MRP II have been past strengths for Steelcase.

• *Strategic positioning of CIM.* At Steelcase CIM is driven by strategy. Steelcase competes by serving its customers better than any of its competitors. CIM, with its emphasis on customer service, product design, product flexibility, and information processes, does nothing more than enable management to achieve its strategic goals. CIM is nothing more than a vehicle; it is a corporate means to a set of strategic ends. This trait is not unique to CIM; it is evident in everything that Steelcase does.

• *Cost-benefit evaluation.* Whenever investing in technology, every investment has been subjected to a cost-benefit evaluation. Every investment must be justified. There must be evidence to show that making that investment makes sense (as discussed in the next trait). However, when evaluating a new investment, it is recognized that the "numbers" don't necessarily tell all. Evaluations reflect both quantitative and qualitative considerations.

• *Doing what makes "sense."* Every change introduced at Steelcase must always go through one critical test: it must show to the naive user that what is being proposed can adequately solve the problem being faced. The emphasis is not on the technical issues. Rather, it is on reducing the problem to basics and showing the fit between the solution proposed and the problem. This is a test more of *common sense* than of economic dollars and cents. That is, for CIM to be successful, it must do the common sense things well.

• *Matching technology and infrastructure.* There are three elements that must be consistent with each other for CIM to succeed at Steelcase. These are technology, information, and organizational structure (see Figure 8-6). All three elements are considered when introducing the changes required by Steelcase. Failure to build the appropriate support infrastructure (information and organization) will result in a drag on technology. The promise offered by the new technology can go unrealized because of problems with infrastructure. This trait also complicates the challenge of implementing CIM.

• *One driving goal: EXCELLENCE IN MANUFACTURING.* This is the ultimate objective for Steelcase. There is a single-mindedness about everything done at Steelcase. Any action taken must ultimately be justified on its ability to maintain Steelcase's reputation for excellence in manufacturing. If an investment or proposed change does not contribute to this goal, it will not be made.

FIGURE 8-6
Balancing Technology and Infrastructure

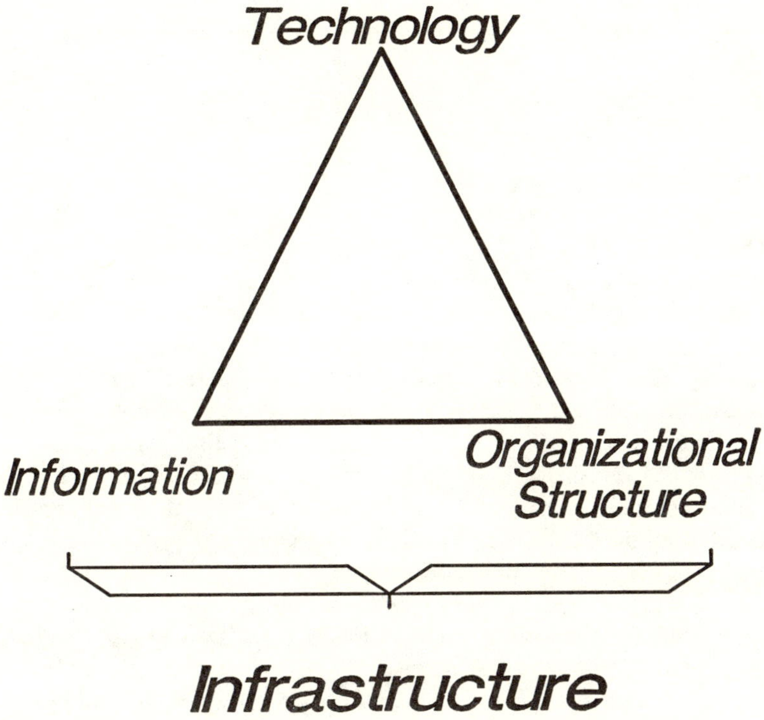

CHAPTER 9

CONTINENTAL CAN (CONTAINER DIVISION)

EXECUTIVE SUMMARY

The Container Division of Continental Can is a major American manufacturer of can-making and can-closing equipment.[1] Its experiences with computer integrated manufacturing are interesting for several reasons. First, it is a company that manufactures a mature product. Its decision to implement began with a very simple but critical question: *Should we stay in the business of making can-making equipment or should we devote our energies and resources to other, potentially more profitable efforts?* At the time that the decision to implement CIM was made, Continental Can was in a crisis. If it was to succeed, CIM had to succeed. Finally, CIM at Continental Can is the most focused and contained of all the CIM systems studied. It deals primarily with the integration and enhancement of the manufacturing system.

For CIM to succeed, the management at Continental Can had to restructure most of the manufacturing system. There was a heavy emphasis on improving the manufacturing database. It was agreed that information was a bottleneck and that it did not support corporate activities. Attention was also devoted to simplifying and improving work flows through such actions as improved shop floor layout. By simplifying the system first, the resulting CIM system could be simple.

As a result of these and other activities, CIM enabled management to make Continental Can a strong and healthy competitor in the can-making industry. However, now Continental Can's future is cloudy. It is faced by several serious challenges to its future. There are doubts about its continued existence in the industry. The reason lies not in CIM. Instead, these doubts are a result of changes in ownership and an associated increase in debt load (incurred to acquire this division).

[1]This case study is based on a presentation done by David Grigsby of Continental Can and has been supplemented by information provided by Richard T. Smith of Ingersoll Engineers.

INTRODUCTION

The Container Division of the Continental Can Corporation, in addition to being a manufacturer of can-making and can-closing equipment, is also a supplier of replacement and spare parts for this same equipment. These products are purchased primarily by bottling and beverage companies (e.g., beer and soft drinks). Continental Can's customers span a wide range, from a local bottler of soft drinks (which has a very small local market for its products) to large regional distributors and bottlers that supply the well-known beers and soft drinks. Its competition comes primarily from two sources: American Can and National Can.

Located in West Chicago, Illinois, Continental Can presents an interesting study in the implementation and use of computer integrated manufacturing. First, Continental Can is a unionized plant. Second, the structure and focus of the resulting CIM system have been very strongly influenced by the mix of products supplied by this division and its strategic stance.

These products are almost a commodity in nature. The market for can-making and can-closing equipment is mature. A set of well-established and widely known criteria dictates how firms compete and win (or lose) market share. To succeed in this market, a firm must be able to do better on these criteria than its competitors.

SUMMARIZING THE KEY SUCCESS FACTORS FOR THE CONTAINER DIVISION

There are three major factors that influence purchases in this market. These are as follows:

Acquisition Cost. A can-making or can-closing machine is a large capital acquisition for most firms. As a result, they are sensitive to the up-front costs of purchasing the machine. Most often, the supplier offering the lowest-cost machine has the highest chance of winning the order (all other things being equal). Factors such as quality and product delivery lead time do not play a major role in the purchasing process. There are no real major differences between the competition on these other dimensions. In general, the quality[2] of a can-making machine supplied by Continental Can is no better or worse than that found in a similar machine supplied by either National Can or American Can.

---

[2]Quality is judged in terms of *defective* cans. A defective can is one that leaks, is not properly sealed, or cannot be opened easily.

There might be slight differences in how equipment is designed and constructed, but the customer is generally unaware of them. In short, a better can-making machine is a cheaper can-making machine.

Speed of Operation. There is only one design and product area in which the competing firms can distinguish themselves from each other. This is operating speed. This term describes the number of cans processed per minute. Operating speed is critical to the customer for several reasons. As the operating speed increases, the operating cost per can falls. Higher speed means more cans processed per hour, which means the availability of more capacity. Finally, since a higher-speed machine offers more capacity, acquisition costs may be affected, since fewer machines have to be purchased.

Availability of Spare Parts. For bottlers and distributors, one of the keys to making money is *uptime*. The can-making and sealing equipment must be available when needed. A machine that is down is not only unable to seal cans but it can also act as a bottleneck to the entire process. Often, bottling operations are line flow processes. A machine breakdown in the can-making and sealing department can quickly affect the other downstream operations. Beverages that cannot be processed must be stored. Once the available storage space is consumed, the process must be stopped.

The key to uptime from the supplier's perspective lies in the effective management of spare parts. The lead time from the moment that the machine breaks down until the spare parts are delivered to the site must be kept as short as possible. There are two approaches that can be used to manage spare parts. The first is to keep an extensive inventory of spare parts available. The second is to have a process in place that can produce the required parts quickly. The shorter the time interval from the moment that the machine goes down until it comes back up again, the better.

Spare parts are critical to the supplier. Since customers are more sensitive to the costs of downtime than to the costs of the spare parts needed to bring the machine back into operation, spare parts tend to be extremely profitable. However, managing spare parts is extremely difficult. It is difficult to forecast what parts to produce when. Given the variety of machines (and parts) supplied, it is difficult to keep all of the needed parts in stock. Often, suppliers are faced by a situation in which they carry high inventories of spare parts. Further complicating this task are the maintenance policies and procedures of the customer. How the customer maintains the equipment affects the demand for spare parts (both urgency and timing). However, these policies are not under the control of the supplier. Since customers demand lead times of less than one week (this figure is constantly falling), this area represents an ever present problem for a supplier such as Continental Can.

From the customer's perspective, the first two factors, acquisition costs and the speed of operation, are order-winning criteria. All other things being equal, the company with the cheapest and fastest machine will win the contracts. Factors such as quality, delivery lead time, and uptime are order-qualifying criteria. It is assumed that each supplier is acceptable in these areas. Uptime is also an order-losing trait. Poor performance on spare parts can seriously affect future purchases.

To a company such as Continental Can, the best you can hope for when selling a can-making or closing machine is either to make a slight profit or to break even. You make your profits on the spare parts.

CONTINENTAL CAN BEFORE COMPUTER INTEGRATED MANUFACTURING

The move to CIM began in 1983. It was triggered primarily by a change in top management. Greg Horrigan was appointed to the position of executive vice-president of the Continental Can Corporation. A person who had advanced quickly through the company, Greg Horrigan was widely recognized as the "heir apparent" to the CEO. One of his actions on assuming this new position was to evaluate the corporation and to assess the performance of all of its divisions. As a result of this review, Mr. Horrigan became aware of the Container Division and its problems.

By 1983 the Container Division was in serious difficulties. It was losing both money and market share. Finally, it had developed a reputation for poor service. To get a spare part from Continental Can would require a wait of months, not weeks. Some of its customers had become so frustrated by the long waits that they had started turning to its competitors for spare parts.

As a result of these and other problems, Mr. Horrigan had decided that something had to be done about the situation at the Container Division. He needed an answer to a simple question:

Should we stay in the business of making cans?

There was pressure to answer this question. Mr. Horrigan had set a time limit of 12 weeks in which to arrive at an answer. It was not enough to simply answer yes or no. If the answer was "Yes, we should stay in this business," a plan for improving operations in the Container Division had to be provided. This plan had to generate some form of distinctive competitive advantage for the division.

Assessing the Problems of the Container Division

To help management answer this question, outside assistance was brought in. Ingersoll Engineers were hired, primarily because of their experience with manufacturing. The consultants, along with selected members of management, were formed together into a team responsible for addressing the question. By the fall of 1983 management was ready to assess the situation in the Container Division. What they found was that the division's lack of competitiveness was simply a symptom of much deeper and more severe problems:

• *A lack of corporate investment.* In the past the corporation had treated the division as a cash cow. As a result, the corporation had invested little money in the division. The newest piece of equipment was seven years old. In walking through the plant, the team was surprised to find still in operation one of the original machines manufactured by Ingersoll Milling. This was a machine built in the 1920s. Not only was the equipment old, but the plants were equally old and outmoded.

• *Poor work flows on the shop floor.* The flow of work on the shop floor was chaotic. The distance between operations was long; the movement of work seemed to follow no logical pattern. The grouping and location of work centers also did not make sense.

• *Excessive capacity.* Another problem facing the division was that it simply had too much capacity. This was not a new problem. Previously, in the mid 1970s, the division consisted of some seven plants located throughout the United States. All production at these plants was then consolidated into two plants located in West Chicago. It was now recognized that a further reduction in capacity was needed. Two plants were still one too many.

• *An inadequate information system.* The information system was not able to support the operation of the division. Since it was primarily manual and relied heavily on the clerical people, the current system was very cumbersome to use and maintain. This factor was evident in a number of ways. Data accuracy was extremely poor. On the average, the accuracy of the bills of materials was only 25 percent. The bills of materials were also incomplete bills. Furthermore, routing information was either bad or incomplete. The only thing that enabled the division to ship products was the informal system. Furthermore, the information system was poorly organized. There was excessive duplication of information in the files. The team agreed that the information system in its current form was a corporate liability.

• *A lack of control over critical activities.* Costs appeared to be out of control. In part, this was a result of the way in which certain activities were managed. The most blatant problem involved the purchasing function. Purchasing was often done by departments other than purchasing. The resulting prices were higher than they should have been.

• *Problems with current management attitudes.* Many of the people in management positions had been educated in the 1940s. They had not kept pace with the changes taking place in manufacturing. They were not aware of the new technologies now available; their views of their work force were very outdated.

- *A lack of critical management skills.* In reviewing the current management group, the team noted that it did not have the right skills in manufacturing and in engineering.
- *A lack of a credible, effective manufacturing planning system.* The current planning system was ineffective. It was basically an *order launch* and *expedite* system.

All of these factors indicated a system that could not do even the basics of manufacturing well. The team concluded that these factors were behind many of the difficulties encountered by the division. Unless these factors could be gotten under control, the team decided that nothing could be done to eliminate the high inventories, long lead times, and high manufacturing costs.

At the end of the 12 weeks, a report was presented to Greg Horrigan. The answer to the question was that Continental Can should stay in the business of building can-making machines. The problems that were currently plaguing the division could be easily identified and costs could be reduced. The factory could be improved.

Beginning the Move to CIM

After assessing the division and its problems, it was decided that CIM offered the most advantages for the Container Division. A proposal was submitted to the executive board of the Continental Can Corporation in March 1984. The primary justification for the proposal was the anticipated cost savings resulting from improved throughput and reduced inventory.

In approving the proposal, top management recognized that what was being asked for was essentially a long-term project. They agreed to commit the corporation to provide the division with access to the required investments. Specifically, top management indicated that they were willing to allocate between $10 to $12 million for the acquisition of any necessary technology. They also agreed to use a long-term perspective when evaluating the division's performance. Finally, they agreed to protect the project and the implementation team from outside pressures. They wanted real change, not "window dressing."

Top management was not the only party interested in seeing a successful implementation of the CIM system. The union, now consisting of some 500 members, was also interested. Everyone knew that the division was in trouble and that the CIM system was the only chance for the division to survive. They decided to support the implementation efforts.

Implementing CIM

Within three months of the project approval, the current vice-president of the division was replaced, primarily for performance reasons, by a new vice-president. David Grigsby, the new vice-president, was from corporate finance. However, he was not unfamiliar with the CIM concept.

Some years previously, he had had the opportunity to observe the changes taking place at Ingersoll Milling. He had developed a long-term relationship with several of the management people from both Ingersoll Milling and Ingersoll Engineers. He was aware of the potential present in CIM. More important, he was convinced that CIM could work very successfully in the Container Division. He was committed to making it work in the division. As a result, Dave Grigsby was to become the CIM champion at this division.

Because of the limited time availability of certain key management personnel, it was decided by everyone involved in the implementation that the Container Division had only one chance to get it "right." This was a constant source of pressure and a constraint with which everyone had to live. To improve the chances of getting it right, top management insisted that certain guidelines be followed in the implementation:

- *Training and education were to be emphasized.* Everyone in the division, from top management to the people on the shipping dock, was to be exposed to the CIM system. Training and education were emphasized for several reasons. These were to be the means of familiarizing people with the new system. These were also to be a conduit for overcoming any resistance to change.
- *Resistance to change was not to be tolerated.* When education and training failed to overcome resistance, the people were to be moved aside. These were viewed as *assassins* and treated as being detrimental to the success of CIM.
- *The CIM implementation was to be focused.* It was decided that the changes demanded by CIM were to be confined to the manufacturing system and the information system supporting its operation (see Figure 9-1). Given the commodity nature of the product and the presence of clear-cut criteria for competition, it was felt that the greatest return on investments would be gained by confining CIM to these two areas.
- *Manufacturing basics were to be improved.* Attention was devoted to the areas of master production scheduling, production planning, materials planning, and shop floor control. These were areas that were currently being done very poorly.
- *The manufacturing system was to be protected as much as possible from external changes.* External changes (such as order changes) were viewed as very disruptive. The system was to be protected in three ways. The first was through the master production schedule. Time fences were introduced. An order

FIGURE 9-1
CIM at Continental Can: The Major Focal Points

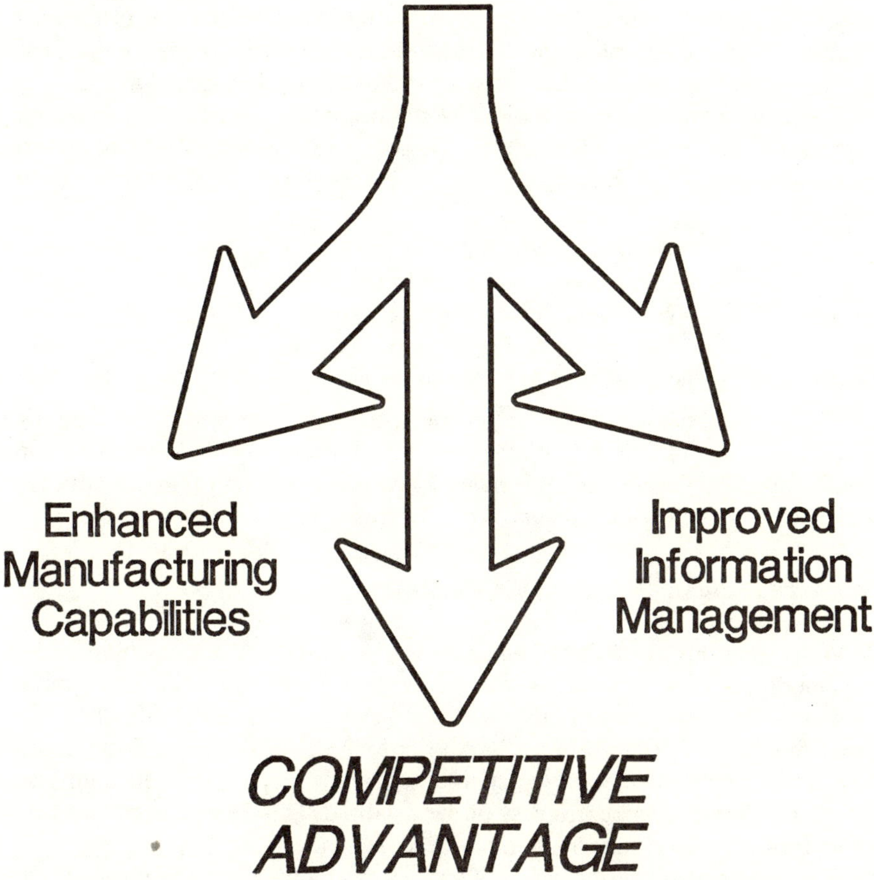

review and release mechanism was introduced to manage the flow of work from the planning system to the shop floor. Finally, extra capacity was added on the shop floor. This final policy reduced the potential impact of any bottlenecks. It also simplified the entire process of capacity planning.

• *System simplification was to precede system automation.* Because of difficulties with work flow and shop layout, management decided that these problems were to be eliminated by changing the layout of the floor and straightening the flows.

• *Manufacturing facilities were to be consolidated.* There was no real need for two facilities. All production was to be concentrated at one plant.

• *Use "plain vanilla" systems where possible.* The systems were to be used as much as possible "as is." There was to be no rewriting of code. This meant that people had to change their practices to conform to the requirements of the new systems. Given the general ineffectiveness of the old system, requiring that people change the way they ran the business was seen as not being all that bad.

There was another reason for this requirement: rewriting code would lengthen the implementation lead time. Given the limited availability of certain critical management personnel, any activity that lengthened the lead time was to be avoided.

• *Get the "right" people involved and committed.* The new CIM system had to be a system that everyone believed in and used. There had to be ownership by the users of this new system. To achieve this goal, systems people (i.e., people from data processing) were treated as support. They were not to lead the implementation effort. That task was to be left to the users.

These guidelines provided the general structure within which the resulting implementation took place. From the perspective offered by hindsight, they also significantly simplified the implementation process by identifying in advance (and resolving) many of the potential pitfalls in the implementation process.

The Steps in the Implementation Described

The implementation of CIM began with two thrusts. The first involved a revision of the corporate database. The redesign was intended to achieve several objectives. The first was to reduce the information lead time (i.e., the time that it took to enter or retrieve information). Prior to the changes now taking place, it took from four to five weeks just to enter data. Second, the redesign was intended to improve the usefulness of the database. This was done by focusing on the format in which information was recorded and stored.

The second was the redesign of the factory. Flows were straightened out; work centers were moved closer together; redundant work centers were eliminated. In addition, work was begun on remedying the problems with tooling. In the past, tooling was a major inventory investment; it was always a major bottleneck. In re-laying out the floor, management tried to improve its control over tooling.

These changes in layout had a significant impact on the operation of the shop floor control. They helped create a simpler system—one in which the need for computers (for data entry) was greatly reduced. By grouping work centers together, the distance traveled between these areas was reduced. Lead times fell as a result. Further reducing the lead times was the order review and release mechanism now in place. This mechanism kept the amount of work on the floor to a minimum. Only that which could be completed at any point of time was now released. Queues decreased, further reducing lead times.

These reductions in lead times meant that an order spent less time on the floor. The need for detailed tracking now fell. Instead of tracking the order at every stage, checkpoints were set up. Information about the order was generated whenever it passed through any one of these checkpoints.

Next the implementation teams began to attack the overhead and material elements of cost. Since labor made up only a very small percentage of total costs, it was essentially overlooked. The major reason for the high-burden costs was ultimately traced to the process nature of the current organizational structure. That is, the division was organized along functional lines. Maintaining the separation between functional areas added to the overhead costs. It also hindered the ability of the firm to work together on various product-oriented tasks. During this time, the division began the process of moving from a process- to a product-oriented focus (with appropriate supporting organizational structure). This was a long-term task.

After the current manufacturing process had been simplified, the division began to implement the various technological elements of the CIM system. Since 1985, seventeen CNC systems have been implemented. The technology embodied in these systems was not very sophisticated. The need for such sophistication had been eliminated by the process of simplifying work flows. Furthermore, simplification had enabled management to eliminate the need for computerized material handling systems.

Technologically, a CAD/CAM system was implemented and made operational in 1985. For planning, an MRP II system was implemented. This was based on the Cullinet system (all eight modules). It was operational by October 1985, some 18 months after its acquisition.

By 1987 the three major modules (CAD/CAM, CNC, MRP II) for the manufacturing side of CIM were in place. However, these were not well integrated. They could be best described as being separate islands of automation (see Figure 9-2). The next stage in the CIM implementation was the merging of these three islands. However, this stage was never completed.

Assessing the Type of CIM System at Continental Can

By 1987 the Container Division had implemented a Type I CIM system. This type of CIM system is appropriate for the business and market in which the Container Division finds itself. Since the basis on which the division competes is well known, there is no real need for extensive strategic integration. Channel integration also does not have to be very extensive. Given that the product is mature and standard, the need for organizational integration is also reduced. By

FIGURE 9-2
The Three Islands of Automation

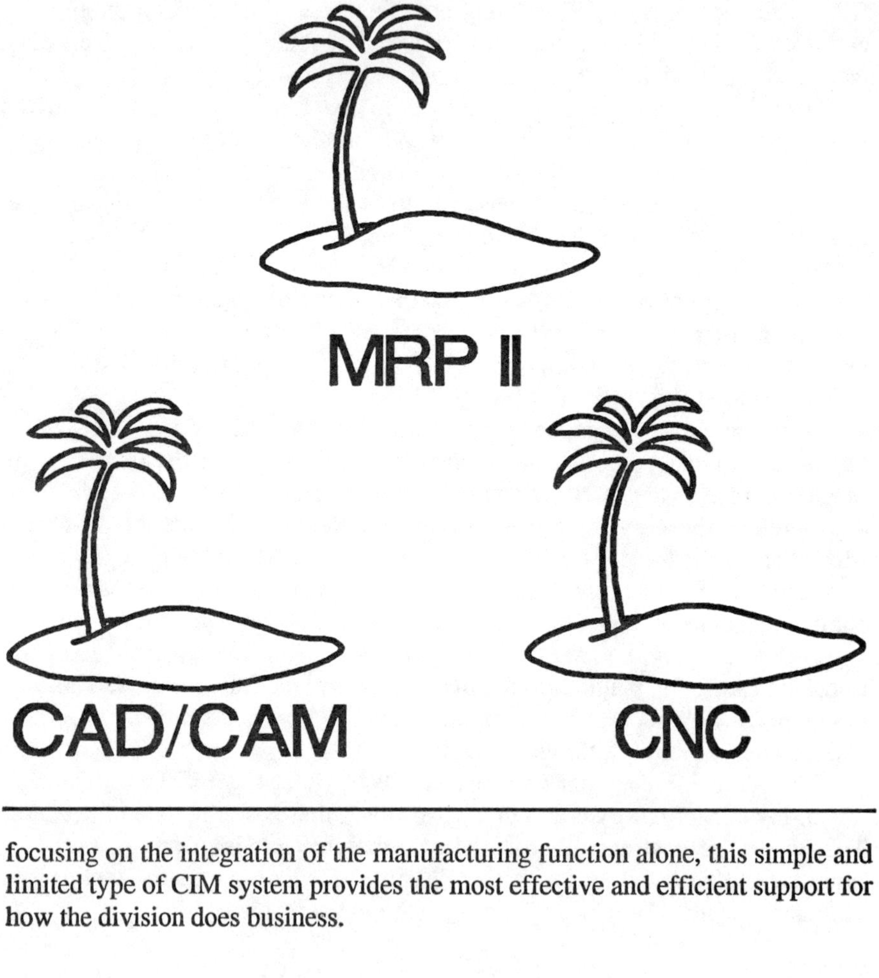

focusing on the integration of the manufacturing function alone, this simple and limited type of CIM system provides the most effective and efficient support for how the division does business.

BENEFITS

By the end of 1987 the CIM system was showing signs of success. The effectiveness of this new system could be found in the following benefits now accruing to the division:

- Lead times had fallen by 40 percent.

- Inventories had fallen by some 33 percent by the middle of 1987, with a 50 percent reduction (on a base of $12 million) being expected by the end of 1987.
- In 1986 only 25 percent of all orders were delivered on time. By August 1987 this had risen to 91 percent, with 95 percent expected by the end of the year.
- Data accuracy had increased significantly. Bills of materials were now 96 percent accurate. Routings were 97 percent accurate. The usual accuracy on cycle counts averaged about 95 percent. Furthermore, errors, whenever encountered, were corrected quickly.
- The lead time on spare parts had dropped from eight to nine months to about three to four days.

The division was now profitable. It was considered to be competitive in the marketplace. Its reputation for service was greatly improved. It had become so successful in the spare-parts segment of its business that management was beginning to consider supplying spare parts for equipment produced by its competitors. Finally, market share was growing to the point that Continental Can could be regarded as being the major competitive force in the market.

In addition to these very tangible benefits, there were other equally important but less tangible benefits:

- People involvement had increased significantly. People were now making decisions and taking responsibility for their actions.
- Cross-functional communication was increasing. The walls between the areas were beginning to fall down.
- The communications with suppliers had improved. The Container Division was now able to give its suppliers better schedules (i.e., the schedules had more visibility and greater stability in terms of timings and requirements). As a result, the division was now able to get better and firmer commitments from its suppliers to the schedules. The Container Division was also able to reduce the number of suppliers to some 500 by the end of 1987.
- Confidence in the CIM system and in the overall ability of the Container Division had improved significantly. The planning was better. People were now able to get good product cost information from the CIM system in hours or days, whereas before it would take weeks or months. Because of this ability to get cost data quickly, the Container Division was now able to compete on its ability to quote rapidly.
- Better data was leading to better actions.

- It was fun to work in the Container Division. The overall operating environment had improved greatly. The level of stress previously experienced was now disappearing. People no longer blamed each other for the problems taking place everywhere in the system.

By every measure, CIM had succeeded at the Container Division. Most of the goals initially set for the CIM project had been achieved. Because it was able to effectively support business activities in this division, everyone agreed that CIM was a strategic project.

LESSONS LEARNED

In reviewing the experiences of the management at the Container Division of Continental Can with computer integrated manufacturing, certain elements can be identified as having a critical influence on the success with which CIM was implemented. These are offered as lessons.

1. *Know what business you are in and use that knowledge to shape the resulting CIM system.* This is the single most important starting point for any system. Before you can undertake the implementation of a CIM system, you must know what capabilities that system has to provide. Identifying these capabilities requires that you closely examine your current business and your customer base. You must understand what you have to do to qualify for orders. You must also understand what you have to do to win orders. Finally, you must grasp how you can lose business. This knowledge is then used to drive the implementation of the resulting CIM system. It affects the shape, content, and structure of this system. It serves to flag what the CIM system must do well and what is not relevant to it.

In the case of the Container Division, this knowledge helped to create a very focused CIM system. Unlike other systems such as those found in Ingersoll Milling, the Container Division did not need a very sophisticated and complex system that linked customers, suppliers, and the division together. It did not need to develop very elaborate linkages between top management (strategy) and manufacturing. Instead, what it had to do was to build its products quickly and cheaply. It also had to ensure that the lead time for supplying spare parts was kept as short as possible. To its customers, the shorter the better. As a result, achieving these objectives required improving the manufacturing and information areas. Why information? Information was a substitute for inventory. By making this side more effective, the firm did not have to keep as much inventory. The more quickly that information was available and the more accurate and complete the data, the less inventory was needed. This lesson can be summarized very simply:

The CIM system must *always* support the business.

2. *Understand the strengths and weaknesses of the current system.* Before investing any money in hardware or software, management must undertake a complete and thorough evaluation of the current system. It must understand where the current system is doing a good job and where it is failing. This evaluation must go beyond symptoms such as high costs, high inventories, and long lead times. It must go to the underlying root causes. These problems must be first eliminated before authorizing any investment in computer-based technology.

For management at the corporate level, this evaluation made them aware of the real magnitude of the problems facing the Container Division. It made them aware that some of the problems facing this division were information based; others were management related. It also made them aware that the current system could not even do the basic things that every manufacturing system should do well (e.g., schedule production, release orders, manage and monitor capacity). Resolving these defects had to precede the move to CIM.

3. *Recognize that the CIM system is ultimately a people system.* A CIM system, if it is to be effective, must be utilized by the user. It must be understood by the user. Ultimately, it must be expanded and built upon by the user. To achieve these objectives, the users must develop a sense of ownership over the CIM system. This sense of ownership was developed in several ways at the Container Division.

First, the implementation team under Dave Grigsby deliberately set out to get the right people involved. The lower-level implementation teams were led by users. Data processing and systems personnel were used primarily to support the users.

Second, an intensive education and training effort was undertaken. This program affected everyone. If you were involved in any way, shape, or form in the building or servicing of a can-making or can-closing machine, you were educated or trained. Everyone from the person working on the dock to the divisional vice-president participated.

Third, the project timetable was long enough so that everyone had a chance to learn about the new system and its operation and to become comfortable with it. This feature helped the users develop a sense of control over the new CIM system.

4. *Understand the importance of a system champion.* One of the factors that ensured the success of CIM at the Container Division was that there were two champions who were interested in seeing it succeed. The first champion was located at the corporate level and was the executive vice-president, Greg Horrigan. The second champion, operating at the division level, was Dave Grigsby. Each was critical to the success of CIM.

Greg Horrigan provided the division with access to those resources needed in the implementation of CIM. He also ensured that the division was protected from external pressures and given the time necessary to make CIM a success. Finally, he backed up the division champion. People in the Container Division knew that Dave Grigsby ultimately had the support of Greg Horrigan.

Just as important was the division champion, David Grigsby. He provided the constant impetus to make CIM work. He believed in CIM. He believed that it could work in the Container Division. He was willing to do whatever had to be done to make it work in the Container Division. He was out convincing people that CIM was right for the division.

5. *Beware of system assassins.* An assassin is anyone who does not believe in the new system and is willing to attack it and its credibility whenever a problem occurs. At times, their attacks are verbal. Other times, they take a more active role in trying to make the system fail. Whenever an assassin is identified, management must be prepared to act quickly. Such persons should be moved out to a position where they cannot harm the system. This might require transferring the person in certain cases. At other times, it might require getting the person out of the system entirely (either through early retirements or terminations).

6. *Simplify before automating.* During the evaluation of the old system, the team became aware that the current shop layout and work flows were too complex and inefficient. As described by one participant, work flows were more like "spaghetti" than anything else.

Before any money was put into automation, the problems on the shop floor were first attacked. Work centers were moved closer together; work flows were straightened out and simplified; distances between operations were reduced. By making these changes, management did not have to buy an automated material handling system. The resulting changes also simplified the operation of the shop floor control system. Tracking and monitoring became very easy. The need for extensive computerization on the shop floor was greatly reduced.

7. *Be aware of the information side of CIM.* CIM systems are "data driven." They require good information to operate effectively. The major difficulty with providing good information is that the process required is long, time-consuming, at times boring, and expensive. However, it cannot and should not be avoided. Without good data CIM systems cannot and will not work well. Their potential will not be realized.

8. *Recognize that the implementation of CIM is a long-term undertaking.* It takes time to build an effective CIM system. After all, CIM cannot be bought off the shelf. It takes time to identify and bring the various components of CIM together. It takes time for the users to learn about CIM and its operation and to become comfortable with it. It also takes time for problems of the old system to be identified and corrected. Finally, it takes time for the organization to change its structure so that it better accommodates the requirements of the new CIM system. The walls that were built up over time have to be brought down. This is not an easy proposition for the firm. People now have to learn to work together with others outside their functional group.

9. *Whenever possible use "plain vanilla" hardware and software.* The goal of the CIM implementation is to enhance manufacturing and corporate competitiveness, not to rewrite code. When acquiring software and hardware, standard systems should be used whenever possible. Such systems are cheaper to purchase, update, and maintain. Using such an approach does mean that the users must change some of their practices so that they are consistent with the software. This can be a source of resistance. However, often the new practices are better than what was done before.

10. *Understand the importance of outside expertise and overcapacity.* There are two types of resources that can benefit the implementation of CIM. The first is outside expertise. Consultants, especially ones who have worked extensively with CIM systems, are extremely useful because they offer the user access to their past experiences and insights. Often, consultants know generally what works and what does not work. Their presence can greatly reduce the implementation lead time and lower overall frustration.

The second type of resource is that of excess production capacity. In the past, firms have tried to run their facilities at as high a level of utilization as possible. While this practice improved utilization, it played havoc with capacity planning and scheduling. Often, very complex capacity planning and scheduling systems were introduced. Meeting due dates in a timely and cost-effective manner became very difficult. To overcome these problems and to allow the users to concentrate on the task of implementing CIM, management decided to overinvest in capacity. The rationale for this decision was very simple. The costs of having excess capacity are far less than the costs of managing a tightly capacitated system. Excess capacity permits the use of simple schedule and capacity planning procedures.

POSTSCRIPT

Since the end of 1987, CIM has reached a plateau stage of development. That is, there have been relatively few new enhancements made in the system. The reason for this state is not that there are no new areas for improvement. Rather, it is a result of changes in corporate ownership.

At the end of 1987 Kewitt-Murdock purchased the parent company, Continental Can Corporation, and all of its assets. This was done through a leveraged buyout (LBO). To reduce the resulting debt load, the management at Kewitt-Murdock began to dismantle Continental Can by selling off its divisions.

In 1988 the Container Division was sold to Figgi International, an American holding company that is located in Minnesota. By 1989 the new owners had decided to focus more tightly the business of the Container Division. The division would no longer manufacture can-making and can-sealing machines. As was noted previously, the profits on this activity are extremely low and the market very competitive. Instead, the division would now concentrate on the highly profitable business of making and supplying spare parts.

As a result of these changes, many of the major players in the CIM implementation left. Greg Horrigan stayed on for a while but left to start his own company. Similarly, the CIM champion, David Grigsby, left in 1988 to work with another company located in the Chicago area. Most of the people who left cited the same reason. They left because it was no longer any fun. They were not enjoying themselves.

CONCLUDING COMMENTS

For the Container Division of Continental Can Corporation, CIM has been a success. It has helped turn a division that was uncompetitive into a strong and aggressive force in the marketplace. It has achieved this transformation by relying on a mixture of actions: the redesign of the shop floor, improved management of information, better management, and the use of appropriate technology. While the resulting CIM system may lack some of the sophistication people typically associate with CIM, there is no doubt that the system works and that it is well suited to the market served by the Container Division. While CIM has been a success, the recent changes in ownership have added an element of uncertainty to its future. Where it will go in the future is anyone's guess.

CHAPTER 10

FROST, INCORPORATED

EXECUTIVE SUMMARY

Frost, Incorporated, is a small privately held firm located in Walker, Michigan (on the outskirts of Grand Rapids), that is engaged in the manufacture of material handling system components and specialty turned metal products.[1] It currently describes itself as the world's best manufacturer of "metal round things." In Frost we see a company whose experiences with computer integrated manufacturing are notable for several reasons.

First, the management at Frost have used CIM as a vehicle not only for survival but also for changing the primary market to be served by the company. In the late 1970s and early 1980s, it became obvious to Chad Frost, the president of Frost, that its current market was reaching the end of the mature phase of the product life cycle and would soon enter the decline stage. As a result, sales would decrease and the competition for this "shrinking" pie would intensify. To survive and to grow in this dying market, top management decided that something had to be done to improve the effectiveness of the company. This something was CIM. However, there were two objectives underlying the move to CIM. The first was that the CIM system had to improve overall operations at Frost. Frost had to become the most cost-efficient manufacturer of material handling system components. Second, since the primary market was dying, a new market was needed if Frost was to survive and grow. The CIM system had to enable Frost to make the transition from simply being a manufacturer of material handling components (the dying market) to being an efficient manufacturer of anything that was metal and round. As part of this second objective, CIM had to help Frost make the transition from being a builder of product to being a seller of process. At the center of this process was Frost's knowledge of how to build round metal items for any need or environment.

Second, with about 150 employees, Frost is a small company. This issue of size has had both its good and bad sides. On the good side, by being small Frost was able to pursue a very aggressive automation strategy. Management has not

[1]This case study is based primarily on a presentation done by Robert McIntyre. Other information was provided by Chad Frost, the president of Frost, Inc., and Amprotech of Grand Rapids, Michigan.

had to cope with all of the organizational and size barriers often encountered in larger firms. Being small, the management has not had the resources available to large firms, however. It has had to undertake a careful evaluation of the various technologies. It has had to justify every action from both a strategic and economic perspective. It has had to show the logic and need of what was being proposed to both its people and to external sources of funding (e.g., banks and government agencies). This has forced management to recognize the importance of justification and performance measurement. Furthermore, being small meant that Frost had to implement CIM right the first time. To do so has meant dealing explicitly with the human and corporate culture aspects of change.

Frost is a Type III CIM firm with the capabilities of becoming a Type IV. However, there is an obstacle preventing Frost from achieving this goal. To become a Type IV, Frost must develop formal ties between itself and its customers. These ties require certain changes in the behavior of its customers, since they will now be buying process from Frost rather than product or capacity. However, being small, Frost has experienced difficulty in convincing its customers of the benefits of change.

In addition, being small has made the overall process of implementation more difficult. Consultants who typically deal with CIM assume that they are dealing with a large company. Similarly, the vendors who supply the various components on which CIM systems are built also assume that they are dealing with large companies. Their solutions and recommendations are often too expensive for a small company to consider. This fact has forced the management of Frost to develop its own expertise in the theory and implementation of CIM. By better understanding the concepts and requirements of CIM, management has been able to implement solutions that are less expensive than those proposed by the so-called experts. These solutions, however, have been extremely effective. As a result, Frost has become a statistical anomaly: a small company that in spite of being small has been able to succeed in making CIM work. Its success with CIM stands in contrast to the experiences of other, large companies that have tried to make CIM work but have failed.

Third, the experiences of Frost illustrate certain principles of successful CIM implementation extremely well. The first is the need for top management support, commitment, and involvement. CIM has succeeded at Frost in large part due to the efforts of Chad Frost, the president. He has been the champion of CIM at Frost. Second, CIM has been used to enhance the productivity of people rather than as a means of eliminating people. Third, before implementing CIM, management standardized operations. They identified the best ways of doing tasks at Frost. Instead of allowing persons to do things in their own individual manner, management required each person to learn this new best way. No deviations were allowed. If there were improvements to be made in these procedures, they were allowed *only* after it could be proved to management that these new procedures did indeed offer a significant improvement. As a result, management was able to control internal variability in the system. Finally, the implementation of CIM was accompanied by a significant change in the organizational structure: it was flattened.

In Frost we see an instance of a small firm's being able to use CIM as a means of ensuring its continued survival and future growth. To the people of Frost, being small has not been a burden. Rather, it has become a strength and a means by which a small firm has been able to successfully pursue an aggressive automation strategy.

The success of CIM at Frost has been recognized in several ways. It has become the subject of a Harvard Business School case study. In addition, its experiences with CIM have been reported in such professional journals as *Production, APICS P&IM Review and APICS News, Business Week, Robotics World*, and *Material Handling Engineering*. It has also been the subject of an industrial research report dealing with manufacturing automation planning services and sponsored by the Yankee Group.

INTRODUCTION

Frost is primarily an industrial supplier of material handling components and systems and currently employs less than 150 employees. Founded in 1913, Frost has undergone two major changes in the products that it offers. When it was first founded, Frost was initially a manufacturer of furniture and builders' hardware. In 1955, it switched markets and began to build metal bearings and overhead conveyor trolleys. These were sold on an OEM basis to conveyor manufacturers located primarily in the United States.

Frost was very successful in this market, with the result that by the 1970s, 80 percent of its product line consisted of trolleys and related components for overhead chain-driven conveyors used in the automotive, appliance, and food-processing industries. Frost had 60 percent of the market. It was considered to be one of the major suppliers. During the 1970s sales in its market had grown by over 600 percent with the returns on sales averaging about 16 percent. However, despite this apparent success, beginning in 1981 Frost has tried to change its marketing focus.

It has tried to wean itself away from this dependency on the overhead chain-driven conveyor market. To the management of Frost, this is a dying market. Instead, Frost is trying to develop the capability of being the world's foremost manufacturer of "metal round things" (especially with grooves) in the 2- to 8-inch size range. The key ingredient to this second transition has been the introduction and successful implementation of computer integrated manufacturing.

Frost has turned to CIM for two major reasons. The first has been as a means of ensuring its continued survival within the overhead chain-driven conveyor market. While everyone recognizes that this market is now in the decline stage, the market is still critical to the continued success of Frost. It is this market that currently provides the bulk of the company's sales (over 60 percent). A small company such as Frost does not have the luxury of being able to turn its back on such a large volume of sales. To succeed in this market, Frost must become the most efficient and lowest-cost supplier. It must become better at reducing costs and delivery lead times than its competitors (which consist of both domestic and foreign suppliers).

Second, CIM is being used as a way of developing new production and process skills for different markets that are not yet in their maturity phases and that are currently very profitable. Here, the capabilities offered by CIM are used to enhance the flexibility of Frost and to enable it to become very adept at building round grooved metal things from casting or bar stock in the 2- to 8-inch range. This type of capability is very appropriate for items using gears or bearings. Currently, 40 percent of total revenue comes from the markets made feasible by these new CIM-generated capabilities.

The CIM implementation, while substantially completed, was still ongoing. Its presence was most evident in the manufacturing and engineering areas. It succeeded in changing the very nature of the company. However, marketing was still learning about how to sell these new capabilities to the marketplace. Frost was still suffering from its position as an industrial supplier. It tended to have a few large customers. While this simplified the task of selling, it increased Frost's dependence on these few customers. Small reductions in their sales would have a major impact on Frost. Finally, Frost was still dealing with the problem of being ahead of its market. Its customers were not as sophisticated as Frost.

To understand the reason for CIM at Frost, it is first necessary to understand the factors that led up to the introduction of CIM. The first of these factors is found in the very nature of Frost in the 1970s.

FROST IN THE 1970S

During the 1970s Frost was extremely successful in its market. This success was a reflection of the extent to which the company structure was consistent with its market and its needs. In the 1970s, Frost had only one major product line: components and parts for the overhead chain-driven conveyor market. Over 80 percent of all its sales came from this market. Its products were extensively used in the automotive, appliance, and food-processing industries.

To service the requirements of this line, Frost had a company of some 185 nonunionized employees organized into 12 departments. Production on the shop floor was organized around 26 screw machines, which were located in 85,000 square feet of plant space. These screw machines were single purpose. They were supported by subsidiary operations such as drilling, deburring, and assembly. All equipment was run 5 days a week and 2 shifts (16 hours) per day.

On this equipment Frost manufactured a range of products for the material handling industry that could be best described as "hourglass" in terms of its product structure. That is, it produced a few high-level assemblies and products (180) that could be configured into over 1,500 various bearing-related special products.

The company, like its manufacturing operations, was very focused. It could best be described as a single-product, single-process company. It was dedicated to serving the needs of its one major market. It developed a resulting inflexibility in both its thinking and its way of doing business. This single-mindedness pervaded the entire company.

For the market that Frost served, this single-mindedness was quite appropriate. After all, the market was mature. The needs of the market were well known throughout the entire organization. Since these needs were well known, there was no real need for continuous interaction between such functions as engineering and marketing. Marketing knew that for Frost to sell to the companies of its market it had to do better than its competitors on the dimensions of cost and lead time. Similarly, engineering knew the technical dimensions of what the customer required. All it depended on marketing for was information regarding order quantities and delivery timings. Manufacturing knew that its major task was to reduce manufacturing costs and lead time.

As a result, the company developed a very rigid hierarchical organizational structure (see Figure 10-1). It consisted of 12 departments. In general, there was little contact between these department heads. Except for weekly meetings, they did not really talk to each other. While each department was fairly autonomous, these departments were effectively organized into four major camps: engineering, sales, manufacturing, and accounting (Figure 10-2). There was little or no crossover of people between these camps. Orders and designs were treated in sequential fashion. They would go from one camp to another (i.e., they were "thrown over the wall").

In the 1970s Frost was a highly structured, very rigid, antagonistic, and segmented company. Everyone had their place; more important, everyone knew their place in this company. It was a 1970s company that had what was described as a "1950s" mentality. These traits are best illustrated by certain incidents.

FIGURE 10-1
Frost Company: Hierarchical Structure of Old Organization

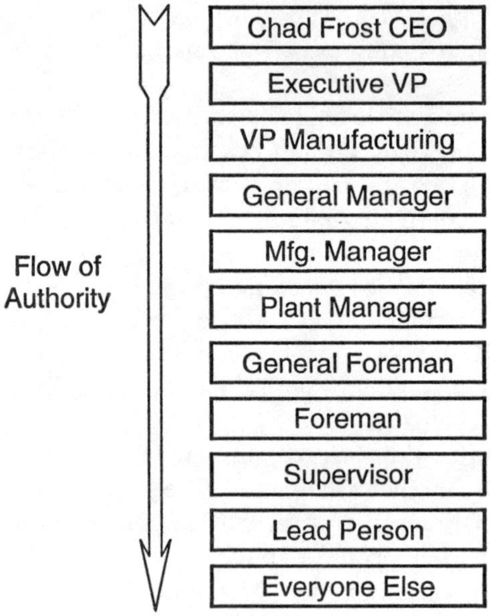

Source: Chad Frost & Frost, Inc.

FIGURE 10-2
FROST, INC.: Four Major Camps

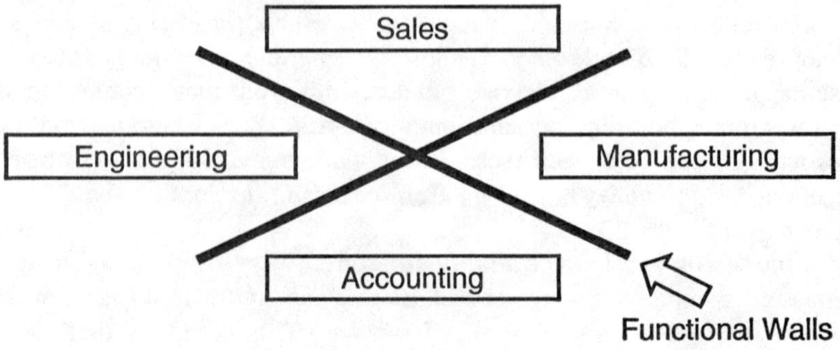

Source: Chad Frost & Frost, Inc.

Parking. The hierarchy of Frost was very evident in the parking arrangements outside the plant. The department heads had the best parking spots reserved for them. These spaces were located in front of the office. Everyone else could take whatever spots were left over. For many employees, parking became a symbol of the things that they did not like about Frost.

Payroll. There were three different payrolls: executive, supervisory, and hourly. The checks for these payrolls came out on different days and were in three different colors. They were handed out by three different people. Each category of payroll had its level of status, with executive being the highest and hourly the lowest. It was commonly felt that anyone who received a handwritten check delivered personally by the controller was at the top. As a result, the payroll became a powerful status symbol to employees at Frost. The importance assigned to payroll is best illustrated by events in 1981. In 1981, in response to lower-than-expected sales, a wage freeze was imposed on everyone in the company. This created a problem, since management could not give the word processing department a raise as a reward for some fine work that they had done. Instead, it was decided to transfer the employees of that department from the hourly to the supervisory payroll. This was meant as a simple token of respect and as a reward (of sorts). The amount paid to the people did not change. Yet, morale went up because the employees saw the change in payroll as a form of promotion.

The Use of Computers and Information. The flow and management of information within Frost mirrored the divisions and segmentation of the firm. The use of computers for information management and control was limited primarily to accounting and data processing. There, they were kept secured under lock and key. It was almost impossible to gain access to the computers if you were outside these two departments. In addition, the controller recalculated all the accounting information on a 10-key adding machine.

Aside from accounting and data processing, the only other computer in the firm was located in engineering. One engineer used an old Wang for job costing.

Establishing the overall accuracy and completeness of the corporate and manufacturing databases was very difficult. No one really knew the overall accuracy of information in the database. Information was fragmented across the various functions. No one had access to all of the information. Database redundancy was present. For many people, information was power. They used their knowledge of certain items of information as a way of influencing the actions of others.

The problems that plagued Frost were in many ways no different from those found in other small companies. More important, Frost was extremely successful. For many people at Frost, there was no real need for change. However, in

1977, change began to take place. The reason for the change was the appointment of Chad Frost as president.

THE BEGINNING STEPS IN THE CIM JOURNEY

The Arrival of a New President

When Chad Frost took over the presidency of the family business in 1977, one of his first actions was to take a closer look at the numbers (specifically sales and profits). What he saw did not excite him. While it was true that sales were growing, the growth was strictly in dollar terms. Unit shipments were flat if not declining. The growth in dollar sales was due to the policy of increasing price to maintain profit margins as costs increased.

The discovery that Frost was covering increased manufacturing costs by adjusting price was a critical discovery, for it indicated to people like Chad Frost that Frost was not really competing on price and lead time.

In terms of unit sales, it became evident to Chad Frost that the market segment currently served by Frost had reached its mature stage and would soon be moving into decline. This realization presented Frost with a bleak future. As demand declined, industry-wide capacity would increase. Competition between firms would increase. This competition would put pressure on prices. Ultimately, this pressure would prevent Frost from relying on its past practice of using price increases as a means of maintaining profit levels. In addition, as the market moved from mature to decline, the demand would change from OEM sales to replacement. In replacement, cost and lead time would become key competitive issues (i.e., order-winning traits). These were dimensions on which Frost, in its current configuration, could not really compete.

There was one final factor that made the future seem very distressing: In a declining market the opportunities for growth are very limited. The only real way to grow is by taking market share away from the competition. This was not a pleasant realization for a company that was accustomed to growth.

The first plan that Chad Frost tried was that of diversification. He decided that if Frost was to grow, new markets had to be uncovered or developed for Frost products. The current market could not easily support continued growth.

Initial Moves: Diversification

The first step in the diversification strategy was to identify a market for which Frost could build products requiring skills similar to those needed in the chain-driven overhead conveyor market and which was not related to this primary market. After studying several options, Frost decided to manufacture metal spindles for lawn mowers. This first attempt was a failure. The reason did not lie in the quality. Instead, Frost failed in this market because its customers had very precise deadline requirements that Frost could not consistently meet.

Next, Frost tried a different tack in its diversification strategy. Management decided that it would be wiser for Frost to "stick to its knitting" and concentrate on building products for the material handling market. This market was examined to determine if there were other areas of growth present. It was subsequently decided to make parts for other types of material handling equipment such as lifts, hoists, and elevators. Again, this attempt at diversification failed. The reason for this failure lay in the organization. The engineers did not know how to design these new items; the salesmen did not know how to sell them. There was no cooperation between these groups. As a result, it took too long to get bids out. When the bids went out, they were often too high.

By 1981 everyone in Frost realized that these initial attempts at diversification had failed. Two primary reasons for these failures were identified. The first was organizational in nature. The very organization that had enabled Frost to succeed in its primary market had also become a hindrance to its attempts to change. Frost was essentially a single-purpose organization. It did not have the flexibility needed for diversification or change. Furthermore, Frost's current set of capabilities and capacities did not match up well with the requirements of these new markets.

Second, there were the constraints imposed by being a small firm. In theory, a company like Frost would attempt to diversify by first assigning several of its key management people full-time to the task of developing these new markets. These people would be relieved of their daily responsibilities so that they could devote all of their attention to this task. However, what should be done and what can be done (and expected) are two different things for a small company such as Frost. It did not have the extra resources available so that it could afford to pull several managers out of day-to-day production. If Frost was to succeed, the existing market had to be served. This was a top-priority task. Frost could not take the chance of taking away scarce resources from this primary task if the removal of resources could affect in any way the servicing of the main market. As a result, no resources could be spared for these new markets.

Automation: Initial Moves (1981 to 1983)

The period between 1981 and 1983 marked a major turning point for Frost. Three major developments took place during this time that made it clear to everyone in Frost (especially Chad Frost) that something had to be done and done quickly. First, there was the fact-finding trip taken by Chad Frost.

In 1981 Chad Frost decided to take a trip to visit major competitors and manufacturers of material handling equipment located in the United States, Japan, and Europe. He visited the John Deere Tractor Works in Waterloo, Iowa. He visited off-shore leading-edge systems. What he saw convinced him even more of the need to decrease his dependence on the chain-driven overhead conveyor market. In Japan and Europe, he saw that firms were increasingly turning to automated equipment and flexible manufacturing systems (FMS). Such new systems demanded a material handling system more flexible than that possible with a chain-driven overhead conveyor. These new systems had to be "smart" and they had to provide the operators with the ability to reroute work quickly and as needed. Because the chain-driven material handling systems were not as flexible, Chad Frost became convinced that this was a product in its decline stage. There was no long-term future in the chain-driven overhead conveyor market.

Second, this period saw the arrival of new competitors from Japan and Europe. These new firms were aggressive competitors interested primarily in establishing a business presence in the American market. They were willing to reduce price in exchange for a larger market share.

Finally, the bottom fell out of the market between 1981 and 1982. The rate of growth in sales fell. Prices also fell owing to the presence of foreign competition and overcapacity in the market. At this point it was generally agreed that something had to be done. This "something" was *system* and not manufacturing automation.

However, before embarking on this automation strategy, two critical requirements were imposed by management. The first was that Frost would not abandon its traditional market. It was Frost's goal, as Chad Frost put it, to be "the last one to turn the lights out." This meant that Frost had to become extremely cost-effective. It had to be the lowest-cost supplier in the market. Second, Frost had to reduce lead time. Since a greater proportion of its sales would be for replacements, lead time would become a critical competitive dimension. The reduction in cost and lead time had to occur while maintaining quality.

Second, Frost would still pursue a strategy of diversification. It would develop new markets so that it could decrease its overall dependence on the chain-driven overhead conveyor market. Based on what Chad Frost had seen in his travels, Frost could survive only by being able to respond quickly to small opportunities whenever and wherever they occurred. This mean that Frost would have to change from a product-oriented firm to a capability-oriented firm.

Learning about Automation

The first step taken by Chad Frost in implementing this automation strategy was to form a small team that would focus on the task of automating the firm while allowing the rest of the company to concentrate on its day-to-day business. This team consisted of three people: Chad Frost, Rob McIntyre, and Randy Nauta.

Rob McIntyre was the marketing director and, as such, was the major source of marketing and strategic planning information. Randy Nauta was the director of new product development. He brought with him a great deal of engineering and computer expertise.

Since the members of this automation committee really did not know much about automation, they decided to learn about it. To do this they hired consultants from several of the country's leading companies in automation. Although this seemed like a good idea at the time, it did not work out for several reasons:

Small-Firm Client, Big-Firm Mentality. Many of the country's leading experts had almost exclusively dealt with large companies. They were used to seeing situations in which very specialized pieces of equipment could be justified and implemented. They were also accustomed to having access to significant resource pools. However, they were not prepared to change their approach and expectations for small firms. They failed to realize that expensive, highly specialized and dedicated equipment was often not a logical alternative for a small firm such as Frost. This problem is best illustrated by two instances.

In the first instance, the automation experts recommended that Frost replace its 26 machines with 4 custom-built machines. These new machines suffered from several problems. They could only make 4 parts; Frost built over 1,600 different parts. Second, they were also very expensive: they cost $12 million. In the early 1980s the annual sales for Frost were about $12 million. Third, they were not flexible. This was not consistent with the image of Frost as a quick-response firm.

In the second instance, another set of automation specialists recommended that Frost create an automation plan that would be strategically driven (by the business plan) and that would affect the entire firm. The consultants even offered to help Frost with the implementation at a cost of $75,000 per month for the next 12 to 18 months. This recommendation came at a time when Frost was faced by downward pressures on price and increasing material and labor costs.

Generic Solutions Often Don't Work. Many so-called experts often draw on a "generic" or common plan for automation. This plan often places a heavy emphasis on automation. It fails to recognize that each firm can reduce the amount of automation needed by simply making the underlying system simpler and by eliminating some of those activities that are to be automated.

In one case, as reported by Chad Frost,[2] a CNC vendor had estimated that Frost would need 12 CNCs to operate at the same capacity levels as then pre-

[2]Chad Frost, "Integrated Automated Program at Frost, Inc.," *P&IM Review and APICS News*, April 1984 (Vol. 4, No. 4), pp. 32–36.

sent. At $250,000 per CNC, this represented an investment of some $3 million. Not included in this estimate were the costs of such support equipment as laser inspectors, robots, parts feeders, part take-away delivery systems, and automatic tool changers (to name a few). After the people at Frost studied the problem and the need for automation, they figured out that by improving integration the same level of capacity could be provided by 5, not 12, machines.

Consultants Only Provide Part of the Solution. Many consultants and vendors do not really sell an integrated automation manufacturing and corporate system. Instead, they sell parts of the solution. They provide CNCs; they sell software planning systems; they manufacture robots. Because they sell only parts of the solution, consultants and vendors often overlook the problem of how to make these various pieces work together as an integrated entity. That task often falls on the shoulders of a client such as Frost.

As a result of these and other problems with these so-called automation experts, the team decided that the best approach would be to have the people at Frost look at the question of automation at Frost. A task force consisting of the 12 department heads was composed. Its mission was to assess the feasibility of automation (specifically CNCs) at Frost.

After six months, the task force returned with its verdict. It recommended that CNCs *not* be used. They were seen as being slower than the current screw machines. They would not be able to justify the levels of investment needed.

After reviewing the report, Chad Frost and the other two members of the implementation committee decided to proceed with the automation project. Their decision, which ran counter to the task force's recommendations, was based on the realization that the task force had not really understood the impact of automation on Frost. Nor did the task force adequately understand the downside of the "do nothing" alternative.

The introduction of CNCs was not simply a matter of replacing one form of process technology by another. This view understated the true impact of the CNCs. The current process was old and suffered from many problems. The machines couldn't hold a tolerance. As a result, the rework rate was about 24 percent. Setups were high; it took about 12 hours to change from one product to another. They needed to be supported by a separate toolroom, rework area, and grinding department. Finally, the existing screw machines were too inflexible. By introducing CNCs, these problems could be eliminated. More important, CNCs would give Frost greater flexibility and responsiveness. These traits were highly prized by management, since they were consistent with the future direction that had been set out for Frost.

Finally, the "do nothing" option proposed by the task force was unacceptable given current market conditions. The entrance of competitors from Japan and Europe had helped instigate a price war. This war put pressure on any increases in price. It also reduced the profit margins, since material and manufacturing costs were still rising. With the market becoming tighter and more competitive, something had to be done and done quickly. Frost did not have the luxury of sitting back and carefully weighing all of its options. It did not have the luxury of time. In 1983 management began the process of making automation (and subsequently CIM) a reality at Frost.

Setting the Ground Rules for the Move to CIM (1982 to 1983)

With the decision to proceed with automation, and ultimately computer integrated manufacturing, at Frost, the automation committee began to map out their strategy for implementing CIM. This decision was accompanied by certain major changes in the company. During this time Chad Frost decided that his chances for success would be greatly increased if he were not only the president of the company but also its owner. He began the process of buying out the company from other members of the Frost family.

This was also a time during which the members of the automation committee began to gather more information about both the company and automation. They talked with people in the firm; they tried to learn as much as they could about both the strengths and weaknesses of Frost. The information gained from this examination phase was subsequently used in shaping a set of implicit guidelines that the committee used in implementing CIM:

The New CIM System Had to Be a TOTAL System. It was not enough to automate manufacturing alone. If Frost was to survive, all aspects of the delivery system had to benefit from the new and improved capabilities offered by automation. This meant that automation had to be extended to such areas as engineering, the toolroom, and the front office.

Whatever Capabilities the New System Offered, These Had to Be Consistent With Current Strengths. With the decision to automate operations at Frost, the automation committee was faced by an interesting decision: The automation project could affect nearly every aspect of production at Frost. It could be tailored to give management whatever capabilities they wanted. Now the question that they had to wrestle with was that of what they really wanted out of the automation project.

Early on the automation committee decided that the new system should give Frost the flexibility to quickly exploit new market niches and needs as they occurred. This decision marked a change in strategy, since in the past Frost was wedded to a specific market.

In reviewing the various capabilities offered by the automation project, management decided that irrespective of the type of automation introduced, it had to be supportive of Frost's distinctive expertises in the market. In reviewing the things that Frost could do well and those things that Frost did poorly, management decided that Frost was really good at working with round metal objects that were grooved or that had bearings. These were typically small items in the 2- to 8-inch range. The resulting CIM system would be structured to enhance the ability of Frost to compete effectively in this marketplace.

These new capabilities would not be ones in search of a market. Based on the demand studies that Chad Frost had carried out earlier, it was estimated that this new market could eventually account for about $4 billion in annual sales in North America. Here was a market that could maintain the growth begun in the 1970s.

The decision to focus on being the best manufacturer of "round metal things" marked a major shift in strategy. In the past, Frost had been associated with a well-defined market niche that was defined by its product. Now Frost would be moving to a market niche that was no longer associated with a well-defined line of products such as chain-driven overhead conveyors. It was now to be a marketer of its new capabilities. Its products would now be a function of its customers' needs. These would change as the customers (or their requirements) changed.

The shift to capability would affect how Frost competed in the marketplace. Price, while important, would become more of an order qualifier. Orders would be won and lost as a result of Frost's flexibility, manufacturing and design capabilities, and lead time performance. The change to capability would also affect how Frost marketed itself to its customers. Marketing could no longer sell a line of products. Instead, salespeople would have to sell Frost and its system. For marketing, this would mean the development of new sales skills.

All of the Capabilities Offered by CIM Had to Be Used. In learning about automation, the automation committee become aware that its introduction would require a change in the way that people looked at capacity. In the past, capacity was determined by labor constraints. With CNCs, production could now take place around the clock. If management could maintain the level at which it was using its CNCs, it could reduce the number of CNCs needed. With only one 8-hour shift per day, 12 CNCs were needed. If the CNCs could be staffed and operated 24 hours a day, 7 days a week, less than 5 CNCs would be needed. Furthermore, by moving to continuous production, Frost could eliminate the need for those subsidiary operations currently present (e.g., the toolroom, grinder).

The CIM System Had to Be Justified. The new system had to be formally justified to two groups. The first consisted of all of the employees at Frost. The second was the business community.

Since Frost was a small company, it could not afford to underwrite all of the costs by itself. Between 1981 and 1983, providing the financing needed for the automation project was becoming increasingly difficult for several reasons. First, sales were beginning to slow. This fact, combined with increased competition, reduced the capability of Frost to generate internally the needed cash flows. Second, in spite of all the work that had been done to reduce the overall costs of automation by the people at Frost, automation was still a major investment. Third, as was previously noted, Chad Frost completed during this period the acquisition of the remaining shares from the other members of the Frost family. This acquisition was seen as being critical by Chad Frost, for it not only gave him control over operations at Frost: if Frost succeeded, he would succeed; if Frost failed, it would be his failure.

In 1982 the management became aware of a program sponsored by the Michigan Economic Development Corporation (EDC). To encourage firms to stay in Michigan and to maintain their long-term viability by automating, the EDC offered to guarantee bonds issued by Michigan cities for automation projects undertaken by local companies. Chad Frost decided to take advantage of this program. To convince the city of Walker to issue the $5.1 million in EDC bonds, Chad had to convince the local government that automation was the most cost-effective option open.

The process of justifying the automation project subsequently had a major impact on Chad Frost. He became convinced that any investment in CIM should not be justified on the basis of gut feel. Investing in CIM should never become a matter of faith. Management had to justify the need to invest in CIM. It had to identify the benefits. This is often difficult in the case of CIM, since the benefits are often not confined to a specific department or area. They tend to be company-wide. However, the numbers must be there if CIM is to succeed. This view is now present whenever Chad Frost evaluates any type of automation project.

In the case of the initial justification to the banks, a conservative approach was taken. Sales were not assumed to grow as a result of the implementation project. Instead, the major benefits would occur from the cost savings gained through inventory reductions and cutbacks in overhead.

The automation committee was successful in convincing the two banks and the city of Walker that the investments in automation were indeed good ones.

The Project Had to Succeed. The automation project was the third major try undertaken by Chad Frost to improve operations at Frost. The first two attempts had failed. Frost could not tolerate one more failure. The investments needed by the new project were very high. If the capabilities were not in place by the time the market turned down, Frost would not be in business over the long run. The company that was started in 1913 would be doomed to a slow and painful death. Also, because Chad Frost was now owner of the company, he would be directly affected by the failure of Frost.

For the CIM project to succeed, those factors that had caused the previous efforts to fail had to be identified and addressed. This meant that the implementation project would have to address simultaneously the technical and human aspects of automation.

CIM and Automation

The automation team initially thought in terms of automation. However, it soon became obvious that automation was not enough. Automation looked at changing the way certain tasks were done at Frost. It sought to replace human muscle and existing equipment by computer-aided systems such as CNC. By itself, automation is not the same thing as computer integrated manufacturing. Automation is too myopic in perspective.

Frost began to make the transition from automation to CIM as soon as the automation committee realized that automation must be linked to corporate strategy and that it must involve the whole company. This realization was reached early on in the implementation process. Even though the term "automation" is used, given how it was implemented, this term is really synonymous with "computer integrated manufacturing."

Defining the Initial Vision of CIM at Frost

Before the implementation of automation (and ultimately CIM) could begin, the automation team had to first address what was meant when the company represented itself in the marketplace as being the lowest-cost and most flexible producer of round metal things. This definition had to be stated in very concrete and operational terms, since it had to be understood by everyone at Frost.

To this end, the automation committee focused on trying to define those changes that would have to take place to lower the break-even volume. By lowering the break-even volume, the minimum scale (utilization) for efficient production would fall. Also, reductions in the break-even were seen as being consistent with increases in overall flexibility. The automation team identified five major changes that would be necessary to lower the break-even point:

- Labor content reduction (both direct and indirect) of the product would be needed.
- Productivity enhancements would have to be achieved (labor that remains has to become more productive and perform more functions).
- The utilization of all assets (human and capital) would have to be increased.
- The integration of all operations through computers would have to be undertaken to minimize "administrativia" and redundant information handling.

- Organization and management for faster decision-making would have to be instituted. Ultimately, this meant a flattening of the organizational hierarchy so that decision-making would be moved closer to the information source.

Every action taken as part of the automation project had to be consistent with these five changes.

Beginning the Implementation of CIM

Having identified the guidelines and changes underlying the CIM project, the automation committee now turned its attention to the task of beginning the implementation of the automation project. Instead of focusing on one area of the project (such as hard or computer automation) to the exclusion of the other areas, the team decided that three different paths had to be followed in parallel. Each path was necessary to the successful completion of the others.

System "Rethink"

The first of the three paths could be best captured by the term "Rethink." Management had agreed that ultimately their firm had to become integrated, well coordinated, and effective. Furthermore, the new Frost would not rely on the old paper backup system. To succeed on these two objectives, the current methods of doing things at Frost would have to be studied, examined, and corrected. This reexamination could not be focused simply on manufacturing and engineering to the exclusion of everyone else. It had to be a reexamination of the *entire* delivery system from the receipt of a request for quotation (RFQ) to the final shipment of the completed order. Areas examined as part of this system rethink stage included order entry, scheduling, and planning. This reexamination was driven by three major questions:

- How are tasks and processes currently organized and carried out (as compared with how we think that they are done)?
- Why are the various tasks and processes present in the delivery system?
- What differences are there between various people carrying out the same or similar task or operation?

The first question established a map of the processes that made up the corporate delivery system. The second forced people to justify the need for the various tasks and operations. This question was included because management realized that processes change over time as people add "temporary" changes that somehow become permanent. The final question reflected management's awareness that the variability present in the processes was not due so much to how the processes were organized as to how different people interpreted and acted on activities.

In addressing this third question, the management became aware that there were no real "standards" at Frost. Most people had introduced their own enhancements and improvements. These changes were rarely if ever shared. As a result, the amount of time it took to do the same task varied from person to person. These differences created variability within the system. Before automating the system, this variability had to be eliminated. There had to be one way of doing tasks, one standard, that was common to everyone assigned to that task. No deviations from this standard could be tolerated.

At this point, management was faced with a fundamental problem: how could they create an environment in which people continuously worked at improving the system without violating the "one standard, one way" guideline? The solution was to make these improvements an "off-line" activity. Under this approach, employees are free to develop new and better ways of doing tasks. However, they cannot implement these improvements without first demonstrating their effectiveness to a supervisor in a setting away from the floor. If the enhancement is accepted, it becomes part of the new standard for everyone in the system. This system allows for the simultaneous existence of creativity and standards.

By using these three questions to structure the review of operations at Frost, management was able to simplify the resulting manual system. The result was also a system that worked and that did not inherently create variability. These improvements simplified the subsequent process of automation.

The process of rethinking had one final impact on operations at Frost: It encouraged everyone at Frost to begin a critical evaluation of the processes and tasks in which they were involved. It made people aware that there were basic system problems present. This awareness and reexamination began to result in the identification and ultimate elimination of those factors that were creating the problems.

Addressing the Human Side of CIM

The members of the automation committee were very aware that one of the major reasons for the failure of past attempts to diversify was organizational, not technological. By ignoring the organizational barriers to change, Frost had failed previously because it was trying to introduce change to a company that was rigid and very hierarchical.

These past failures had two effects on the resulting implementation. The first was that it made the committee aware of the need to change the organizational structure at Frost so that it was consistent with the new directions planned for the company. Second, people had seen initiatives tried twice in the past and they had seen these initiatives fail. People were now somewhat leery of this new initiative. They had to be convinced that management was serious about changing things at Frost.

To bring about the human changes required to make automation and CIM successful at Frost, the automation committee mapped out a three-pronged strategy. This strategy consisted of the reshaping of the Frost organization, the developing of system ownership on the part of the employees, and extensive training and education.

Removing Organizational Barriers. The Frost company of 1983 was not consistent with the Frost company of 1987. Frost in 1983 was a rigid company that was organized into 12 different departments and that had close to 17 levels present between the shop floor and Chad Frost. It was very much a divided camp—something that was unacceptable in a company that competed by identifying and quickly responding to market needs as they became apparent. It was a company in which decision makers on the shop floor did not have timely and ready access to all of the information needed in carrying out their tasks. Finally, it was a company in which morale was low.

To make Frost more flexible and responsive, the automation committee began to attack those traits that were resented by the people who worked at Frost. The three payrolls were consolidated into one payroll, which was now handed out by Chad Frost on Wednesdays (the old supervisors' pay day). By handing out the checks, Chad Frost got to know his employees and to become aware of the problems and difficulties that they were facing. Next, the reserved parking spaces were removed. Those who got to Frost the earliest got the best spaces.

The organization was seen as being too top heavy and too hierarchical. Something had to be done to streamline Frost. The automation team began by studying information flows and needs. They quickly found out that the people who needed information were not getting the right information at the right time. The current organizational structure was identified as being the major barrier to the effective flow of information.

The automation team decided to restructure the organization so that it was consistent with and supportive of the effective flow of information. The number of organizational layers at Frost was reduced from 17 to 4 (see Figure 10-3). The 12 departments were reorganized into 4 areas that reported directly to Chad Frost:

- *External operations.* This department was responsible for sales, marketing, engineering, and manufacturing engineering.
- *Internal operations.* The second major department was responsible for manufacturing, material control, purchasing, and data processing. The plant manager now fell under this area's control.

FIGURE 10-3
The "New" Flattened Frost, Inc., Organization

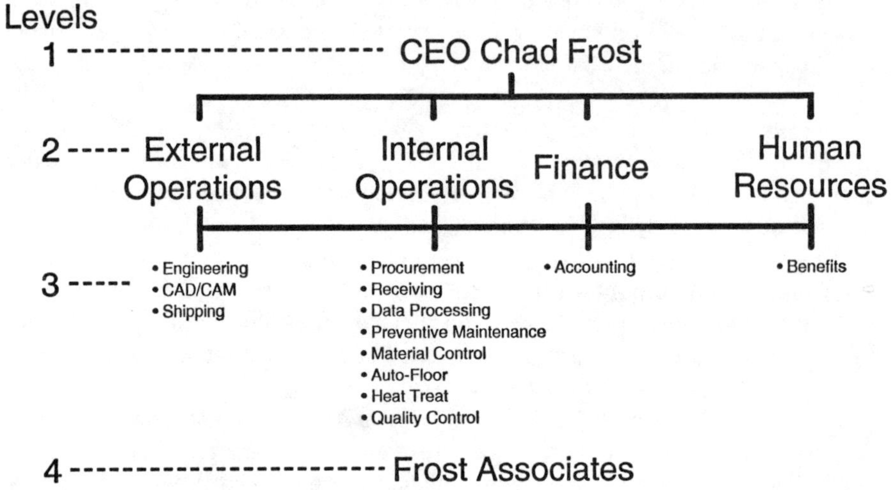

Source: Chad Frost & Frost, Inc.

- *Human resources.* The major responsibility of this department was for training, recruiting, and outplacement counseling.
- *Finance.*

To further emphasize the changes that were taking place at this time, many of the head office arrangements were changed. Many of the offices now had no doors (including that of the president, Chad Frost). Any employee now has direct access to the president.

These changes were not received enthusiastically by everyone in the organization. Several people were not impressed by what was being done. They did not believe in automation or CIM. In some cases they wanted to be left alone. Ultimately, these people were told that if they did not want to become part of CIM, they should leave Frost. For those that decided to leave, Frost had set up an outplacement service.

By October 1984 employment had dropped from 185 (1981) to 152. Among the departed were the controller, the managers of data processing, quality, manufacturing, material control, and purchasing, and the plant manager.

Developing System Ownership. Changing the organizational structure to remove barriers to change was not enough. The people who worked at Frost had to have a direct stake in the success or failure of automation. To develop this sense of involvement and ownership, management implemented certain changes.

In October 1984, one year after the decision to automate, an open house was held at Frost. Employees and their families were invited. The various pieces of automation were shown off to everyone. The time clock was dismantled indicating that the role of the employee was changing at Frost. At a table, Chad Frost greeted each employee and gave that employee 10 shares in Frost. That person was no longer simply an employee. When that person took that stock, he or she became a *shareholder-employee*. In addition, a 401(k) plan was also established. This plan made available the opportunity for employees to increase their stake in the company by acquiring more shares.

Second, a new compensation and incentive system was introduced. Everyone became salaried (to simplify compensation). To encourage employees to take a longer-term perspective, a new profit-sharing plan was implemented. This plan is based on the concept of equal shares. Everyone, including Chad Frost, gets the same number of shares. There is also a quarterly bonus program that is dependent on corporate productivity. Along with the bonus, there is also a "celebrity fund" that can be used by managers as they need it. One manager was granted his request to draw on this fund for money with which to hire a belly dancer to perform in the office.

Finally, the employees were given more say in the implementation of the automation project. This new role was most evident in the scheduling decisions. Management explained to its employees that the new equipment would now be run 24 hours a day, 7 days a week. The resulting worker schedules had to be changed to ensure that there were enough workers on the shop floor. The employees were asked to develop a new schedule. The solution that they came up with was to have people work three 12-hour shifts and be off four days. It was the employees' decision and they made it work.

Training and Education. Early on in the implementation process, the automation team realized that training and education would play a very critical role in the success or failure of automation and CIM at Frost. The people at Frost would have to take the lead in implementing and extending automation. The reason was that there was no real outside professional help available.

The resulting vacuum would have to be filled by the employees. If these people were to have any chance, they would need access to training and education. This training and education focused on two elements: computers and general business skills.

In the new Frost, computers would play a central role. However, the Frost of 1983 was poorly positioned to take advantage of the power of the computer. One reason was that the company was generally computer illiterate. Prior to 1983, only 5 people out of the 165 employees (i.e., about 3 percent) had any exposure to computers. For most employees, the introduction of computers would result in FUD (fear, uncertainty, and dread). To avoid this, the employees would have to feel comfortable with the new computer-based tools. The automation team decided that this comfortableness would be best developed through a combination of classes, hands-on experience, and experimentation.

• *Classes.* In October 1983 the automation committee held a day-long computer literacy class and meeting that was open to all Frost employees. This first class was a test of the potential interest and commitment of the employees. It was held on a Saturday. Attendance was voluntary. While the meeting was eight hours, Frost would only pay for four hours of time. Everyone from Frost was there.

By the end of the meeting, it was evident to the automation committee that more classes would be needed. This class marked the start of a policy by which all aspects of the new computer systems would be explained in classes sponsored by Frost. As computers have been added to the Frost system (e.g., IBM terminals), there have been classes designed to familiarize the employees with these new additions. This policy is ongoing.

• *Hands-on experience.* The automation committee decided that one way of overcoming any resistance to computers would be for the employees to have their own personal computers. As a result, Frost arranged for a local computer retailer to offer a company discount on the purchase of Commodore 64 computers. In addition, Frost offered to pick up 50 percent of the cost with the remaining half being deducted from the employee's paycheck ($10 per week).

The program resulted in the sale of 104 Commodore computers. All of these computers went home in time for Christmas 1983.

• *Experimentation.* The final tactic was most evident in the approach taken by management in introducing robots to the shop floor. Before the robots were introduced into daily operation, the robot vendor (Roberts Corporation of Lansing, Michigan) delivered a demo system three months early. This robot was installed in the company cafeteria. Employees were allowed and encouraged to experiment with the robot. Before long, nearly everyone at Frost had learned how to control the robot. They had also learned what the robot could do well and what the robot could not do well. People were no longer intimidated by the introduction of robots at Frost.

For Frost to succeed, computer literacy, while important, was not enough by itself. The employees had to become familiar with the general skills that they would now be using. They would have to understand what an MRP system was and how it operated. They would have to know what statistical process control was. Furthermore, with the general reduction in the number of organizational levels at Frost, the employees each had to know what they were doing. They could not expect to have their work checked by a supervisor.

The management approached several local universities, colleges, and junior colleges in the Grand Rapids area and had them offer courses for Frost employees. The management even structured a course dealing with skills and areas necessary for operations at Frost. This course, offered through the Grand Rapids Junior College, is now known as Frost 101.

As a result of these initiatives in education and training, certain critical benefits were gained for Frost. First, familiarity with the new tools was developed. The people understood the tools and they were not intimidated. More important, they had developed the confidence necessary to make the most of these new tools.

Resolving the Automation Issues

At the outset the focus of the automation team was not on getting the most sophisticated or comprehensive hardware or software available. Instead, it lay in getting technology that was consistent with the needs of Frost. To simplify the process of buying computer hardware and software, the committee decided to pick one vendor as the sole supplier for hardware and one vendor as the sole supplier for software. In both instances the same vendor (IBM) was selected.

IBM was selected for one reason: support. IBM could readily provide such support, since it was the only vendor with a fully staffed field support office located in the Grand Rapids area. Second, the people at IBM were interested in supporting a small company such as Frost.

In the case of software, the automation team decided to go with standard off-the-shelf packages. The use of such packages was attractive for several reasons. First, they were relatively inexpensive to acquire. Second, they required a minimum amount of custom software. They were also easier to maintain and upgrade. Finally, the lead time needed to install and debug them would be significantly reduced.

For engineering, the committee selected IBM's CADAM system running on an IBM 4361. This package, while not the most sophisticated, offered Frost the features that they were most interested in. It was also relatively easy to use. For production and inventory control, IBM's MAPICS system was selected.

In automating manufacturing on the shop floor, standard hardware was used as much as possible. The screw machines were replaced by five Mazak CNCs. To feed these machines, robots manufactured by the Roberts Corporation were installed. These robots were installed upside down to facilitate access to the CNCs. With the robots mounted to the floor, moving the arm up still left the pedestal on the floor in front of the machine. To mount the robot on the ceiling required reversing the thrust bearings.

The last problem to be addressed involved the interfacing of the robots and the CNCs. Initially, these did not "talk" to each other. There was also no standard interface available for linking these two pieces of technology. Randy Nauta and Charles Cope took over the task of developing a general-purpose device-to-device interface. This interface was based on the Motorola 68000 processor. It is now used for interfacing all devices within Frost.

The Founding of Amprotech

The automation committee soon realized that the experiences they had gained and the lessons learned in automating Frost would be of interest to other small firms. This fact combined with a realization that the members of the automation committee were essentially acting as full-time consultants to Frost led to the formation of Amprotech of Grand Rapids, Michigan, in October 1983.

With the establishment of Amprotech (consisting of the three members of the automation committee), the managers who were engaged in daily production were freed to focus on getting product out the door (their full-time tasks). Amprotech (mainly McIntyre and Nauta) took over full-time responsibility for managing the implementation process. The only way that an intensive and extensive undertaking such as CIM could succeed at Frost was by having someone responsible for its implementation on a *full-time* basis.

Other Implementation Issues

From 1982 to 1985 the automation committee met with various members of the four new management areas. During these discussions, seven guidelines were developed that were to influence the shape and practice of CIM at Frost:

- Enter data once, close to its source of origin.
- Eliminate double paper handling by entering data correctly the first time.
- The person entering data is responsible for its accuracy.
- Problems are best solved once, at their source.
- Operate the business plan with the smallest number of people required to fulfill the assigned function.
- Remain open to change, listen to others, and be willing to help others.

- Working together at Frost is based on mutual respect.

In hindsight, these guidelines reflect several just-in-time perspectives. However, they were developed because they made sense to the people involved and they accurately reflected the new organizational environment now present at Frost.

As the implementation progressed, management began to eliminate supervisory people. Their jobs were essentially taken over by the machine operators. There were several reasons for this action. First, the elimination of the supervisors was consistent with the new *flattened* organizational structure. Second, and more important, the members of the automation committee and the four new management areas soon recognized that the operators were the real experts in the system. They knew better than anyone else how to make things run better the first time. It was only logical that the people who knew what they were doing should be made responsible for managing and controlling their own actions. The guidelines were used to help them on the floor.

Assessing the CIM System in 1985

By 1985 most of the work required to begin the implementation of automation at Frost was completed. At this time the CIM system was primarily a manufacturing and engineering development. Automation was most obvious on the shop floor and in the engineering areas. The changes in organizational structure and the emphasis on system ownership were taken to improve the chances of a successful implementation.

At this time Frost became a Type III CIM system. It had strong functional integration (especially within the manufacturing and engineering areas). It was developing better organizational and strategic integration. The next challenges for Frost were to better "fine-tune" the existing CIM system and to develop the external linkages needed to better support the operation of this new corporate way of life.

GROWING THE CIM SYSTEM (1985 TO 1987)

The implementation of CIM at Frost began to change the way in which the other support functions operated. These changes were most evident in the purchasing and the front office areas.

CIM and Purchasing

In 1985 Frost was faced by a problem. It had implemented the foundations of its CIM system. However, it was now faced by 21,000 square feet of accumulated inventory. This inventory was the result of a purchasing system that used the economic order quantity for ordering parts and that tried to reduce total purchasing costs by relying on price breaks.

To a large extent the large stockpile of inventory was a temporary problem. It was consumed quickly by the CNCs now in place. When this stockpile was eliminated sometime in 1986, management was now faced by a new, *hidden* problem: past purchasing practices were no longer acceptable for the new Frost. In the past, purchasing had bought large quantities of bars and tubes. The presence of CIM now required the frequent delivery of small quantities of forged blanks. To make this transition, both purchasing and Frost's suppliers had to change the way that they did business.

- *Impact on purchasing.* Purchasing now had to think in terms of the continuous flow of raw materials from supplier to manufacturing. The emphasis now had to be on better scheduling and ongoing communication between purchasing and its suppliers.
- *Impact on suppliers.* For the suppliers, the effect of CIM was much greater. They were now under pressure to meet Frost's needs for

  Tolerance control

  Timeliness of delivery

  A commitment to deliver material in containers that could be readily used by Frost's material handling system

  A reasonable price for a product that was delivered to Frost in small shipments.

The problem for Frost was that there were no existing suppliers that could meet these requirements. The purchasing department at Frost now had to engage in supplier development. They had to go out to potential suppliers and convince them that in making the changes needed by Frost both Frost and the suppliers would ultimately benefit. This was a task that required Frost to sell its suppliers on the benefits of CIM and on the notion that what was happening at Frost was a precursor of future developments. By working with Frost, the new suppliers would learn those skills needed to service not only Frost but other CIM-type customers. Furthermore, since the Frost requirements were very similar to those encountered under just-in-time, these suppliers would also learn how to work with JIT buyers.

This was not an easy selling job. After all, Frost was a small company. For many suppliers, it did not represent a great deal of volume. CIM at Frost was still at an early stage. It had only been two years since Frost seriously started working at implementing CIM. In the interim Frost had to rely on its old screw machines to make the blanks needed by the CNCs. It took eight months for an appropriate supplier to be found for Frost. This supplier was not only able to meet Frost's requirements but to do so at a cost lower than those paid in the last two years.

CIM and the Rest of Frost

The last area affected by the introduction of CIM was the front office. How paperwork was now managed was changed. No longer were typewriters the primary vehicle for transcribing work. Instead, all paperwork was done on a combination of mainframe word processing (using PROFS on an IBM System/36) and microcomputers. All microcomputers and computer terminals were tied together so that information could be quickly passed and shared.

CIM AT FROST: 1987 TO 1988

By 1987, the implementation of the CIM system was mostly completed. The resulting system was one that seemed to be greater than the image of CIM typically found in the relevant literature. This was a system that provided a *glue* to the organization. It tied everyone in the organization together by the real-time computer links now present. Information, instead of being hoarded, was now a resource widely shared by all and available to all.

For Chad Frost and Frost, Inc., the key to the future lies in developing and growing the capabilities side of the market. That is, Frost must decrease its reliance on the chain-driven overhead conveyor market. It must become better at building "metal round things." This focus is evident in the orientation of the CIM system.

As Frost has become better at building round metal things, the resulting improvements have had an impact on the traditional side of the market. Techniques and shortcuts learned in the new market have often found ready use in the traditional market. As a result, a unique symbiotic relationship has developed between the two markets served by Frost.

The Key Features of the System Summarized

Currently, CIM at Frost is a corporate-wide system. It is present everywhere and is used by everyone in the firm, from the person on the shop floor to the president. The CIM system is so pervasive that some at Frost believe that it should be called *CIC* (computer integrated corporation) or *CIB* (computer integrated business).

The computer system that supports and facilitates this corporate-wide system is broken down as follows:

Corporate Database. There is one corporate-wide database that supports everyone in the firm. This database is maintained on an IBM System/36 through the IBM MAPICS system. Except for one instance, this information is available to everyone in the firm without restriction. The only restriction is that of payroll, which is available on a view-only basis.

By making the information readily available to everyone in the firm, management has further emphasized the need for trust.

General Business. All general business activities such as word processing and messaging are managed through a mixture of terminals and personal computers. The terminals are linked to the IBM System/36, which has PROFS. This system is used for word processing, general business tasks, and messaging. Personal computers are used for local tasks such as tool crib management and spreadsheet (e.g., Lotus).

In all cases there is a network linking all terminals (over 40) and personal computers to the IBM System/36. This network serves several purposes. First, it allows for uploading and storage of data collected at the personal computers by the mainframe. Once stored at the mainframe, this data can then be viewed and backed up. Second, by linking terminals and personal computers together, users located at a terminal can "look" inside the personal computers.

Manufacturing Planning and Control. The major vehicle for manufacturing planning is again IBM's MAPICS. This system has been modified by dropping its MRP module. It was decided that, given the high rate of routing changes, MRP would not be apropriate for Frost. In its place, management introduced three more focused and appropriate modules:

- Customer order servicing (provided by GMD).
- A stock locator package (provided by GMD).
- A purchasing package (provided by Marcam).

The entire system is interactive and operates in real time.

Engineering Design. To enhance the efficiency of the engineering area, a coding and classification system based on the D-Class processing system provided by Brigham Young University is used. The major control device is the part number. Every part receives a unique part number that describes the major design features of the product. When beginning to design a new part, the existing part numbers are checked to see if there are any identical or similar existing parts currently in the system. This check reduces design lead time and its associated costs. All designing is done via a computer terminal using IBM's CADAM program. The system is implemented on an IBM 4361 system.

The engineering information generated is stored in one of two databases. MAPICS stores all corporate information. It also contains manufacturing notes and instructions. CADAM is used to store geometry and tolerances.

Classifying Frost's CIM System

As of 1989 Frost was a Type III CIM system. It was a company that had all three forms of internal integration: functional, organizational (across functions), and strategic (between top management and the shop floor). What it lacks is channel integration. The type of integration between suppliers and customers, on the one hand, and the firm, on the other, found in other CIM systems such as Ingersoll Milling is lacking. This lack is not so much a planned consequence as it is the result of Frost's size and the lack of experience compatibility.

• *Size.* Frost is a small company. It is often very difficult for any small company to convince its customers or its suppliers to change how they do business so as to facilitate integration. Frost simply does not have the influence or power necessary. It does not have the volume needed to bring about the desired changes.

• *Lack of Experience Compatibility.* By any standard Frost is a leading-edge company. It has been actively working with the cutting edge of technology since 1983. As a result, many of the practices and changes that Frost has instituted are now considered commonplace and mundane by the people at Frost. These same practices and changes would be considered radical and extremely unusual by its customers or suppliers.

For channel integration to take place, the customers and suppliers must develop to the extent that they are at a level of development similar to that of Frost. At this point channel integration can become a meaningful goal. Without this level of development, the people at Frost must educate its suppliers and customers. This is an expensive and time-consuming process—more so for a small firm such as Frost.

Benefits

There is no doubt in the minds of people at Frost that the CIM journey has borne a great deal of fruit for the company. It has had a significant positive impact on the firm. The benefits gained from the implementation and use of CIM have been both quantitative and qualitative. The following are the most significant of these benefits:

• *Better utilization of resources.* CIM has influenced the utilization of both machinery and manpower. Machines now spend more time cutting and processing parts instead of waiting. More important, with the elimination of supervisors, each worker does more. The elimination of supervisors has reduced overhead (an important objective) and increased employee productivity.

- *Reductions in floor space.* As equipment and people become more efficient and as the need for inventory is reduced, the amount of floor space needed for production is also diminished. In 1983 production took place on 85,000 square feet; it now takes only 28,000 square feet for a higher level of output. This benefit has simplified the problem of capacity expansion. Should the need ever arise for more floor space, manufacturing can simply grow into the floor space abandoned through consolidation and improved utilization.
- *Increased flexibility.* The new equipment now allows for rapid expansion at a lower cost. Management estimates that they can increase the capacity and capabilities of their current system with a minimum amount of investment.
- *Reduced setup times.* Setups that used to take hours now take only minutes as a result of CNC capabilities.
- *Reduced lead time.* Delivery lead time to the customer has gone from 12 weeks in 1983 to less than 4 weeks in 1989.
- *Increased output per person.* In 1983 each employee only produced $80,000; in 1987 each employee produced $164,000. As productivity has increased, so has the overall profitability of Frost.
- *The survival of Frost.* The biggest benefit of CIM for Frost has been that it has allowed Frost to survive. The increases in output per person have occurred during a time when prices fell by more than 27 percent. Without CIM, Frost would not have still been in business in 1989. Everyone at Frost believes this to be a fact.

LESSONS LEARNED

After over seven years with CIM, people like Chad Frost have developed certain strong views about what should and should not be done. These views, based primarily on the experiences of Frost, are the lessons that the firm can offer the reader:

1. *Know where you are going.* CIM by itself does not provide direction for the firm. It does not present the firm with those skills and capabilities that are needed to survive in the marketplace. These issues, as well as the guidance needed by CIM, are set by top management in the strategic planning process. Strategic planning drives CIM; CIM is nothing more than a mirror of the strategic objectives set by the firm. Unless there is a clear and unambiguous under standing of how the firm competes, who it services, and the key skills needed by its market, there will be confusion. The implementation of CIM only serves to magnify these strategic shortcomings.

2. *Know who your customers are.* CIM is customer driven. However, any firm faces three different types of customers: those to whom you think that you sell; those to whom you really sell and those to whom you really want to sell. These customers may not be the same. When planning for CIM, management must first eliminate any confusion between the first two types of customers. It must also decide whether the CIM system is intended to serve the existing customer base or a new customer base.

Frost had to address this issue in its implementation. Management responded by building a CIM system aimed directly at serving its new customers: those interested in "metal round things." This market was a growth market, unlike the current chain-driven overhead market, which is in its mature and early decline stages.

3. *Justify all investments.* No investment, especially one as significant as that required by CIM, should be accepted on faith alone. It should be justified. Benefits must be identified. With CIM these benefits are often company-wide and can be found in the areas of overhead reduction, inventory reduction, quality improvements, and lead time reductions.

Justification sells the CIM investment at many levels. It sells CIM to outside sources of capital; it sells CIM to top management; it sells CIM to the people on the floor. Such "sells" help to create acceptance.

4. *Rethink before automating.* Automation should never be the first step in any implementation plan. Instead, it should be viewed as being the last step. One of the first steps is that of reviewing every function in the firm to see if it is really being done properly and efficiently. To guide in this evaluation, management must be prepared to address the following critical questions:

Can we make better use of our people's time?

Can we remove obstacles to doing a better job?

Can we reduce waste?

Can we focus our people on value-added work?

Can we computerize routine activities?

Can we eliminate obsolete reports?

Can we eliminate paperwork?

The purpose of "rethink" is to get people to work smarter, eliminating procedures that are not needed and implementing more efficient and productive methods and procedures.

5. *Get people involved early in the process and from all levels of the company.* For any firm to succeed, it cannot rely on technology alone. It must rely on its people. It is the people who will implement the CIM system; it is the people who will make the CIM system ultimately work. As a result, the challenge for management is that of securing user involvement and commitment.

6. *Use closed-loop systems for the effective involvement of people.* The management at Frost learned this lesson from their experiences in justifying the automation project to the external financial institutions. When examining the reasons for automation, the management formed a committee. One of the members of the committee was a banker. Having a banker present in the committee from the outset had a number of benefits. The banker could tell the other committee members what had to be done to improve the presentation. The banker also served as a *sounding board* for committee ideas. More important, however, the banker was able to learn quickly about what automation was and was not. The banker was also there to respond immediately to issues and concerns raised in discussions. If the banker had not been there, the committee would have had to note down the ideas and issues, wait until the end of the meeting, approach the banker, and wait for his response. The resulting process would have been wasteful and frustrating for the people involved.

As a result of this experience, management recognized that when forming committees, all of the people essential to the success of that committee or its solution must be involved in the committee from the outset. They must be involved in a closed-loop system: The issues that are raised are responded to immediately in committee. For every problem raised in the committee, there is a response that comes from the committee. This helps drive the committee to success.

7. *Understand and eliminate organizational obstacles to CIM.* Often, organizations cannot take advantage of CIM because their existing structures are too rigid and contribute to a myopic perspective. This was the case in Frost. Before CIM could even be considered, these organizational obstacles had to be identified and removed.

This lesson has a significant implication for management. It requires that managers examine any recent failures in attempts at bringing about similar changes. They must determine whether these failures were merely symptoms of deeper organizational problems. If they are, management must deal with these problems. If they fail to do so, the attempt to implement CIM will simply become another expensive failure.

8. *Use outside "change" managers.* Whenever possible, outside consultants who are willing to be involved in a full-time capacity should be secured. First, consultants have seen other implementations. They have experience. They know what works and what does not. Finally, by being involved in a full-time capacity, they can devote the necessary time and effort to making CIM a reality. For most firms, the implementation of CIM, while critical, is not the only activity taking place. There is another activity that is paying the bills for the implementation of CIM: that of meeting the daily production schedule. Since most employees will be involved in this task, they will be limited in the time and effort they can devote to the implementation of CIM. This "vacuum" is best filled by an outside consultant.

9. *Eliminate jargon.* CIM and many of its techniques are surrounded by a lot of very technical jargon. For most people who are not CIM experts, jargon tends to confuse and intimidate. If people are to be involved, then they must feel comfortable with the concepts underlying CIM. Many of the terms associated with CIM can be restated in less threatening and more commonly used phrases.

10. *Use education and training.* Education and training are necessary. They help the users develop a sense of ownership over the new project. They familiarize people with the terms and concepts underlying the new CIM system. They give users an opportunity to experiment and explore the new systems. Finally, they help users develop confidence. All of these traits encourage user involvement in the implementation of the new CIM system. They help make people excited about CIM.

11. *Work with vendors that can and will support the implementation.* An implementation is never easy. Inevitably something will go wrong on either the hardware or software side. If the vendor will not support the implementation process, it is doomed to failure. It will take too long to correct necessary problems. Without support, the user is left in the position of either solving the problem alone or hiring expensive outside help. In either case the result is an increase in the implementation lead time (which may doom the implementation).

The vendor is the expert. If he or she is willing to work with the implementing firm, problems that are identified can be quickly resolved. The vendor, by acting as the technology expert, allows the users within the firm to focus on the other aspects of the implementation problem.

12. *Senior management involvement is critical to success.* People in any firm take many of their cues from the senior management. If the senior management is not interested in CIM, it is likely that very few people in the firm will be. On the other hand, if the CEO is very involved and his or her involvement is highly visible, many people will rethink their views.

Another reason for senior management involvement lies in their control over resources. It is the senior managers who will approve or disapprove the investments in equipment, software, consulting, and training. It is also the senior managers who will release key people to work on the implementation team.

Without visible and active senior management involvement, the CIM implementation will fail. People will tell you that one of the reasons that CIM succeeded at Frost was the continual presence of Chad Frost. Chad Frost, by virtue of his ownership of the firm, had a very sizable stake in the success of the CIM project. He was also out every week giving paychecks to the various employees. As a result, everyone knew that Chad Frost wanted the project to succeed.

THE NEXT STAGE: MEETING THE CHALLENGE OF THE 90S

With the coming of the 1990s, the challenges for Frost have not ended with the introduction of CIM. Frost must still address the following issues if it wants to maintain its continued survival in the marketplace:

• *Size.* As has been pointed out previously, Frost is a small firm often dealing with larger firms. This fact has affected its ability to get outside support. It also influences the ability of Frost to change the behavior of either its suppliers or its buyers.

• *Frost as an industrial supplier.* Frost sells primarily to industrial buyers. A sale, if successful, can simplify planning and scheduling, since the supplier can devote a portion of its capacity to the new buyer. Industrial sales are attractive because they are large and they cover long periods of time. There is a downside, however: An industrial supplier deals with few buyers. Should one of the buyers have a problem with sales, there is a direct and, in some cases, immediate impact on the supplier. A large portion of its sales is now gone. The result may be an immediate crisis.

Since Frost is an industrial supplier, it must live with this reality. It is an aspect that affects the entire firm. For the 1990s, the challenge facing management is that of how to reduce this dependence on a few buyers.

• *The risk of market development.* Frost has used its CIM system as a method of building a capability that can be sold in turn to a new market. This capability is that of making the best "metal round things" in the world. Successfully building this activity does not necessarily mean immediate success in the targeted market. The reason is that it takes time to develop the new market. There is always the risk that the market will not meet anticipations. New marketing skills (on the part of Frost) must be learned and applied. As of 1989, it was not certain whether the new market targeted by Frost would really assure the continued survival and growth of Frost.

CONCLUDING COMMENTS

Frost is in many ways an interesting study. It is a company that has been able to succeed with CIM in spite of the odds stacked against it. It is a small firm that has succeeded with CIM because its management decided to view CIM as a corporate, not a manufacturing, system, and because they provided CIM with a strong strategic focus. Frost has used CIM not simply as a means of ensuring survival but also as a method of generating growth.

With Frost we are faced with a company that is still developing its CIM system. The foundations that it has laid in the past should serve to help Frost succeed in the future. Irrespective of what the future holds for Frost, it can be said that for this firm, *small is beautiful.*

Frost, Inc.: A Postscript

In 1991, Chad Frost reviewed the progress of his company with CIM. CIM had forced Frost, Inc. to carefully redefine its strategic focus. Initially, Frost saw itself as being the world's foremost builder of round, grooved metal things from casting or bar stock in the 2- to 8-inch range. This definition, which seems fairly well focused at first, is still far too vague. There are other firms that also define themselves as the world's foremost manufacturers of round, grooved metal things from casting or bar stock in the 2- to 8-inch range. However, what they build is very different from what is provided by Frost. Most of these competitors excel at long runs. They can produce the same items over and over again at low cost and with little or no scrap. That is not, however, what Frost excels at. Rather, Frost excels at building round metal things in which flexibility is critical. Flexibility is defined in terms of the ability of Frost to cope with many part numbers, which are subject to frequent change due to either variation in demand or variation in order quantity. Furthermore, Frost excels at dealing with order quantities that are less than 22 pounds.

Frosts's CIM structure is well designed to deal with such variability. Organizationally, Frost emphasizes group technology principles that enable it to take advantage of commonality. The firm is good at fitting together work loads that are similar either in design or processing. In addition, Frost has become proficient at material and capacity planning and scheduling. It can simultaneously satisfy customer needs and manpower requirements. Frost can take highly variable demands (that are often changing) and successfully smooth out the variations on the floor. Finally, it has also been successful in reducing nonproductive uses of capacity. All of these capabilities build flexibility.

It has taken Frost a long time to realize that its initial strategic position was too vague. Orders were accepted in the past that, while involving round metal things, did not require flexibility. Frost did poorly on these orders. It found itself unable to compete with those other firms that were good at long runs. The reason was not CIM-related but strategically based. As a result, both Chad Frost and Frost, Inc. have learned that success with CIM requires a clear, well-defined, specific strategic statement. To succeed, you must know exactly what you can do well and what you cannot do well.

CHAPTER 11

COMPUTER INTEGRATED MANUFACTURING: PAST PRESENT, AND FUTURE

There is a fundamental paradox underlying any study dealing with CIM CIM, by its very nature, is dynamic. It is constantly evolving. These change: reflect developments in technology, organizational structures, strategi initiatives, and consumer expectations. As a result, a CIM system that i: considered state of the art in 1985 is not state of the art in 1990. Given thi continuously evolving nature of CIM, a premium is placed on the study o current CIM systems. Both researchers and managers must be kept aware o the latest developments in CIM. However, these studies, no matter hov quickly they are completed and published, are doomed to be obsolete. The: are but a snapshot of CIM at a given point in time. By the time their finding are published, the evolution of CIM will have carried it past the characterization contained in the study. Hence, we are faced by a paradox: Becaus CIM is dynamic, we must continually study its operation so that we ca better understand it. However, these studies are doomed to be obsolete a soon as they are completed. This paradox raises the basic question, *Wh study CIM systems at all?*

The answer to this question lies in another trait shared by all CIM systems CIM systems, while different, do possess some commonalities owing to the fac that the development of CIM is *evolutionary,* not *revolutionary.* The CIM system of the 1990s, while different from the CIM system of 1985, is influenced b developments present in 1985. In studying "snapshots" of CIM systems at different points in time, we can identify those trends and developments that wil influence the future shape of CIM.

This chapter summarizes and presents the authors' view of the *futur CIM*—the CIM of the future. We use the term "future" to denote the next fiv years (from 1991 to 1995). The discussion of future CIM consists of three basi sections:

- *The "shapers" of future CIM.* The first section examines those factor that will influence the shape of future CIM.

- *The obstacles to future CIM.* These are the obstacles that will hinder the development of any CIM system in the future. How these obstacles are overcome will affect the shape of future CIM.
- *The traits of future CIM.* The final section presents the major characteristics present in future CIM.

Our view of future CIM as presented in this chapter represents extrapolations of major CIM-related trends observed in the case studies. Where possible, we have discussed these views with industrial representatives from the companies involved. The reason for these discussions was to get their feedback on the extent to which these views are realistic.

THE SHAPERS OF FUTURE CIM

In general, manufacturing tends to be evolutionary in its development, not revolutionary. The manufacturing systems of today reflect influences of the past. In turn, the manufacturing system of the future will be very much a product of developments now evident. Duane Shull of GE FANUC North America best summarized this view when he noted that:

> **One does not need to gaze into a crystal ball looking for new technologies that do not yet exist. The building blocks of the future are scattered around us right now....Our task is to sort them out and put them together.**[1]

As a result, if we are to predict the shape of future CIM with any degree of success, we must identify and examine those factors now present that will influence the development of future CIM. The following lists many of these *shapers*. This list (as summarized in Figure 11-l) is not intended to be exhaustive. Rather, it is intended to highlight a few of the most important factors.

Many Paths, One Common Goal

Any manager interested in improving the overall competitiveness of a firm is now faced with many alternatives. These include developments such as JIT manufacturing, TQC systems, MRP II, OPT, and, of course, CIM. Each development shares a common goal, that of ensuring corporate survival by enhancing

---

[1]Hamid Noori, *Managing the Dynamics of New Technology* (Englewood Cliffs, N.J.: Prentice-Hall, 1990).

FIGURE 11-1
The Shapers of Future CIM

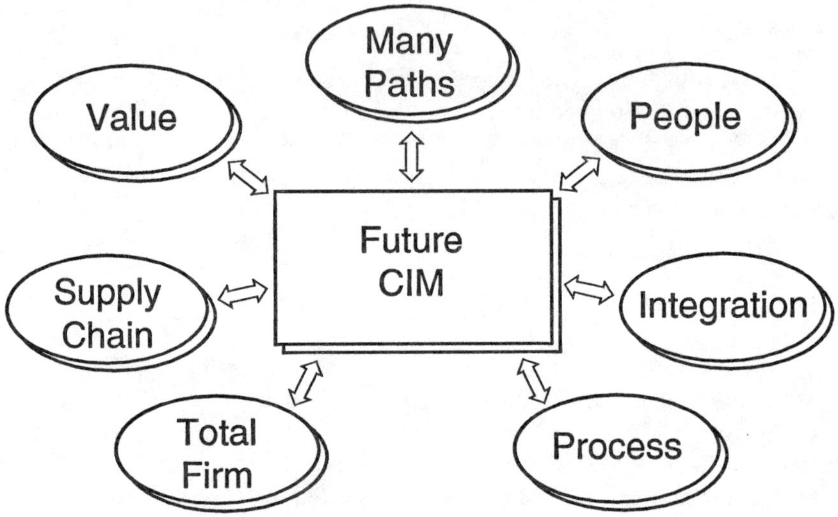

the ability of the firm to develop a distinctive competitive advantage in the marketplace. However, how each tries to achieve this objective could be somewhat different.

Just-in-time, for example, focuses attention on *waste*. Waste is any activity that adds cost without adding any offsetting value to the product. Classic examples of waste include excessive inventory, scrap, salvage, rework, illogical paper flows, incoming inspections, bad data, and management by crisis. Having identified waste within the firm, just-in-time attempts to eliminate it primarily by uncovering and eliminating the underlying causes. Scrap, for example, is only a symptom. Its presence can indicate a number of problems such as poor housekeeping, inadequate incoming materials, poor operator training, inadequate specifications, the lack of adequate capacity, and bad data. Unless these underlying problems are identified and corrected, any action aimed at reducing the scrap levels will be doomed to failure.

Often, this process of waste elimination draws heavily on simplifying the underlying processes. Instead of using a real-time, computer-based data tracking and control system, visual control is used. On-floor inventory stacks are kept low (less than 4 feet) so that managers can get a "feel" for the condition of the shop floor quickly by simply looking at the floor. Another example of just-in-time's emphasis on simplification can be found in the *kanban* system.

The term "kanban" is used to denote the system of visual triggers that are used to control production and inventory flows on the shop floor. These triggers can take a variety of forms. In Toyota, cards are used. At Hewlett-Packard, squares are used. The simplicity introduced by just-in-time can best be understood by examining the kanban squares used at Hewlett-Packard's Greely, Colorado, plant. The various work centers are linked together by kanban squares. The status of the square controls production. That is, when the square is full, the supplying point cannot produce any inventory for replenishment. However, once the square is empty (i.e., the inventory contained within the square has been consumed by the upstream work center), the supplying work center can produce. The result is a simple system to operate.

Total quality systems focus attention on quality. Manufacturing resources planning, in contrast, emphasizes the need for a comprehensive, integrated, and effective manufacturing planning system that manages all of the resources (manpower, tooling, material, capacity, and financial capacity) of the firm. With OPT, attention is drawn to the need to identify and closely manage the constraints or bottlenecks of the firm. It is these bottlenecks that ultimately determine the level of output that the firm can generate as well as the resulting cost levels. Finally, there is CIM, which draws on the capabilities offered by computers and computer-based technologies to enhance corporate competitiveness.

In the past, these developments were viewed as competitors and substitutes. For example, the argument was often over whether a firm should implement CIM or JIT. Implicit in this argument was the view that JIT and CIM were competing alternatives and, therefore, mutually exclusive.

This view has now changed. Managers, consultants, and researchers see these various developments as complements. Ultimately, the solution to the problem of how to improve corporate competitiveness will be a matching process, whereby management draws on those elements of each development that are appropriate to the specific problems facing the firm. As a result, the distinctions between CIM, JIT, TQC, MRP II, and OPT are now beginning to blur. We now recognize that while there is one goal, there are many ways of reaching this goal. No one path contains all of the answers. Rather, each path contains part of the answer.

A Changing View of Value

All effective manufacturing systems are value driven. They are designed to generate value for their customers. They attempt to reduce the gap between what the system produces and what the customer desires. Whenever a system is value driven, it is forced to become dynamic. The very nature of value forces this fact on both management and the firm. Value is defined not by the firm but by the customer. Customer needs change over time—often as a result of such factors as customer expectations, customer education, the level of income, competitive actions, and governmental actions. For any manufacturing system and for a technology such as CIM, the challenge is that of keeping pace with these changes.

Because of significant lead times (required to add new capacity or to change existing capacity), these changes must be anticipated if the manufacturing system is to keep pace. If we are to predict the shape of future CIM, we must first identify those trends in customer needs that the new system will be expected to satisfy.

Stanley Davis, in his book *Future Perfect*,[2] presented his views on the future shape of the business enterprise. He saw the firm of the future as being shaped by several critical influences:

Any Time. At the heart of this first influence is time. Time has value for the customer. It influences the actions of the customer. It is a resource, not a constraint. As a result, the challenge for the supplying firm is to reduce the lead time faced by the customer.

Lead time reduction has several different dimensions. First, there is a reduction in the average lead time faced by the customer. This dimension of lead time is often referred to as speed. Second, there is the reduction in the variability of lead times faced by the customer. This second aspect ultimately manifests itself in uncertainty reduction for the consumer. Third, the concept of lead time not only encompasses the supply of goods but also the flow of information. Fourth, in reducing lead times, attention cannot simply be focused on one isolated aspect of lead time (e.g., manufacturing lead time, delivery lead time). Rather, it must deal with all aspects. It must examine the process of bringing a product to market. Areas such as engineering and design lead time must be considered. Fifth, significant lead time reduction cannot be attained by simply doing the same things better. Rather, the entire process of designing and delivering a product must be reevaluated and changed where necessary. Finally and most importantly, all discussions of lead time must be based not on the supplying firm's perspective but on the customer's view of lead time. It is the customer's view of lead time that should dominate the firm's view of lead time and drive its lead time reduction activities.

Certain innovative firms are now wrestling with the implications of "any time." As reported in a 1989 *Fortune* article,[3] these firms are beginning to see the fruits of their efforts to reduce lead times and to compete on time (see Tables).

---

[2]Stanley M. Davis, *Future Perfect* (Reading, Mass.: Addison-Wesley, 1987).

[3]Brian Dumaine, "How Managers Can Succeed Through Speed," *Fortune*, February 13, 1989, pp. 54–59.

TABLE 11-1
Superfast Innovators

Company	*Product*	*Development Time*	
		Old	*New*
Honda	cars	5 years	3 years
AT&T	phones	2 years	1 year
Navistar	trucks	5 years	2.5 years
Hewlett-Packard	computer printers	4.5 years	22 months

Source: Brian Dumaine, "How Managers Can Succeed Through Speed," *Fortune*, February 13, 1989, p. 56.

With "any time" we are forced to recognize that, for many customers, time is far more important than the issue of cost. Some customers are willing to pay a premium for products quickly designed and delivered. We are also forced to recognize that, for the suppliers, the benefits of speed are very attractive. As has been noted by George Stalk and Thomas Hout,[4] as time is compressed, the firm experiences the following changes:

- Productivity increases.
- Costs are reduced.
- Prices can be increased.
- Risks are reduced.
- Market share is increased.

Finally, managers are realizing that with "any time," slow firms can never effectively compete with fast firms.

Any Place. This second influence emphasizes the need to *reduce the distance* between customer and supplier. In the future, successful firms will be able to bring together in time and distance customers and suppliers. This closing of the gap will take place in several different ways. First, an increasing amount of production will take place at the customer's location. Customers will also play a more active role in the design, manufacturing, and delivery activities of the suppliers. Second, bringing together suppliers and customers will demand a reduction in the number of stages between customers and suppliers. For many firms, the reduction of these stages will take the form of the elimination of unneces-

[4]George Stalk Jr. and Thomas M. Hout, *Competing Against Time* (New York: The Free Press, 1990).

TABLE 11-2
Superfast Producers

Company	*Product*	*Order-to-Finished-Goods Time*	
		Old	*New*
GE	circuit breaker boxes	3 weeks	3 days
Motorola	pagers	3 weeks	2 hours
Hewlett-Packard	electronic testing equipment	4 weeks	5 days
Brunswick	fishing reels	3 weeks	1 week

Source: Brian Dumaine, "How Managers Can Succeed Through Speed," *Fortune,* February 13, 1989, p. 56.

sary steps in the design and delivery process. Finally, the basic customer-supplier relationship must move from one of competition to one of cooperation.

No Matter. The third factor describes a change in the way customers view the outputs of the firm and the product that the supplier actually delivers to the customer.

In the past, the buyer-seller relationship has been built around the flow of physical, tangible products. This relationship is now changing. Suppliers no longer sell physical goods alone; customers no longer buy physical goods alone. Rather, intangibles are beginning to play a greater role in this relationship. These intangibles take many forms, such as the increasing importance of security, information, goodwill, and service. Suppliers are now paying more attention to the development of closer buyer-supplier relationships.

Mass Customization. Simply described, this final influence describes the importance of a system that is able to customize on a mass scale. This trait is becoming more desirable as firms move from focusing on broadly (and often poorly) defined markets to specific customers. The challenge now facing the supplier is that of offering these customers more options (resulting in customized goods) while not increasing lead time and costs.

We will face a customer who is no longer influenced primarily by issues of cost and quality alone. Instead, we will deal with a customer who evaluates the firm along a number of different dimensions simultaneously. We will be faced by a customer who is interested in issues of lead time, cost, quality, and flexibility. As a result, the firm must think in terms of customers, not markets. This trend has been recognized recently by marketing professionals and has been described as *micromarketing* by *Business Week*.[5]

The Increasing Importance of People

People play an important role in the manufacturing system. On the one hand, they are the ultimate consumers of the outputs of this system. Simultaneously, they are also inputs into the process. However, the expectations of what is needed of workers as inputs is changing. We are now seeing a shift from the use of labor. Employees are no longer hired for their manual strength or dexterity but rather for their knowledge and intelligence. Increasingly, people in their dual roles of consumers and resources are placing greater demands on the manufacturing system. These views are summarized in Figure 11-2.

As consumers, they have greater expectations of the overall quality of products and services that are now available to them. People can now afford to be choosy. For any good or service that they might want, there are multiple suppliers (both domestic and foreign). In addition, in some industries, such as automobiles, competition for sales dollars has been heightened by the presence of overcapacity. As a result it is now harder to sell to these customers. It is also very difficult to get repeat business should a customer experience any problems with the product.

As inputs, workers have had several effects on the manufacturing system. First there is the general lack of skilled trades. It is becoming more difficult to find workers with such skills as tool and die making. Firms and managers must now work much harder at recruiting such employees and at keeping them within the firm after they are hired.

Second, managers now recognize that people are sources of major improvements. People who work at any job for a period of time become familiar with that job. They often discover shortcuts; they develop better ways of doing things. As a result, the task for management is that of trying to bring these items of information into a formal system. This realization has changed how workers are viewed. A firm's employees are no longer laborers but, rather, problem solvers.

In addition, workers now expect to play a greater role in their work settings. They want a more attractive work setting. They want a workplace in which they will be challenged and in which their abilities as problem solvers are used and their contributions to the firm and its success recognized. Finally, workers want greater control over their jobs.

---

[5]Zachary Schiller, "Stalking the New Consumer," *Business Week*, August 28, 1989, pp. 54–62.

FIGURE 11-2
Increasing Importance of People

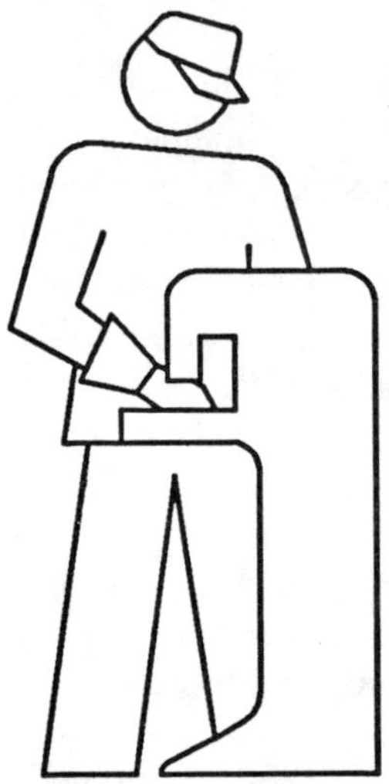

* As consumers
* As scarce; skilled workers
* As source of knowledge

Workers can now make such demands because they now have alternatives. If a firm will not provide the "right" work environment, the best employees can leave and look for work elsewhere. Alternatively, they can upgrade their positions (and enhance their mobility) by taking advantage of the education system.

These views have a parallel in the experiences of countries such as Sweden with its own work force. In a recent *Business Week* article, the decision of Volvo of Sweden to change its mode of building cars from the traditional assembly line to a teamwork concept was described:

> It wasn't a lack of success in the car business that drove Volvo to design such a radically different system. Sales totaled $16.1 billion in 1988, a 50% increase since 1983. Last year, operating profits totaled $1.2 billion, and Volvo has no net debt. But it has a problem that haunts all Swedish manufacturers: *The country's highly educated, well-trained labor force doesn't like to work in factories* [italics the authors'].
>
> Volvo's Swedish plants suffer absenteeism of 20%, and almost one-third of its workers quit yearly . . . . With unemployment at 1.6%, there is no lack of jobs. Says Per-Olov Bergstrom, a national bargainer for the Swedish Metal Workers' Union: "The main problem in the '90s simply will be to keep a work force" [italics the authors'].[6]

---

[6]Jonathan Kapstein and John Hoerr, "Volvo's Radical New Plant: 'The Death of the Assembly Line?"' *Business Week*, August 28,1989, pp. 92–93.

For many workers, the two key words of the 1990s will be "enfranchisement" and "empowerment." With labor (especially skilled) becoming scarcer, managers must recognize and be prepared to cope with these demands.

An Awareness of the "Total" Supply Chain

For most managers, CIM has been viewed primarily from an *internal dimension.* That is, its primary goal has been to improve operations within the firm. Implicit in this approach has been the assumption that by improving operations within the firm, the overall effectiveness and efficiency of the firm is improved. In turn, these developments are expected to improve profitability. In focusing on the firm in *isolation,* this approach has ignored the *external dimension.*

The external dimension considers how actions taken by a firm's suppliers and customers affect that firm's operation. By studying these external actions taken outside the firm, we recognize the presence of *interdependence.* That is, actions taken by suppliers or buyers can have a profound impact on the firm. Poor-quality materials that are not delivered on time and in the right quantity can bring to a halt any manufacturing system (no matter how well managed it is). Similarly, a buyer who provides the firm with schedules of requirements that lack adequate stability and visibility can adversely affect operations. By considering this external dimension, we are looking at the firm within the context of the *supply chain.*

Every firm is an integral element of the supply chain. This chain consists of all the stages that exist between the raw material (e.g., iron ore in the ground) and the ultimate consumer (e.g., the person who buys a Ford car from a Ford dealership). At each step in the supply chain, there are linked buyers and sellers. Often, firms involved in this chain act simultaneously as both buyers and sellers. A firm buys iron ore from a mining company that has extracted the ore from the ground. In turn, the buying firm sells the steel or pig iron that has been made out of that iron ore to another customer.

Each stage is influenced by actions taken elsewhere within the chain. As a result, for the firm to succeed, it must not only effectively manage operations within its own company; it must also ensure that there is some form of coordination and cooperation between itself and its buyers and suppliers. The purpose of this coordination and cooperation is to ensure that, at least, actions taken outside the firm do not adversely affect internal operations and that, preferably, external actions are supportive of internal operations. For most firms, it is easier for management to control internal operations than it is to control the external dimension. After all, the firm has direct access to its manufacturing planning and control system. It can replace ineffective employees. It can use new technology as needed. The challenge facing most managers is that of controlling the external dimension.

In general, there are two options open to those interested in exercising some form of control over this external dimension. The first is that of *vertical integration*. The second is that of *active persuasion*.

With vertical integration the firm attempts to establish control over the supply chain by purchasing controlling interests in its suppliers and its customers. Control is established through ownership. One of the best illustrations of vertical integration in practice can be found in the activities of the Ford Motor Company in the 1910s and 1920s.

In a 1926 book entitled *Today and Tomorrow*, Henry Ford summarized many of the experiences and lessons he had learned in managing the Ford Motor Company.[7] One of the lessons described how Ford had gained control over this chain. To Ford, this chain began with the mining of the raw material (i.e., coal and iron ore) and ended with the delivery of the final product (the Model T car) to the customer at a Ford dealership. He recognized that problems anywhere within this chain could have severe repercussions on operations at Ford in Michigan. The method that Ford used to gain control over this chain was the physical acquisition of most of the intermediate stages in the chain. The result was a supply chain almost totally owned by Ford and managed by Ford-appointed managers.

Iron ore was mined from Ford mines. It was then hauled by Fordson tractors to Ford docks. There, it was loaded onto Ford ships and delivered to River Rouge, Michigan. There, it was unloaded by Ford material moving equipment and moved to a Ford steel plant (where it was converted into iron and steel). From the steel plant, the steel was moved into the Ford assembly plants, where it was used to build Ford cars and trucks and Fordson tractors. The finished Ford cars were then shipped to a Ford dealership for final delivery. Ultimately, this entire supply chain could be controlled by decisions made at Ford headquarters in Dearborn, Michigan.

However, this system did not survive. The Ford system, like many other instances of vertical integration, could not overcome two flaws inherent in it. The first is that vertical integration requires that the firm be able to commit the financial resources needed to acquire control over the downstream (from it down to its suppliers) and upstream (from it to its customers) stages in the supply chain. Second, vertical integration requires the owning firm to be able to control and coordinate activities in the supply chain. Achieving this control often requires the presence of a large bureaucracy. In many cases the costs of maintaining and managing this bureaucracy are far greater than the resulting benefits. The bureaucracy becomes too large and unwieldy. It cannot respond

---

[7]Henry Ford, *Today and Tomorrow* (Cambridge, Mass.: Productivity Press, 1987), reprint.

fast enough to the requirements of a changing market. Such was the case with Ford. As a result, this system of control was ultimately abandoned.

In contrast, active persuasion is a very different approach to the problem of coordinating and controlling the supply-chain linkages. Instead of imposing control through ownership, active persuasion imposes control by making the potential benefits to either the supplier or the customer very attractive. Implicit in the practice of active persuasion is the *carrot and the stick*. The management of the firm goes out and actively convinces the management of the other stages of the supply chain that by acting in certain ways, there are advantages (the carrot) to be gained. For the supplier, these advantages might take the form of increased business. For the customer, the benefits might include better or more cost-effective deliveries. However, for those suppliers who do not choose to heed these enticements, there is the stick of the potential loss of business in the future. For the buyer, the stick may consist of higher costs or longer lead times.

In all instances active persuasion attempts to create a situation in which the suppliers, buyers, and the firm share the same set of relevant values. What the suppliers view as desirable is identical to those requirements of the firm that best support the operation of its internal system.

In this study, we have seen several instances of active persuasion in practice. These practices have been most noticeable in the Steelcase and Ingersoll Milling cases. However, for the best example of active persuasion, we must look outside of these firms to General Motors, another American company that is struggling with the problem of controlling the supply chain.

Recently, General Motors implemented the *Target for Excellence* program. One of the major purposes of this program is to examine all aspects of the systems that General Motors' suppliers have developed to receive, process, and deliver orders and to determine the extent to which this process can operate in a timely and cost-effective manner.

As part of the Target for Excellence program, General Motors is now sending out teams consisting of representatives from such areas as purchasing, engineering, and quality. One of the purposes of these teams is to determine the extent to which the suppliers' processes are operating in a timely and cost-effective manner. All assessments are to be done by a three-day on-site plant visit.

The management at General Motors has tried to ensure that all of its suppliers understand the importance of these assessments. For the suppliers who either have developed, or are in the process of developing, effective delivery systems, there is the promise of increased future business from customers such as General Motors. After all, such suppliers represent the least amount of risk for General Motors. These are suppliers that have delivery processes that are under control. For the suppliers that have not developed such systems and that are not interested in improving the capabilities of their current systems, there is the risk that their share of future work from General Motors will decline. These suppliers represent the greatest amount of risk for General Motors, since their processes are not under control. For both General Motors and its suppliers, the Target for Excellence program represents a major step in gaining greater control over the supply chain and its activities.

Currently, management of the supply chain is limited to one level out (i.e., to the immediate suppliers and the immediate customers). In the future, the management of the supply chain will broaden. The major reason is again that of *interdependence*. It will soon become evident to the supplier, for example, that for the firm to do well, there must be greater control over the activities of the supplier's supplier. As this realization begins to filter through the supply chain, each stage will try to get greater control of its suppliers. This realization will affect the management of both suppliers and buyers.

CIM systems in the future must recognize the importance of the external dimension. They must affect interactions within the firm and, more important, between firms. They must address not only issues of how to improve operations internally but also the question of how to better enhance interorganizational coordination (see Figure 11-3).

An Increased Awareness of the Importance of Integration

A key element in any CIM system is integration. The notion of integration has been at thc heart of such CIM-related developments as MRP II. With MRP II the emphasis has been on the need for integration or cooperation between manufacturing and the other functions within the firm (e.g., accounting, finance, and marketing). In this discussion, manufacturing has been at the center.

In studying the role of integration in CIM, there has been some confusion or disagreement about whether integration is the result of the successful implementation of CIM or whether it is a prerequisite to the implementation. Many managers have implicitly assumed that the presence of the "right" technology will help make integration a reality. For example, the presence of a communications network can help managers in departments that are geographically separated communicate more quickly and efficiently. A company-wide database should help the firm to integrate its various functions. The problem with relying on these computer-based developments is that they do not necessarily affect the *practice* of integration. In most cases all they do is to enhance the *structure* of integration.

The practice of integration refers to the penchant of managers located in different departments to work together as needed to resolve problems (real or potential) of mutual interest. It also refers to a set of views held by people working for the company. Typical of these views are the following:

- It is a rare problem that can be effectively solved by one group or function working in isolation.

FIGURE 11-3
Increasing Awareness of the Supply Chain

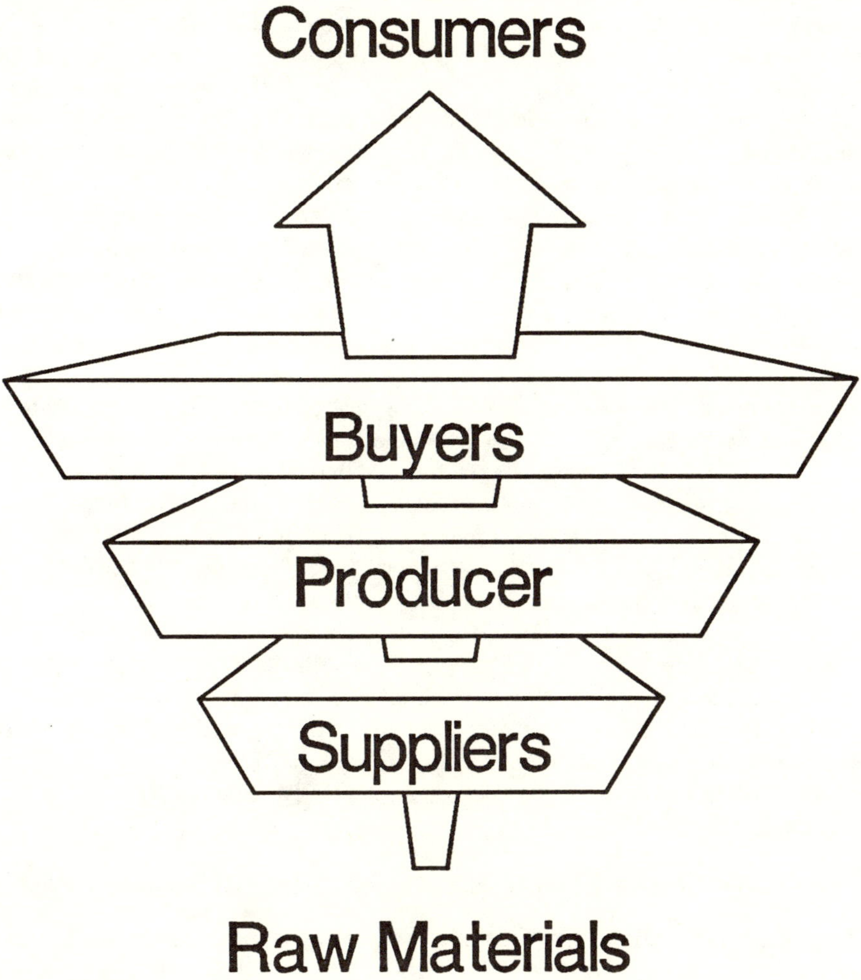

- Most problems, when they appear, have multiple causes (attributable to actions taken by one or more departments) and they affect each department differently.
- When possible, actions should be taken in parallel and not sequentially (i.e., avoid the "over the wall" syndrome).

- Working together with people from other departments on issues of mutual concern is both effective and desirable.

When the practice of integration is present, the barriers imposed on the firm by functional groups tend to be weak or nonexistent. The organization is fluid and flexible. When the practice of integration is present, the solutions that are generated may not necessarily be the best (i.e., optimal). However, the resulting solutions tend to be effective in that they do address the problem. More important, they are also effective because the people involved in the generation of that solution accept the solution and try to make it work.

Building the practice of integration where none was present previously is difficult to accomplish. It requires that people in the firm become accustomed to working together. The structure of integration can help management build this practice. A company-wide communications network can replace the need for face-to-face contact. However, such a network, by itself, cannot cause this practice to appear.

As a result, managers are now realizing that integration is a *prerequisite* to CIM and not a result of CIM. The practice of integration must be present in the manual system before the introduction of CIM can take place.

Firms must emphasize the use of teams and cross-functional groups. There is evidence of this trend in the actions taken by such companies as Procter and Gamble, General Mills, Digital Equipment, Federal Express, and 3M.[8] These are not computer-aided developments but people-oriented developments.

For future CIM, the implications of this trend are important. Future CIM systems will increasingly be built around manually integrated systems. They will be used to enhance the current methods of integration and to facilitate integration. They will support the operation of these integrated systems. In short, in future CIM, islands of automation will no longer be regarded as an indication of partial success with CIM but rather as proof that the current CIM implementation has failed.

Management of the Total Firm, Not Manufacturing and Engineering Alone

At present, CIM efforts have focused primarily on the activities of manufacturing and engineering. However, these are not the only activities that take place in most companies. There are activities such as order entry, shipping, costing, invoicing, and employee training and recruitment. These activities involve such areas as accounting, sales, finance, personnel, quality, and office groups. In most companies these areas make up the indirect cost element of total costs and are frequently assigned to overhead. With direct labor costs now representing

[8]Brian Dumaine, "Who Needs a Boss?" *Fortune*, May 7, 1990, pp. 52–60.

less than 10 percent of total costs, the challenge facing most firms is that of reducing the overhead burden.

To reduce this burden rate, management must improve the overall productivity of these other groups. It must examine the activities of the office to identify weaknesses and areas where automation can be used to enhance output. It must identify and eliminate all forms of waste present in these indirect support activities. Finally, management must also try to close the gap between these indirect activities and the operation of the manufacturing process. For example, consider the interaction between cost accounting and manufacturing. Under ideal conditions, the process that generates the product should also generate the resulting costs incurred in building that product. These two activities must be integrated and not sequential in nature (i.e., the generation of the cost data should not be a separate task that occurs only after the completion of production). Under this ideal scenario, the accounting of cost has essentially been transformed from an indirect to a direct task.

In short, to enhance the overall competitiveness of the firm, we must be prepared to enhance the overall effectiveness of all areas of the firm. Bottlenecks need not occur on the shop floor; they can be found in the front office or in order entry. Such bottlenecks can be as limiting as ones found on the floor.

The Increasing Importance of Process

In the past, we have tended to regard manufacturing firms as suppliers of goods and services. As long as these firms were able to meet their schedules, we had no reason to be interested in the delivery systems that these firms had developed to process and meet the orders issued by their customers. As a result, we treated the firms as "black boxes." This view, however, is now changing.

Managers are now beginning to appreciate the need to make the delivery system highly visible, easily understandable and accepted. The term "delivery system" includes all of the activities that take place in receiving, processing, and completing the customer's order. It includes activities such as order entry, order confirmation, engineering design and review time (if needed), scheduling, manufacturing, shipping and receiving, warehousing, transportation, and purchasing. It is a company-wide system and not simply a manufacturing and engineering system. Furthermore, the delivery system must be integrated and under control. It must be a system that is structured to ensure that schedules are met in a timely and cost-effective manner (and not because of excessive expediting). Both suppliers and customers are now coming to appreciate the importance of a visible and simple delivery system.

For the supplier, a visible and simple delivery system is a highly effective marketing tool. It is a way of demonstrating to a potential customer that the supplier can deliver the goods when and as needed. The supplier has a process in place that makes such deliveries "inevitable." By having the customer accept the validity and integrity of the supplier's delivery system, the customer accepts the supplier as being "willing and able." In this study, we have seen four instances of companies that marketed their processes (and the associated capabilities) to customers. These were Steelcase, Frost, Ingersoll Milling, and Cone Drive.

For the customer, a supplier who has a visible, easily understood, and effective delivery system is to be prized. After all, such a supplier exposes the customer to less risk. Lower risk ultimately manifests itself in the form of lower costs.

As managers accept the importance of process, this acceptance will affect the way in which future CIM systems develop. It will no longer be enough to add technology in hopes that the resulting changes will improve manufacturing's effectiveness. Instead, when introducing computer-aided techniques, managers will try to ensure that these changes enhance the extent to which the process (i.e., the delivery system) is visible, integrated, complete, under control, and easily understood.

THE OBSTACLES TO FUTURE CIM

Working against the preceding shapers are obstacles. These obstacles, as summarized in Figure 11-4, hinder the growth and development of future CIM. The extent to which the firm is able to take advantage of the capabilities and power of future CIM will be dependent on its ability to identify and overcome these obstacles.

Organizational Barriers

In the previous section we argued that, organizationally, firms must become more integrated while retaining a strong element of flexibility. Organizational structures must change when and if required. A company that is organized along functional lines must be able to restructure itself along product lines. Instead of having all of the people who work in the same function together, we now have all of the people who are involved in the delivery of a specific product or part family working together (i.e., product-oriented groups versus functional groupings). In addition, structures must be flexible enough to permit cross-functional integration.

However, if the organizational barriers are too strong and too well entrenched, the changes required by future CIM may not occur. Managers of functional groups may refuse to permit the existence of product-organized groups. They may see these new groups as threats to their power. They may not

FIGURE 11-4
Barriers to Future CIM

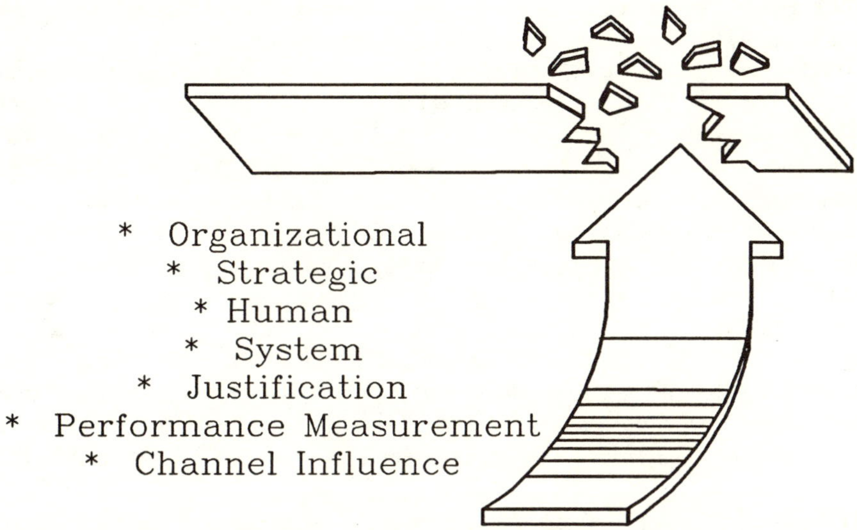

be willing to accept the threats created by change. More important, they may not see or accept the benefits resulting from having organizational flexibility.

Without this flexibility, future CIM systems cannot develop, since the emphasis on process and delivery systems called for by future CIM will be hard to achieve. These firms will continue to be functional (i.e., manufacturing and engineering) endeavors rather than corporate systems.

Strategic Barriers

As was noted in Chapter 4, CIM and corporate strategy are tightly bound. For CIM to be used effectively, there must first be a corporate strategy that is designed to generate a *sustainable* competitive advantage for the firm. The CIM system that is implemented is then an expression of this corporate strategy. Its structure, orientation, and objectives are derived from the corporate strategy. CIM becomes a means by which strategic goals are achieved.

However, there are two instances in which strategy can act as a barrier to the development of CIM. The first is one in which the firm lacks a well-developed, comprehensive, complete, and feasible corporate strategy. In these cases top management may not know what it is that they wish to do with the firm.

They may have no clear idea of the role to be played by CIM (except that they know that they want to have a CIM system). They may see CIM as the end rather than as a means to an end. The result is a poorly focused and ineffective CIM system. Often, it cannot grow beyond islands of automation because no one has a clear idea of how to use CIM as a corporate system.

The second instance is one in which corporate strategy is inflexible. The linkage between CIM and corporate strategy is not simply one way. It is not enough to realize that strategy drives CIM. Managers must also recognize that CIM can influence corporate strategy (see Figure 11-5). This second linkage from the CIM system to strategy is often a result of experience with CIM.

The implementation and use of CIM is very much a learning experience. It is somewhat different for every firm. Because CIM affects the operation of the entire corporate system, it is often difficult to generalize one firm's experiences with CIM. As each firm begins to implement CIM, its people undertake a process of discovering the capabilities present in the new system. Many of the capabilities that are uncovered are unanticipated. They reflect the interaction between the users and the system (i.e., the users discovering new and better ways of using the system). Information about these new capabilities must be fed back to management so that it can assess the strategic implications of these capabilities. Corporate strategy must be changed, if necessary, to take advantage of these newfound strengths. If strategic objectives cannot or will not be

FIGURE 11-5
The Strategy–CIM Interplay

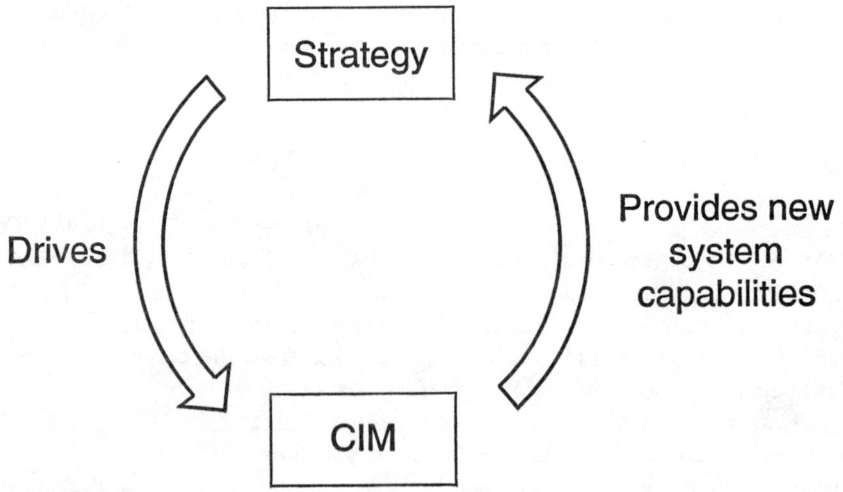

changed, the potential offered by these new capabilities will go largely untapped.

Human Barriers

If CIM is to succeed and to evolve over time, it must be accepted and supported by the people who will use it. This is not an easy condition to satisfy, since for most people, CIM represents change. Change is often seen as a threat (often with good reason).

First, there is the uncertainty associated with the changes introduced by CIM. People may not know how CIM will affect the firm, themselves, and their jobs. Second, CIM will change job requirements and skills. In many cases it will require new skills and make several of the current skills obsolete or unnecessary. It will often change the very nature of the jobs and tasks being carried out by people. In some cases CIM, by changing skills and information requirements, may affect the power structure of the firm. People who were experts under the old system may find themselves no longer experts under the new system. The prestige and position that these people enjoyed in the old system is no longer present in the CIM system.

Sometimes people may object to CIM because they associate it with a reduction in the firm's work force. This situation is encountered when the demand for the firm's product is stable. The implementation of CIM will make each employee more productive. However, since the demand is not changing, the firm now has too many workers. Reduction through attrition may not be enough. As a result, management must reduce its work force through dismissals or long-term layoffs.

If people see CIM as a threat, they will not accept or support it. More important, they may work to ensure that the system fails. As a result, CIM is doomed. It will not develop and evolve. Its impact on the firm will be limited at best.

System Barriers

CIM should be regarded as a magnifying glass in that it makes good systems better and bad systems worse. CIM is not a replacement for an ineffective, incomplete, or inefficient formal manufacturing planning and control system. Because of this fact, firms must have effective manual systems in place prior to implementing CIM.

These manual systems must be able to carry out the basic tasks of manufacturing. That is, they must be able to

- Generate feasible and valid master production schedules.
- Ensure that adequate levels of resources (i.e., workers, machines, material, and tooling) are present for given production schedules.

- Ensure consistency between corporate objectives and actions taken on the shop floor.
- Maintain good control over orders once they are released to the shop floor.
- Ensure that information generated on the shop floor is fed back to the planning system in a timely and effective manner.

Without these capabilities the CIM system is primarily an exercise in automation. It is not an effort designed to enhance the formal system through the application of appropriate computer-aided technologies. After all, there is no formal manual system to enhance.

Evaluation and Justification Barriers

Traditionally, CIM systems have suffered from problems with cost justification. It has always been difficult to justify the introduction of CIM systems using such traditional investment justification procedures as discounted cash flows (DCF) or payback. Much of these difficulties are related to the differences between the CIM investments and the assumptions that underlie the traditional justification procedures.

Implementing a CIM system is a major undertaking for any firm. It requires a major investment of time, money, equipment, and manpower. Much of this investment has to be made "upfront" with the result that, initially, CIM results in a cash outflow. It isoonly later that there is an offsetting resource inflow. The benefits, when they occur, are a mixture of quantitative and qualitative elements. Reductions in cost (quantitative) are often accompanied by improvements in lead times, quality, shop floor congestion, and performance predictability—improvements that are qualitative. The benefits are also company-wide. One department may make the necessary investments, yet the entire firm often benefits. For example, consider the implementation of a coding and classification scheme. Such a scheme helps users better manage parts within the system. It enables engineers to simplify the design of new parts by drawing on information pertaining to past similar designs. Purchasing can enhance its effectiveness by using the scheme to identify similar parts. Costs can be reduced by placing orders for such similar parts with one vendor. Because the parts are similar, the vendor should be able to take advantage of the similarities to reduce cost. It is important to note that if engineering makes the necessary investment in this scheme, it does not gain all of the advantages.

In addition, CIM implementations are long-term undertakings, often requiring in excess of five years. It is difficult to generalize experiences with CIM systems because each firm is different. Finally, it is difficult to identify at the outset all of the benefits to be gained from the implementation of a CIM system.

Benefits are often uncovered as users learn more and more about the new system, its operation, and its capabilities.

These traits adversely affect the ease with which investments in CIM can be justified using more traditional procedures. Justification techniques such as the discounted cash flow are heavily oriented toward the use of quantitative costs. They have difficulty in incorporating qualitative benefits. They are also short-term in orientation. Projects that can generate benefits early on are rated more highly when compared with projects with benefits occurring late in the implementation process. Finally, these techniques ignore *externalities*. It is assumed that the department or area that makes the investment stands to gain most of the benefits.

Unless more appropriate justification procedures are developed, the process of justifying the implementation of a CIM system will seldom support the decision to proceed (even though it may be strategically in the company's best interest).

Justification is a critical test that must be applied whenever the decision to implement CIM is being considered. However, it should be viewed as a test of logic rather than numbers. As evident from the cases contained in this book, every firm has justified the implementation of a CIM system. However, the justification in many cases was a mixture of numbers and *gut feel*. It was also a logical test, one in which management tried to answer the question, Does it make sense?

Performance Measurement Barriers

Performance measurement plays a very important role in how CIM systems are used and developed. People respond to the rewards and punishments present in the performance measurement system. They change their behavior in response to the stimuli present in the performance measurement system. How they view the new CIM system and its intended uses is often influenced by the performance measures used in evaluating their performance. As David Packard, one of the two founders of Hewlett-Packard, once observed:

Tell me how a person's measured and I'll tell you how he behaves.

For example, if the performance measurement system focuses on cost control, people will use the CIM system to control and reduce cost. The behavior encouraged by the performance measurement system will always win out over the behavior desired by management (especially if the two types of behavior are not consistent). If the performance measurement system focuses on efficiency and the management wants quality, the users will give management improved efficiency.

There must be consistency between management's objectives for the new CIM system and the performance measurement system in place. More important, the performance measures that are present must reinforce management's goals for CIM. With such consistency, the users tend to develop CIM in directions desired by the firm. Without such consistency, the performance measurement system becomes an effective and ever present barrier to the use and future development of CIM.

Channel Influence Barriers

As firms become increasingly aware of the importance of the total supply channel and of their dependence on the acceptable performance of every link in this chain, they must address the question of whether they can really assess performance and influence behavior within this chain. At issue is the relative influence of buyers and suppliers in this chain. Relative influence identifies the extent to which a specific party (supplier or buyer) within the supply chain can affect the actions of the other involved groups. One party's relative influence is often based on *volume*, where volume consists of the *quantity* or size of orders placed and the *frequency* or the extent to which business is transacted between a buyer and a supplier during a given period of time (e.g., a year).

One party may try to change the practices of the other group for a number of reasons. A supplier may request that the buyer present information in a certain format so as to simplify order entry and processing and order scheduling. A buyer may want the supplier to use bar codes on all shipments made to the buyer's plant. The buyer may also want the supplier to ensure that all parts in the shipment be oriented in a particular fashion so that the parts can be used immediately in production. In each case one of the parties is being asked to change one or more aspects of the way that they do business.

A supplier can successfully resist these requests in those cases in which relative influence of the buyer is limited. The buyer's relative influence can be limited for several reasons: the buyer accounts for a small percentage of the supplier's total sales; sales to the buyer generate little profit; the costs of accommodating the requested changes may be seen as outweighing any potential benefits; the supplier sees few, if any, gains in the changes requested by the buyer; sales to the buyer are seen as highly unpredictable (i.e., the time between buys is too long or variable). Similar reasons can be identified that can reduce the relative influence of the supplier.

When the relative influence of either the buyer or supplier is limited, CIM tends to evolve into an internal development. It is something that takes place within the firm. Its growth becomes a function of those activities that management within the firm can affect and change. In those situations in which no buyer or supplier can exercise a great deal of relative influence, the implementation of CIM in the supply chain can be described as a series of CIM systems that are poorly integrated with each other.

THE TRAITS OF FUTURE CIM

Future CIM, or the face of CIM systems in the future, will ultimately be a product of the manner in which firms and management respond to the two sets of factors described in the preceding sections. The shapers encourage changes in the future form and structure of CIM, while the obstacles act to retard change taking place in CIM. Given that the relative power of these shapers and obstacles varies by firm, future CIM will tend to be different from firm to firm. However, there are certain general traits that can be expected to be present in any future CIM system (irrespective of the firm-specific settings). The following is a list (summarized in Figure 11-6) of the most important of these future CIM traits.

FIGURE 11-6
Future CIM

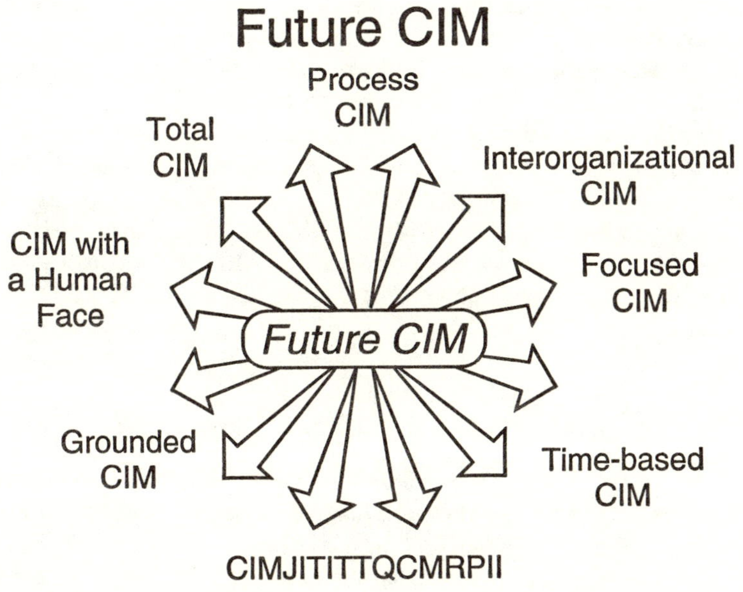

CIM with a Human Face

The hallmark of the future CIM system will no longer be its emphasis and reliance on computer-aided technology. Instead, future CIM will become a system that supports the user and that enhances his or her effectiveness through the use of appropriate technology. This marks a change in perspective from CIM as a technical system to CIM as a human system. Under this perspective, the role of CIM is not to replace the human but to complement the user.

This change in perspective is based on a greater appreciation of the role played by the user in the operation of the firm. The user is no longer being viewed as a bottleneck and as a source of inefficiency. Instead, we now see the user as someone who brings certain critical skills to the system. People are good at identifying problems. They are well suited to act as problem solvers. Because of their knowledge of the system and its operation, they are the source of major improvements. In short, they generate efficiency.

The task for future CIM is that of supporting the activities of the user. This task will be achieved in a number of different ways. First, the computer-aided technologies present in CIM can be used to take over those activities for which people are poorly or ill suited. Computers can be used to do bookkeeping, data recording and tracking, and data summarization. They can also be used to do work that is repetitive, boring, physically demanding, or potentially dangerous.

Second, computer systems can be used to replace people in tasks that are routine and not very demanding. It is here that artificial intelligence and expert systems are particularly appropriate. By taking over these tasks, the users are freed to concentrate their attention on more demanding or less routine problems.

Third, computer systems can be used to enhance decision-making by the user. They provide the user with ready access to that information needed for decision-making. They can analyze any data pertaining to a problem and summarize it graphically. They allow "what if" analysis.

This change in CIM, a change from CIM with a machine visage to CIM with a human face, has been best summarized by ABB of Sweden. The management of ABB no longer talks about CIM. Instead, the term used is *MIC* or "Man In the Center " (see Figure 11-7).

Grounded CIM

In the past, many CIM implementations failed because management suffered from the "Grapefruit 45" syndrome. This syndrome gets its name from a diet product that was heavily advertised on television in the late1980s. The advertisements for this product implied that the user did not have to reduce calorie consumption or exercise to lose weight. Instead, all that was required was

FIGURE 11-7
The Change in CIM

simply taking two of the diet tablets a day. The problem with this product was that it implied that the user did not have to correct poor eating or exercise habits (the true cause of weight gain for many people). These problems could be corrected by the use of the "right" pill. Ultimately, the Food and Drug Administration (FDA) succeeded in getting this product pulled from television.

Many managers view the technology component of CIM in very much the same manner as dieters saw the diet product. The right technology, if selected and applied appropriately, could correct any problems created by poor data, ineffective planning, or other similar problems. What these managers have learned, after the fact, is that "good" technology can never replace bad systems. Rather, technology is more like a *magnifying glass* in that it makes bad systems worse and good systems better.

To gain the maximum advantage from CIM, managers must first build effective and efficient manual systems. They must make the current systems better. They must ensure that their databases are complete, current, and accurate. They must also be capable of *DTDTR* (see Figure 11-8). Their planning systems must generate feasible and valid plans; they must ensure that shop floor operations are tightly controlled and monitored.

It is not enough to build effective manual systems. These systems must be simplified. Unnecessary or excessive steps must be identified and removed. Where possible, operations should be combined. By simplifying the manual system, managers assure themselves of a system that can be computerized relatively inexpensively and quickly. They also guarantee themselves a system that is easy to understand and to use. Such systems do not intimidate people.

FIGURE 11-8
The Key To Manufacturing Success

DTDTR*

(* - Doing The Dumb Things Right)

In the future, managers will view these enhancements and improvements to the manual system as being necessary first steps in the CIM implementation process. As a result, the resulting CIM systems can be viewed as being *grounded* in that they are based on simple, effective, and efficient manual systems.

CIMJITITTQCMRPII

In the future, we can expect to see the distinctions between such systems as JIT and CIM blur. As implementers of CIM become familiar with techniques used in JIT, they will borrow those JIT procedures that are attractive (i.e., that work and are relatively cost-effective). As a result, future CIM systems will incorporate elements of JIT, IT, TQC, and MRP II. As a result, it will become increasingly more difficult to differentiate, for example, between a JIT and a CIM system. For many managers, the real issue is that of implementing systems that work and that make sense. There is no real gain to be had from being dogmatic about whether the firm has a pure JIT or CIM system.

Time-Based, Risk-Reduced CIM

As firms become aware of the need to compete on speed and risk, these considerations will become very evident in the structure and orientation of future CIM systems. Managers will try to incorporate into their CIM systems features and systems that can help the firm reduce lead time (either in terms of time from design to market or time from order to delivery). Furthermore, managers will try to ensure that the resulting CIM system reduces the customer's exposure to risk. This is an important consideration when dealing with industrial customers. A late delivery or a shipment of defective parts creates critical problems for the buyer. In some cases, it may succeed in shutting down production. As a result, the CIM system that is implemented must help the manager be able to identify and eliminate problems within his or her system. It should help enhance effective control over the system.

In short, the benefits of speed and reduced risk are too real to be ignored. They must become incorporated into the resulting form of future CIM.

Focused CIM

CIM systems are not homogeneous. There is no such thing as a "plain vanilla" or generic CIM system. Instead, CIM systems can and do take on a number of different forms. This fact, evident in the case studies presented in this book, will become an accepted trait of all future CIM systems. There are two reasons for this diversity. First, the environment in which each firm operates is somewhat different. Second, how each firm tries to establish a sustainable competitive advantage in its market also differs. As a result, these differences will be reflected in the shape and structure of the resulting CIM systems.

The challenge for managers is to develop that type of CIM system most appropriate for their market and their strategic objectives. When such systems are identified and implemented, they can be described as being focused. The CIM system that is implemented by a firm competing in a mature market involving a well-defined and commodity-like product will be very different from the CIM system that is imposed by a firm that must deliver a highly engineered product in which the requirements are not completely known at the outset.

Increasingly, future CIM systems will bear the hallmark of focus. Such systems will always outperform any CIM system that is either generic in nature or unfocused. With focused CIM systems, there is direct linkage between corporate strategy, the market, and the resulting CIM system.

Interorganizational CIM

Today's CIM system is focused on improving operations within the firm. It often ignores the linkages between the firm and its suppliers as well as its customers. Future CIM will explicitly recognize the importance of the supply chain. It will attempt not simply to coordinate activities within the firm; it will also try to coordinate activities across the chain. It will span corporate boundaries.

With the spread of EDI and the increased use of integrated engineering design systems, customers will download requirements and specifications electronically to suppliers. Suppliers will generate the necessary designs and upload the designs to their customers using one of the available graphics standards.

Suppliers and customers can now access each other's systems and obtain information in real time. More important, suppliers and customers are now linked in real time. This linkage of systems is facilitated by the CIM systems. In some cases the two systems are electronically linked.

The advantages presented by such linked CIM systems are very attractive. Communication time and product development lead time are reduced. Friction and potential problems between suppliers and customers are avoided. Better designs are generated much more quickly and at lower costs. Suppliers and customers will regard each other as partners rather than as adversaries. There will be less confusion present in the supply chain.

George J. Hess of Ingersoll Milling has developed a term to describe this situation. He refers to it as Interorganizational CIM. In the future, more and more firms will extend their CIM systems to include either their suppliers or their customers (or both).

Process CIM

With more and more firms selling process (i.e., the capabilities of their delivery system) rather than simply product, it can be expected that future CIM systems will become more process oriented. That is, the changes that accompany the implementation of these CIM systems will be intended to enhance the overall process by making it easier to use, faster, clearer (more visible to the user), and more under control. Future CIM will support management's attempts to sell process to the firm's customer.

Total CIM

In the future, for the firm to succeed, it will not be enough for manufacturing and engineering to be effective and efficient. Rather, the firm as a whole must become more effective and efficient. Every aspect of the firm must be better managed. The front office with its secretaries must be as efficient in the execution of the tasks assigned to it as are the operators on the shop floor. This trait will become very apparent in future CIM systems.

Future CIM systems will apply computer-augmented systems to every aspect of the firm. Future CIM will not only affect the operation of such direct activities as manufacturing, engineering, and quality assurance, but it will also influence the operation of such support activities as accounting and finance. Increasingly, these support activities will become more closely linked to direct activities in both time and place. For example, as soon as a part is produced at a given workstation at a given point in time, the accounting system will be able to identify the actual costs associated with that part. Finance will be able to determine the impact of that item's production on the firm's financial position.

Future CIM will also affect in a similar fashion the operation of other support areas such as customer service, the front office, order processing, field service, and personnel. Each area affects the manner in which the firm serves its customers. The operation of each area can be improved through the capabilities offered by the computer technology present in CIM.

With the move of CIM from a manufacturing and engineering system to a corporate-wide system, it may no longer be appropriate to describe future CIM as *computer integrated manufacturing*. Rather, future CIM may be better described as the *computer integrated business* (CIB) or the *computer integrated enterprise* (CIE). These terms, which are currently being used by some to describe CIM, better represent the broader perspective of this new corporate-wide CIM system.

THE FALLACY OF THE "LIGHTS OUT" FACTORY

It is 10:30 at night and production on the shop floor is in full swing. The sound and noise of production is in the air. Orders are being released to the shop floor; carts are routed to the appropriate workstations; parts are being loaded, processed, checked, and unloaded. Problems such as machine breakdowns or scheduling conflicts are addressed as soon as they occur. Carts are moving between workstations. As soon as a job is completed, information about that job is summarized, recorded, and transmitted to the company's database. Everywhere, there is activity. Everywhere, work is being started and completed.

This would seem to be a typical, well-run shop—typical except for two important traits. The first is that the shop is cloaked in almost complete darkness. Except for the occasional overhead light or the flashing of a red warning light, the floor is dark. Second, except for a handful of people in the control room, there is no one on the shop floor. All of the work for that night is being managed without the interference of any human. Instead, it is computer-aided automation in the form of expert systems and artificial intelligence systems, automated storage and retrieval systems, NC machines, robots, and automated tool changers that is managing the transformation of work from start to finish. What we have described is the *"lights out" factory*.

For many managers and researchers, this factory represents the ultimate goal of any computer integrated manufacturing system. It is a factory in which automation, not the human, is the key to success. It is also a factory in which the variability and unpredictability of the human has been replaced by the certainty and regularity of technology. After all, machines do not get tired; machines do not get thirsty; machines do not have hangovers or marital problems; machines do not strike; machines do not require double time for working holidays or overtime. It is a system that is under total control.

However, the lights out factory does not represent the utopia of CIM. It represents a fundamentally flawed view of CIM. Why? The goal of a lights out factory suffers from the following shortcomings:

• *The lack of a linkage with corporate strategy.* In most discussions of the lights out factory, one critical element is often overlooked or ignored. This is the linkage between the new factory and how the firm competes in the marketplace. The automation of a factory, while exciting as a technological advance, does not necessarily help the firm compete more effectively in the market. It may not help the firm gain or maintain market share. It may not create more value for the customer.

The goal of all effective CIM systems is the creation and maintenance of a *sustainable* competitive advantage. It is not the implementation of technology or automation.

• *The view of the human as the weak link in manufacturing.* Implicit in the lights out factory is an important assumption regarding the human in the system. The human is viewed and treated as a bottleneck and a source of uncertainty. People are treated as if they have a neutral impact on the system (under the best of condition) or, at worst, an adverse effect on operations. There are several major shortcomings in this view.

First, this view ignores the advantages that people bring to any manufacturing system. As was pointed out previously, people can learn about systems. They have the ability to offer new ways of improving operations on the shop floor. As was noted in one study of leading-edge shop floor control systems,[9] people are the major source of efficiency in the manufacturing system.

Second, the savings generated by eliminating people from the shop floor are not that great. In general, labor costs now account for less than 10 percent of total costs. To attack this 10 percent, the lights out factory now introduces an increased level of overhead costs. It is also ignoring the 90 percent of costs incurred elsewhere in the firm. In short, attacking the human component does not result in the "biggest bang for the buck."

Finally, there may not be a direct link between reducing the human element in the manufacturing system and improved competitiveness. The task for most CIM systems, is not one of labor replacement but rather one of resource enhancement. That is, the challenge is to make existing employees more productive by eliminating those activities to which people are ill suited (e.g., bookkeeping) and by supporting those activities for which people are well suited (e.g., problem-solving).

---

[9]Steven A. Melnyk and Phillip L. Carter, *Shop Floor Control: Principles, Practices and Case Studies* (Falls Church, Va.: American Production and Inventory Control Society, 1987).

• *CIM as a shop floor development.* The lights out factory deals primarily with the manufacturing and engineering aspects of manufacturing. It also focuses attention on the shop floor. The problem is that this perspective is too narrow. It ignores the other elements of the firm that can affect operations. These include the office, marketing, accounting, finance, quality, and strategic planning. These elements can and do have a significant impact on corporate actions and performance.

• *Failure to address the major operating cost elements.* As was pointed out previously, labor is no longer the major component of the cost of goods sold in most manufacturing plants. Instead, the major cost component now comes from overhead costs. Overhead costs are the result of such factors as investments in automation and the presence of support activities such as cost accounting, finance, and the front office. The lights out factory does not address the question of how to reduce these sources of overhead. It does not deal with the problem of how to make the front office more productive or of how to make cost accounting an integral element of the manufacturing process. Instead, by emphasizing automation, the lights out factory increases rather than decreases the overhead burden rate.

• *A myopic view of CIM.* Finally, the lights out factory focuses attention on the activities of the firm in isolation. Ignored are the linkages between the firm and its suppliers, on the one hand, and between the firm and its customers, on the other. As a result, the impact of the lights out factory on the supply chain is minimal. Yet as was shown elsewhere in this chapter, this chain is becoming increasingly more important.

At the heart of the lights out factory is automation; at the heart of computer integrated manufacturing is the intelligent marriage of corporate strategy, organizational structure, and computer-aided technology. For the lights out factory, the goal is automation; for CIM, the goal is corporate survival and the development of a sustainable competitive advantage. As a result, the pursuit of the lights out factory will not result in the achievement of CIM. In short, the lights out factory is nothing more than the false prophet of CIM.

The fallacy of the lights out factory was best summarized by James T. Clark of IBM, who once commented that:

CIM is not robots dancing in the dark.

PROMISING DIRECTIONS FOR FUTURE CIM RESEARCH

In many ways this book has summarized the results of a limited study of computer integrated manufacturing. For various reasons the authors have decided to confine their attention to the study of CIM as it is being practiced in one type of specific, well-defined, and limited environment. The environment selected for this study was that of discrete batch manufacturing firms located in the continental United States.

By limiting the scope of the study, the result was a manageable project that could be successfully executed by the authors. However, the resulting study was and is not all-inclusive. CIM occurs in a wide range of different settings. By studying these other settings, we can obtain a better and more complete idea of how CIM is used within firms and how it affects the resulting operation of companies.

There are certain settings that offer researchers great potential for investigation. These include the following:

CIM in the Process Industry. Manufacturing takes place in basically two types of environments. The first involves the processing of items in discrete units. The output of this process is naturally divisible. Examples of such output include cars, pills, chairs, bearings, and beverage containers.

The second type of process involves items that are not naturally divisible. Examples of such output include oil, paint, and fruit juice. The output occurs in a stream that is often broken up into discrete batches (e.g., cans of orange juice) when the output is packaged. This type of manufacturing environment is commonly referred to as process.

It is recognized that each type of manufacturing environment is managed differently. Practices that work well in one environment would not necessarily work well in the other. As a result, a study is needed that examines the implementation and use of CIM within the process industry. This study would examine how CIM is modified to better accommodate the unique requirements and needs of firms located in this industry.

A Comparison of CIM in Small and Large Companies. This study brought together both large and small firms. At one extreme, we examined Steelcase of Grand Rapids. This is a firm with over 8,000 employees in the Grand Rapids area alone. At the other extreme, we studied Frost of Grand Rapids. Here was a firm that had about 150 employees. This study did not attempt to treat large and small firms as separate populations. Instead, they were aggregated into one population.

However, in the course of the study it became evident to the authors that the CIM experience in small firms is different from that encountered in large firms. Some of the differences are obvious. Small firms do not have the resources found in large firms. Often, the risks of failure and the benefits of success are much greater. Once committed to CIM, the process of implementation is often very quick. In contrast, large firms have to contend with organizational inertia.

Because of their size, it is often more difficult to start the process of change. By their very size, the educational and training process is far more complex and demanding. The obstacles to CIM are also more prominent (and often more difficult to overcome).

As a result, what is needed is a comparative study of CIM in both large and small firms. Such a study would be a three-stage undertaking. In the first stage a sample of small firms that have successfully implemented CIM would be studied. In the second stage the process would be repeated except that the target of the study would be large firms. The first and second stages could easily be made into parallel activities. Finally, after these two stages have been completed, the results would be compared and differences and similarities identified and examined.

The Study of Interorganizational CIM in Practice. CIM is no longer something that takes place within the firm. With the advent of interorganizational CIM we see CIM crossing corporate boundaries. It has become a means of linking suppliers and buyers and for coordinating the flow of information. Ingersoll Milling provides us with one instance of a firm actively engaged in the use and development of interorganizational CIM.

Since interorganizational CIM is a relatively new development, it is still relatively unknown. There are a number of questions about it for which we currently do not have any answers:

- Under what conditions is interorganizational CIM an appropriate and attractive option for a firm to pursue?
- What are the benefits and costs associated with interorganizational CIM?
- What are the prerequisites to interorganizational CIM?
- What are the major steps of implementing interorganizational CIM?
- How does the presence of interorganizational CIM affect the activities of the supplier? Of the producer? Of the customer?

To answer these and other questions, a study of the practice and principles of interorganizational CIM is needed. This study should study interorganizational CIM as it is practiced domestically (i.e., American suppliers working with American producers who satisfy the needs of American consumers). If possible, this study should also try to examine cases in which interorganizational CIM is applied internationally as a means of coordinating the activities of either non-American suppliers or consumers with those of American firms.

A Comparison of CIM Systems from Around the World. In this study we focused our attention on the experiences of American CIM users. CIM is not just an American system. It is being used and implemented in other countries. There are good examples of CIM systems present in countries such as Sweden, West Germany, Great Britain, France, Japan, and Canada. It would be useful for both researchers and managers to have access to a comparative study that examines CIM systems in selected countries. This study would try to identify and explain major similarities and differences in the CIM systems found in different countries. It would also try to determine how the managers of these various systems see the future of CIM (and why).

CIM, as a concept, has been with us since the early 1970s. In spite of all the research that has been done to this point, we have not addressed all of the questions associated with the theory and practice of CIM. As soon as we succeed in addressing one set of questions, we are faced by another set. As is evident from the brief discussion found in this section, the study of CIM is still very fruitful. Our knowledge of CIM is fragmentary. Only with more research can we develop a better understanding of CIM.

CONCLUDING COMMENTS

In this the final chapter of this book, we tried to look into the future and to predict the form of CIM in the future. Why? The reason is found in the very nature of CIM. CIM is dynamic. It is ever changing. The manner in which it changes is a result of the interaction of shapers and barriers. Shapers promote the evolution of CIM; barriers retard its development. Managers and researchers must try to understand these forces, since the CIM system that is begun today must be ready to function in tomorrow's, not today's, environment.

With this chapter, we have completed our study of the principles and practices of CIM at the leading edge. In examining CIM at the leading edge, we have discovered the following themes.

- CIM can and does work.
- CIM is most effective when it is linked to and driven by corporate strategy. In CIM we see a very concrete expression of how the firm intends to compete in the marketplace.
- CIM is heterogeneous.
- CIM is one but not the only approach available for a company interested in developing a sustainable competitive advantage.
- CIM is not for everyone.
- At the center of CIM are the elements of people, strategy, and integration. Technology, while a critical component, is not the major key to success with CIM.

- CIM cannot and will not work in a manufacturing vacuum. For CIM to succeed, the manual system that precedes it must first operate.
- CIM is dynamic. It is now beginning to expand beyond the boundaries of the firm and to link together suppliers and buyers. It is also moving beyond the confines of manufacturing and engineering to the firm as a whole. Its influences can now be seen in such areas as the front office.
- The ultimate goal of CIM is not CIM but the presence of a sustainable competitive advantage in the marketplace.

CIM is a viable weapon in a manufacturing firm's arsenal. If it is to be used most effectively, it must be understood completely. This study has tried to help both managers and researchers develop a better understanding of this system known as computer integrated manufacturing.

APPENDIX 1

THE 1987 CIM CONFERENCE

THE MEMBERS OF THE CIM STEERING COMMITTEE

To a large extent, this study greatly benefited from the insights and contributions made by the members of the CIM Steering Committee. The steering committee was a group consisting of six executives from the business community and two academic representatives. The six executives were selected for their knowledge of computer integrated manufacturing, both in theory and practice. Each of the executives can be viewed as an expert in the field of CIM.

The corporate members of the steering committee played an important role in guiding and shaping this CIM study. They were constantly consulted regarding the structure and format of the CIM conference. They were solicited for information about firms to approach for the conference. At times they were asked to help in securing the cooperation of certain firms. They were always willing to share their experiences with the authors of this study. They were also ready to act as teachers. They taught the authors a great deal about what CIM is and what it is not. They saw CIM as a working system, not as a theoretical entity—a perspective very evident in this study. Their insights, comments, suggestions, knowledge, and industrial perspectives were greatly appreciated. To them goes a good deal of the credit tor this project.

In this appendix we recognize the members of the steering committee and publicly express our gratitude to them.

CIM Steering Committee

Steering Committee Members	*Supporting Corporations*
Franklin Becker (Corporate)	John Deere and Company Moline, Illinois
George J. Hess (Corporate)	Ingersoll Milling Machine Company Rockford, Illinois
Les Kelly (Corporate)	Parke-Davis Rochester, Michigan
Steven A. Melnyk (Academic)	Michigan State University East Lansing, Michigan

Ram Narasimhan (Academic)	Michigan State University East Lansing, Michigan
Frank E. Plonka, Ph.D. (Corporate)	Chrysler Motors Corporation Detroit, Michigan
Richard T. Smith (Corporate)	Ingersoll Engineers, Incorporated Southfield, Michigan
Sam Tomas (Corporate)	Motorola Corporation Scottsdale, Arizona

CIM CONFERENCE PARTICIPANTS

In addition to the members of the steering committee, the writers also wish to publicly acknowledge the contributions of the seven companies that participated in the CIM conference. This conference was held at Michigan State University in the fall of 1987. Without their insights, inputs, and willingness to discuss their experiences with CIM, this book would not have been possible. We thank them for their involvement.

CIM Conference Participants

Company	*Company Representative(s)*
Allen-Bradley Co. 747 Alpha Drive Highland Heights, OH 44143	Ed Braxton
Cone Drive P.O. Box 272 Traverse City, MI 49684	Paul Brauninger
Continental Can Co. 1700 Harvester Road West Chicago, IL 60184	David C. Grigsby
Digital Equipment Co. 2500 West Union Hills Drive Phoenix, AZ 85027	David Richards
Frost, Inc. Represented by Amprotech Inc. 660 Cascade West Parkway Grand Rapids, MI 49506	Robert McIntyre

Ingersoll Milling Represented by Ingersoll Engineers First Center Office Plaza Suite 550 26913 Northwestern Highway Southfield, MI 48034	Richard T. Smith
Steelcase P.O. Box 167 Grand Rapids, MI 49501	Jim Allan Cecil Bradshaw

APPENDIX 2

GLOSSARY

The following terms are important in developing a better understanding of the principles and practices of computer integrated manufacturing. These terms have been used previously in this study. They are collected and defined in this section for clarification.

ADS Automated design system. See *computer-aided design.*

AGVS See *automated guided vehicle system.*

AI See *artificial intelligence.*

AMH See *automated materials handling.*

artificial intelligence The study of the application of various computer-based procedures and techniques with the ultimate aim of enabling computers to emulate and eventually improve on the reasoning abilities of humans.

AS/RS See *automated storage and retrieval system.*

assemble-to-order A manufacturing scheduling practice whereby the manufacturing planning system builds inventories of high-level components and assemblies (i.e., levels 1 and 2). No end item orders are built in advance because of uncertainty regarding the final configuration of the product (most often a result of the presence of options). Instead, the process of satisfying is managed through the final assembly schedule. That is, the order is placed by the customer. Provided that there is sufficient capacity available and the materials needed are present, the final product is then built to order. This practice is commonly referred to by its abbreviation, ATO.

ATO See *assemble-to-order.*

automated design system See *computer-aided design.*

automated guided vehicle system Vehicles equipped with automatic guidance equipment that follow a prescribed guide path that interfaces with workstations for automatic or manual loading and unloading. In an FMS environment, the guided vehicles generally operate under computer control.

automated handling system System(s) used to automatically move and store parts and raw materials throughout the manufacturing process and to integrate the flow of workpieces and tools with the manufacturing processes. In an FMS environment, the *automated materials handling system* operates under computer control.

automated identification A broad term used to describe any information-gathering procedure in which information is encoded in a form that can be transmitted and recorded using an electronic device. Examples of automated identification include bar coding, optical character recognition, magnetic strips, and voice recording.

automated inspection and testing The automation of one or more steps in the inspection and testing procedure developed at a company.

automated materials handling The use of computers to direct the movement of raw materials, work-in-process, or finished-good inventories. These systems are often integrated with *automated identification* systems to facilitate movement and handling.

automated storage and retrieval system A high-density rack storage system with rail-running vehicles servicing the rack structure for automatic loading and unloading. Vehicles interface with AGVS, car-on-track, towing, or other conveyor systems for automatic storage and retrieval of loads. In many manufacturing environments, the automated storage and retrieval system operates under computer control. The storage and location information generated by the automated storage and retrieval system is used by the computer system to track, monitor, and control material.

automation 1. The technology concerned with the application of complex mechanical, electrical, electronic, and computer-based systems to the operation and control of production. This technology includes automatic machine tools for processing parts, *automatic materials handling systems*, automatic assembly machines, continuous flow processes, feedback control systems, computer process control systems, and computerized systems for data collection, planning, and decision-making to support manufacturing activities (Groover, 1980, pp. 3–4). 2. Semiautomatic or automatic material handling, automatic loaders and unloaders, and other labor saving devices (TMEH, 1987, p. 8). 3. The application of machinery and equipment to perform and control automatically and continuously all operations in a manufacturing plant (TMEH, 1987, p. 8). 4. The theory, art, or technique of making a process automatic, self-moving, or self-controlling.

automation, hard The use of specialized machines and machine lines to manufacture and assemble items. Typically hard automation involves dedicated equipment in that each machine or line can only do one function or task or a very limited range of functions and tasks. Hard automation is typically most appropriate for the high-volume manufacture of mature products.

automation, soft The use of flexible, reprogrammable machines, equipment, or *work centers* to manufacture and assemble items. Integral to the concept of soft automation is not only the use of equipment but also the structure, management, and use of information. In contrast to *hard automation*, soft automation involves equipment that can be reprogrammed quickly to do a variety of tasks or functions. Soft automation is typically most appropriate for the low- to medium-volume manufacture of products. It is also appropriate for products in which the design is still fluid and subject to change.

bar code An array of rectangular bars and spaces arranged in a predetermined pattern to represent characters for machine reading.

CAD See *computer-aided design*

CADD See *computer-aided drafting and design.*

CAE See *computer-aided engineering.*

CAIT See *computer-aided inspection and testing.*

CAM See *computer-aided manufacturing.*

CAPP See *computer-aided process planning.*

cellular layout As contrasted with *functional layout,* a physical rearrangement of the shop floor in which the various machines are grouped into *manufacturing cells.* A physical implementation of *cellular manufacturing.*

cellular manufacturing An approach to the production of parts in which each part is assigned to a specific group of machines for production. The machines assigned to the cells are restricted to the production of a limited number of parts, typically members of a *part family.* The group can be limited to the planning stage alone (with the assigned machines being physically located in functional *work centers*) or it can be physically implemented on the shop floor. Cellular manufacturing is most frequently a result of the application of *group technology.*

CIB Computer integrated business. See *computer integrated manufacturing.*

CIM See *computer integrated manufacturing.*

closed-loop material requirements planning A planning system built around *material requirements planning* that includes the additional production-oriented planning functions of production planning, master production scheduling, and capacity requirements planning. Once the planning phase is complete and the plans have been accepted as realistic (i.e., feasible) and attainable, the execution functions (i.e., *shop floor control*) come into play. The term "closed-loop" implies that each of these elements is included in the overall system and that there is feedback from each lower level of the system to the higher levels in order to ensure that the plans are kept valid at all times (APICS Dictionary, 1980, p. 5). More commonly referred to as close-loop MRP.

CMM See *coordinate measuring machine.*

CNC See *computer numerical control.*

common database An approach to the management of databases that stresses that there should be only one set of files present (i.e., no redundancy of data), which are continuously maintained and made available to all those who have a legitimate need to access the database. Other traits of a common database include high accuracy (i.e., in excess of 95 percent), completeness, and currency of information.

computer-aided design Any design activity that involves the use of computer or computer-based technology to facilitate the creation and evaluation (through such procedures as simulation) of new engineering designs and the modification of existing designs (Groover, 1980, p. 262).

computer-aided drafting and design Any drafting activity that involves the use of computers and computer-based technology to facilitate the creation and modification of product or process designs. In contrast to *computer-aided design*, computer-aided drafting and design focuses on the actual process of creating designs. It can be regarded as a subsystem of computer-aided design and the complement of *computer-aided engineering*.

computer-aided engineering Any designing or engineering activity that draws on the use of computer-based technology during the creation and maintenance of engineering designs. In contrast to *computer-aided drafting and design*, computer-aided engineering deals primarily with the process of generating and evaluating the designs. It is a subsystem of *computer-aided design*.

computer-aided inspection and testing The effective use of computer technology in the inspection and testing procedure. Typically, it incorporates such technology as sensors, *machine vision*, *coordinate measuring machines*, *automated identification*, and statistical process control.

computer-aided manufacturing The effective utilization of computer manufacturing technology in the management, control, and operation of the manufacturing facility through either direct or indirect computer interface with the physical or human resources of the firm. Typically built around the following technologies: *NC* technology, production or process planning, *robotics*, the *flexible manufacturing system*, and factory management systems.

computer-aided process planning Any activity that draws extensively on the use of computers and computer-based technology in the preparation, maintenance, and modification of *routing sheets*. The development of these sheets or plans takes advantage of standard manufacturing operation sequences that are stored in a computer. These sequences can be developed as a result of the implementation and use of a *parts coding and classification system.* Computer-aided process planning acts as the link between the design and manufacturing activities of a CAD/CAM system.

computer integrated manufacturing 1. A corporate system designed to enable the firm either to develop a distinctive competitive advantage through manufacturing or to achieve its strategic objectives by using computer technology to enhance manufacturing (and ultimately corporate) integration. 2. A corporate system that draws extensively on computer-aided technology to help management organize the company to focus on a market or specific market niche with the goal of obtaining and maintaining a *sustainable* competitive advantage.

computer numerical control A numerical control system that uses a dedicated, stored-program computer to perform the basic NC functions, tasks, and operations.

continuous manufacturing A major classification of production involving the high-volume production of a nondiscrete item (e.g., petroleum) in a process characterized by a sequence of operations common to most items, short lead times, and very small queues.

coordinate measuring machine A machine that measures in three-dimensional space the surfaces and features of a workpart. This measurement is done by means of a contact probe and a positioning mechanism.

corporate database A database that consists of all of the information that the firm needs for planning, execution, and control activities. Typically, the corporate database consists of two major sets of data. The first is the financial database, which contains primarily financial data (e.g., cost data, balance sheets, expense accounts). The second is the operational database, which contains all of the information needed for the manufacture, design, and delivery of goods (e.g., bills of materials, inventory records, routings). For the corporate database to be effective, the two sets of data must be linked on an ongoing basis.

database A collection of data or information maintained and stored in some organized way so that stored data can be quickly accessed when needed.

design attributes Similarities between parts that are reflected in geometric shape or size. Often used as a means of developing *part families* and of assigning membership to such families.

design for manufacture See *integrated engineering design system.*

DFM Design for manufacture. See *integrated engineering design system.*

direct numerical control A system in which a large central computer is used to direct the operations of a number of separate NC machines.

discrete batch manufacturing A classification of production processes that involves the manufacture of discrete units in small batches. Each batch does not necessarily follow the same sequence of operations (as contrasted with continuous or repetitive manufacturing). Lead times tend to be relatively long due to large queues. Also referred to as job shop manufacturing.

DNC See *direct numerical control.*

EDI See *electronic data interchange.*

electronic data interchange The exchange of business transactions electronically. The resulting "paperless" flow of information creates a situation characterized by increased data accuracy, reduced inventories, reduced lead times, and increased productivity. Electronic data interchange is often facilitated by the use of *automated identification.*

engineer-to-order A manufacturing schedule practice most often encountered with products that the firm has never built before. These are either new products or existing products that are undergoing extensive customization. Under this practice, nothing is done until the order (and initial set of product specifications) is received from the customer. At this point, the firm starts to work, beginning with engineering. Typically, engineer-to-order products experience the longest lead times because they incur not only the manufacturing lead time but also the time required by engineering for initial design, customer review and revision, and design revision. This scheduling practice is commonly referred to by its abbreviation, ETO.

ETO See *engineer-to-order.*

expert systems An intelligent computer program that uses knowledge and inference procedures to solve problems that are difficult enough to require significant human expertise for their solution. Knowledge necessary to perform at such a level can be thought of as a model of the expertise of the best practitioners of the field (Harmon and King, 1985, p. 5).

flexible manufacturing system A flexible manufacturing system consists of a group of processing stations (such as CNC machines or machining centers) that are interconnected by an *automated materials handling system* and controlled by a central computer. Flexible manufacturing systems can be broken down into five different types: 1. Flexible manufacturing modules (FMMs), which are appropriate for production where there are large numbers of different parts to be produced but the associated volume is relatively low. 2. Flexible manufacturing cells (FMCs), which are appropriate in situations in which the number of different parts to be produced is less than in an FMM and the volume of parts is higher. They tend to be self-contained in that all of the operations needed to be done on the part are done within the cell. 3 Flexible manufacturing groups (FMGs), which, in contrast to the FMC, are appropriate for situations in which parts variety is lower and volume higher. 4. Flexible machining systems (FMS). 5. Flexible manufacturing lines (FMLs) are the most comprehensive form of FMS. They are appropriate for those manufacturing settings in which volume is higher and parts variety low.

FMC See flexible manufacturing cells, under *flexible manufacturing system.*

FMG See flexible manufacturing groups, under *flexible manufacturing system.*

FML See flexible manufacturing lines, under *flexible manufacturing system.*

FMM See flexible manufacturing modules, under *flexible manufacturing system.*

FMS See flexible machining systems, under *flexible manufacturing system.*

FMS See *flexible manufacturing system.*

functional layout A physical arrangment of the shop floor in which the various machines are grouped together based on the similarity of the function(s) performed. The opposite of a functional layout is a c*ellular layout.*

group technology A systematic methodology under which component similarity (in terms of product, processing, or both) is used to form *part families*, plan common production processes, and establish *manufacturing cells* so that items typically made in short-run discrete batch runs can be manufactured as efficiently as those produced in larger runs.

hard automation See *automation, hard.*

information technology The collective means used to assemble and electronically store, transmit, process, and retrieve words, numbers, images, and sounds, as well as the electronic means to control machines of all kinds, from everyday appliances to vast automated factories. Also referred to by its abbreviation, IT.

integrated engineering design system A system designed to improve and enhance the process of designing products. One of the key traits of this system is that it brings together into groups (often referred to as product design teams) representatives from those functions that affect the product. Typically, these representatives come from such areas as design engineering, manufacturing engineering, quality assurance, sales and marketing, accounting/finance, production and inventory control (PIC), top management, and, in some instances, the customer and the suppliers. These groups are charged with reviewing and assessing the product while it is still in an early stage of development. The purpose of this review is to identify any potential problems attributable to design and to uncover any improvements. The changes resulting from this review can be made very quickly and relatively inexpensively at this early stage of product development. The ultimate result is a better product that is cheaper to develop and build and that is able to get to the market in a much shorter period of time. This system is also referred to as design for manufacture, simultaneous engineering, and parallel engineering.

interorganizational CIM A CIM system that extends beyond one firm's boundaries. With interorganizational CIM, the focus is on building closer ties between suppliers and customers. These ties are facilitated by means of such technologies as *electronic data interchange* and *integrated engineering design systems*. The CIM systems of the suppliers and the customers are now tightly linked and tend to operate as a coordinated unit. Under interorganizational CIM, customers can download requirements and specifications electronically to suppliers. Suppliers can generate the necessary designs and upload the designs to their customers using one of the available graphics standards. Suppliers and customers can now access each other's systems and obtain information in real time. More important, suppliers and customers are now linked in real time. The term "interorganizational CIM" was first coined by George Hess of Ingersoll Milling.

islands of automation Describes a situation created typically when applying automation to a manufacturing problem. The goal in applying automation is to solve the problem (which has been myopically defined) or "optimize" system performance. However, the impact of the new technology on the rest of the system and its operation is often ignored. In addition, issues of integrating or interfacing the new system with the other elements of the overall manufacturing system are not considered in great detail. The result is an "island" that works well by itself but that is unable to work well with the rest of the system. It is a situation in which the operation of one element of the total manufacturing system has been optimized, but the operation of the entire system has not been optimized.

islands of expertise This describes a situation very similar to that encountered with *islands of automation*. With islands of expertise, there exist within a company or organization experts who have, at best, a local impact in that the rest of the organization may not be aware of their presence. Even if the rest of the organization is aware, access to them may be very limited. As a result, organizational learning is adversely affected. Errors previously made are unnecessarily repeated. Solutions uncovered elsewhere in the company are not available to the rest of the organization. As a result, there is a great deal of "reinventing of the wheel."

IT See *information technology*.

JIT See *just-in-time manufacturing*.

job shop control See *shop floor control.*

just-in-time manufacturing A corporate-wide system (philosophy) intended to produce goods and services having the highest *value* for the firm's customers by continuously identifying and eliminating all forms of corporate *waste.* Also called zero inventory.

machine vision The acquisition by computer and mechanical means of image data followed by the processing and interpretation of this data by computer. Machine vision is used primarily for (1) inspection, (2) recognition, (3) gauging, and (4) robot guidance and position control.

make-to-order The manufacturing scheduling practice whereby production is initiated by the receipt of a customer order (either actual or anticipated). Until the order is received, no inventory (either raw, work-in-process, or end item) is kept in stock. Once the order is received, ordering of inventory takes place and production begins. This type of practice is appropriate for items that the firm has previously manufacturing but that cannot be kept in stock (i.e., too expensive to keep in stock, not possible to keep the items in stock, or the demand is either too low or too sporadic to justify the presence of inventory). The lead time under this schedule is long, since the buyer must wait for all purchasing, manufacturing, assembly, and delivery activities to be completed. This scheduling practice is commonly referred to by its abbreviation, MTO.

make-to-stock The manufacturing scheduling practice whereby all customer orders are filled from a finished-goods inventory. This inventory is built in advance of actual demand and must be present when the customer's order is placed. The basis for this production is a forecast, not actual orders. Typically, this practice is appropriate for mature products that are fairly standard (i.e., the configuration desired by the customer is known in advance). This scheduling practice is commonly referred to by its abbreviation, MTS.

manufacturing activity planning See *shop floor control.*

manufacturing attributes Similarities between parts that are reflected in the sequence of processing steps required to build the part. Often used as a means of developing *part families* and of assigning membership to such families.

manufacturing cell The manufacturing cell consists of all the machines and equipment required for the production of a *part family.* The manufacturing cell, most often the result of *group technology* activities, can be found in a range of configurations. In its simplest form, the manufacturing cell can consist of only one machine. At the other end of the spectrum, a manufacturing cell can consist of a group of machines connected by a conveyor system.

manufacturing database The manufacturing database describes a set of files that contain the information base needed by the manufacturing system. Ideally, this set of files should be centrally located and part of a *corporate database*. The manufacturing database consists of two broad sets of files: planning files (those needed during planning) and control files (those needed to control work and resources once work has begun on the shop floor).

manufacturing resources planning Also referred to as MRP II. A method for the effective planning of all the resources of a manufacturing company. Ideally, this method addresses operation planning in units and financial planning in dollars. It also has a simulation capability to answer "what if" questions. It links together a variety of functions: business planning, production planning, capacity requirements planning, and the execution systems for capacity and priority control. Outputs from these functions are integrated with financial reports, such as the business plan, the purchase commitment report, the shipping budget, and inventory projections. Manufacturing resources planning is a direct outgrowth and extension of MRP (APICS Dictionary, 1980, p. 16).

manufacturing technology The use of technology in the process of taking the design created by *product technology* and implementing it in the manufacturing planning and control system. Included as part of this type of technology are manufacturing planning, scheduling, the development of standards, and the development and acquisition of necessary equipment and tooling.

material requirements planning A system that uses bills of materials, inventory records, planned lead times, open order data, and master production scheduling information to calculate requirements for materials. The system makes recommendations to release replenishment orders for materials and to reschedule open orders when due dates and need dates are not in phase. Originally seen as merely a better way to order inventory, materials requirement planning is now viewed primarily as a scheduling technique—as a method for establishing and maintaining valid due dates on orders. (APICS Dictionary, 1980, p. 18). Commonly referred to as MRP.

MICLASS A parts classification and coding system frequently found in *group technology* implementations. The term MICLASS stands for the Metal Institute Classification System. MICLASS uses a classification number of 12 to 30 digits, in which the first 12 digits are a universal code and the other 18 digits can be used to code data specific to the firm.

MRP See *material requirements planning.*

MRP II See *manufacturing resources planning.*

MTO See *make-to-order.*

MTS See *make-to-stock.*

NC See *numerical control.*

numerical control A form of programmable automation in which the processing equipment is controlled by means of externally written and stored programs. An NC machine consists of three major components: a program, a machine control unit, and processing equipment.

Opitz classification system A parts classification and coding scheme found in *group technology* implementations.

PAC Production activity control. See *shop floor control.*

parallel engineering See *integrated engineering design system.*

part family A collection of parts that are similar because of geometric shape or size (i.e., *design attributes*) or because of similarities in the processing steps required to build them (i.e., *manufacturing attributes*).

parts coding and classification system A system designed to identify similarities between parts and to describe these similarities by means of a coding system. These similarities can be described in terms of either *manufacturing attributes* or *design attributes.* The resulting coding system can be used to simplify the design and manufacture of parts by taking advantage of the lessons learned from designing or building similar parts in the past. Two of the most commonly used coding and classification schemes are *Opitz* and *MICLASS.*

product technology The use of technology in support of all product design activities (from conceptualization to layout, evaluation, revision, maintenance, storage, and retrieval).

production activity control See *shop floor control.*

robot A reprogrammable multifunctional manipulator designed to move material, parts, tools, or specialized devices through variable programmed motions for the performance of a variety of tasks.

robotics The study and use of computer-controlled manipulators or *robots* to automate such manufacturing tasks as welding, material handling, painting, and assembly.

routing sheet The listing of steps, *work centers,* processing times, activities, and their sequencing required in the completion of a given part.

shop floor control Also referred to as manufacturing activity planning (MAP), production activity control (PAC), and job shop control. A major subsystem within the manufacturing system. Shop floor control is that group of activities directly responsible for managing the transformation of planned orders (i.e., orders related to the shop floor by the planning system) into a set of outputs that conform to some set of prespecified evaluation criteria. Shop floor control governs the *very short-term* detailed planning, execution, and monitoring activities needed to control the flow of an order from the time that the order is released to the time that the order is filled and its disposition completed. The shop floor control system is responsible for making the detailed and final allocation of labor, machine capacity, tooling, and materials to the various competing orders. It collects information on the shop floor activities involving the progress of orders and the status of resources and makes this information available to the planning system. Shop floor control consists of five major activities: (1) order review and release; (2) detailed assignment; (3) data collection and monitoring; (4) control and feedback; and (5) order disposition (Melnyk and Carter, 1987, p. 12).

simultaneous engineering See *integrated engineering design system.*

soft automation See *automation, soft.*

technology 1. The physical processes combined with the intellectual or knowledge processes by which materials in some form are transformed into outputs used by another organization or subsystem. 2. The body of knowledge about the means by which we work on the world, our arts and our methods. Essentially, it is knowledge about the cause and effect relations of our actions. It is knowledge that can be studied, codified, and taught to others. 3. A system of components involved in acting on or changing an object from one state to another. 4. The set of physical and informational elements that are used to enhance current products or capabilities or to substitute improved products or capabilities for existing items. 5. The system (consisting of knowledge and physical methods) by which the firm satisfies its needs and desires.

tooling All equipment and special features that the system can draw on and use during the setup and operation of a machine, *work center*, or assembly process. Included in this definition is such equipment as jigs, features, drills, pallets, cutting tools, rollers, brushes, tool holders, laps, chucks, mandrels, collets, centers, arbors, step-blocks, angle-plates, templates, molds, dies, cams, and patterns. These items can be broken down into three major categories: transportation tooling, setup tooling, and production tooling.

tooling requirements planning Tooling requirements planning is a planning system that uses the principles of *material requirements planning* to plan and schedule the *tooling* resources of a company. It is intended to answer the following questions:

When is tooling required?

How much tooling is required?

What tooling is required?

It is a relatively new technique. Tooling requirements planning is also referred to by its abbreviation, TRP.

TRP See *tooling requirements planning.*

value The assessment of the usefulness or utility of a particular item, good, or service. This assessment is ultimately made by the customer, not the firm. The purpose of any effective system is to provide goods and services that generate the highest level of value for the customer by means of a process that has identified and eliminated any activities or steps that do not directly contribute to value.

vision, machine See *machine vision.*

waste Any action taken within the entire firm that adds cost without increasing *value* (as perceived by the customer). There are seven major categories of waste (as identified by Suzaki, 1988): (1) waste from overproduction; (2) waste from waiting time; (3) transportation waste (i.e., overhandling of the product); (4) process waste (i.e., doing more than is required to meet the specifications, for example by excessive setups); (5) waste of inventory; (6) waste of motion; and (7) waste from product defects.

work center A production facility consisting of identical or similar machines that can be grouped together and considered as one for the purposes of planning or scheduling.

zero inventory See *just-in-time manufacturing.*

APPENDIX 3

BIBLIOGRAPHY

The following is a list containing all of the articles, papers, books, and presentations reviewed in the preparation of this book. It should be noted that given the large body of literature pertaining to the concepts, components, elements, principles, and practices of computer integrated manufacturing, this list should not be regarded as comprehensive. It is not. Rather, it is a compilation of some of the major works pertaining to CIM.

Each article is presented in standard reference format. Furthermore, each of the articles denoted by a § indicates a citation that is summarized in the CIM Source Book. Information about this reference can be obtained by contacting either the Education and Research Foundation of the American Production and Inventory Control Society or the authors of this study.

Abdel-Malek, Layek, and Boucher, Thomas O. "A Framework for the Economic Evaluation of Production System and Product Design Alternatives for Robot Assembly." *International Journal of Production Research* 23, no. 1 (1985), pp. 197–208.§

Abou-Zeid, M.R. "Manufacturing System Analysis and Evaluation for Group Technology Application." *Spring Annual Conference Proceedings of the AIIE.* Washington, D.C., 1975, pp. 279–286.

Accampo, Paul W. "MAP Pilots: Promises and Pitfalls." *CIM Technology* 5, no. 1 (Spring 1986), pp. 19–23.§

Aceti, John, and Carroll, Charles. "Robotic Gear Assembly." *Manufacturing Engineering* (April 1985), pp. 92–94.§

Adler, Paul S. "Managing Flexible Automation." *California Management Review* 30, no. 3 (Spring 1988), pp. 34–56.§

"AGV's at GM Provide Untended Material Handling." *Manufacturing Engineering* (December 1985), p. 51.§

Aiello, Joseph L. "CIM: The Factory of the Future...NOW." *Computer Integrated Manufacturing and Flexible Manufacturing Systems Seminar Proceedings.* New Orleans, April 24–26, 1985, pp. 41–43.§

"Allen-Bradley Unveils CIM System." *High Technology Business* 8, no. 10 (October 1988), pp. 62–64.§

Anderson, E.P. "Bar Coding Technology Enables Integration of Information and Control Systems and Manufacturing." *Industrial Engineering* (November 1984), pp. 40–47.§

Appleton, Daniel S. "Integration Technology." *CIM Technology* 5, no. 1 (Spring 1986), pp. 16–18.§

Arcuri, Gerald. "Chart a New MAP." *Computerworld* 21, no. 22 (May 30, 1988), p. 52.§

Argote, Linda, Goodman, Paul S., and Schkade, David. "The Human Side of Robotics: How Workers React to a Robot." *Sloan Management Review* (Spring 1983), pp. 31–41.§

Armstrong, Robert E. "Storage Systems of the Future." *P&IM Review and APICS News* 2, no. 4 (April 1982), pp. 22–23, 33.§

Arnold, John C., and Johansson, Hank J. "Establishing a Framework for the Analysis and Justification of Computer Integrated Manufacturing." *Computer Integrated Manufacturing and Flexible Manufacturing Systems Seminar Proceedings*. New Orleans, April 24–26, 1987, pp. 58–66.§

Arnold, K. "Implementing Robotic Deburring." *Manufacturing Engineering* 99, no. 1 (1987), pp. 35–37.§

Aseo, J. "Protocols Set Course for Next-Generation Networks." *Digital Design* 15, no. 12 (December 1985), pp. 36–38,40–41.§

"AS/RS—A Better Way to Handle Minimum Inventory." *Production Engineering* 32, no. 6 (June 1985), pp. 86–94.§

"Automated Factory Looks Out for Itself." *Production Engineering* 31, no. 12 (December 1984), pp. 22–32.§

"Automatic Identification—Now It's Really Taking Off!" *Modern Material Handling* 37, no. 14 (September 21, 1982), pp. 32–43.

The Automation News Factory Management Glossary. New York, N.Y.: Grant Publications, 1984.§

Avishai, Bernard. "A CEO's Common Sense of CIM: An Interview with J. Tracy O'Rourke." *Harvard Business Review* 67, no. 1 (January–February 1989), pp. 110–117.§

Bailey, Harald A. "The Drive Toward Computer Integrated Manufacturing." *P&IM Review and APICS News* 5, no. 3 (March 1985), pp. 48–52.§

Baker, J. A. "Winning Your Case for Automation." *Manufacturing Engineering* (July 1984), pp. 72–73.§

Barger, B., and Gould, L. "Computers—Where We're Going Tomorrow." *Modern Material Handling* (November 21, 1983), pp. 42–50.§

Barrett, D. "Brick Wall Crumbles into Flowing Channel." *Computerworld* 20, no. 14 (April 3, 1986), p. 20.§

Bartik, J. "MAP's Complement: TOP." *Data Communications* 14, no. 3 (December 1985), p. 152.§

Beaver, J.E. "TOP: New Standards for the Office?" *Computer Decisions* 18, no. 3 (January 28, 1986), pp. 46–47, 80.§

Benjamin, Robert J., Rockart, John F., Scott Morton, Michael S., and Wyman, John. "Information Technology: A Strategic Opportunity." *Sloan Management Review* (Spring 1984), pp. 3–10.§

Bennett, Earle D. "NTC Flow Management Program." *Computer Integrated Manufacturing and Flexible Manufacturing Systems Seminar Proceedings.* New Orleans, April 24–26, 1985, pp. 159–161.§

Benoit, C. "It Already Is Bar Coding Technology." *P&IM Review and APICS News* 2, no. 10 (October 1982), pp. 50–52, 69–70.

Benson, P. George, Hill, Arthur V., and Hoffmann, Thomas R. "Manufacturing Systems of the Future—A Delphi Study." *Production and Inventory Management* (Third Quarter 1982), pp. 87–106.§

Bergstrom, Robin P. "AI: Fad with a Future?" *Manufacturing Engineering* (April 1985), p.65.§

———. "Computer-aided SQC Makes Impact at Pontiac." *Manufacturing Engineering* (May 1985), pp. 71–72.§

———. "FMS Users and Vendors: Where We Are Today" *Manufacturing Engineering* 96, no. 6 (March 1986), pp. 48–53.§

Bessant, John, and Haywood, William. "Experiences with FMS in the UK." *International Journal of Production and Production Management* 6, no. 5 (1986), pp. 44–56.

Boaden, R.J., and Dale, B.G. "Development of a Model for Use in Planning for Computer-Integrated Manufacturing." *International Journal of Operations and Production Management* 6, no. 5 (1986) pp. 8–19.

Bowman, Daniel J., and Bowman, Annette, C. *Understanding CAD/CAM.* Indianapolis: Howard W. Sams, 1987.§

Bozman, Jean S. "Deere Reaps MAP, CIM Benefits." *Computerworld* 22, no. 23 (June 6, 1988), pp. 43–44.§

Brandt, Richard, and Treece, James B. "Retool or Die: Job Shops Get a Fix for the Future." *Business Week.* No. 2951 (June 16, 1986), pp. 106, 108.§

Brauninger, Paul W. "Building Towards Computer Integrated Manufacturing." *Computer Integrated Manufacturing and Flexible Manufacturing Systems Seminar Proceedings.* New Orleans, April 24–26, 1985, pp. 84–86.§

Bray, Olin H. *Computer Integrated Manufacturing: The Data Management Strategy.* Waltham, Mass.: Digital Press, 1988.§

Browne, Jim, Harhen, John, and Shivnan, James. *Production Management Systems: A CIM Perspective.* Wokingham, England: Addison-Wesley, 1989.

Buchholtz, Robert F. "How to Automate (Without Losing Your Mind)." *Production Engineering* 29, no. 9 (September 1982), pp. 78–81.§

Burbidge, J.L. "AIDA and Group Technology." *International Journal of Production Research* 11, no. 4 (1973), pp. 315–324.

———. *The Introduction of Group Technology.* New York: John Wiley & Sons, 1975.

———. *Group Technology.* London: Mechanical Engineering Publications, 1979.

Burgam, Patrick M. "Programmable Robots Accelerate Arc Welding Automation." *Manufacturing Engineering* (December 1984), pp. 64–65.§

Burgon, Claron D., Miller, Karen S., and Neeley, Parley S. "MRP II Software Flex-Out...The Forgotten Step (Part 1)." *P&IM Review and APICS News* 5, no. 8 (August 1985), pp. 40–45, 70.§

———. "How to Proceed with MRP II Software Flex-Out (Part 2)." *P&IM Review and APICS News* 5, no. 9 (September 1985), pp. 56, 67–69.§

Buss, Martin D.J. "Making it Electronically." *Harvard Business Review* (January–February 1982), pp. 89–90.§

———. "Managing International Information Systems." *Harvard Business Review* (September–October 1982), pp. 153–162.§

———. "How to Rank Computer Projects." *Harvard Business Review.* (January-February 1983), pp. 118–125.§

Buzacott, J.A. "Flexible Manufacturing Systems: A Review of Models." *TIMS/ORSA Thirteenth Joint National Meeting*, Detroit, p. 77.

Buzacott, J.A., and Shanthikumar, J.G. "Models for Understanding Flexible Manufacturing Systems." *AIIE Transactions* 12, no. 4 (December 1980), pp. 339–350.

Bylinsky, G. "The Race to the Automatic Future." *Fortune.* Vol. 107, No. 4 (February 21, 1983), pp. 52–64.

Callagher, Gerald R., and Gullo, John J. "Developing a Closed Loop MRP System—What's In It for the Design Engineer?" *Production and Inventory Management* (Fourth Quarter 1980), pp. 21–37.§

Calligan, Stephen C., and Hazarika, Panna. "Shop Floor Control in a Continuous Production Flow Environment." *Production and Inventory Management* (Fourth Quarter 1982), pp. 87–119.§

Campbell, K.L. "A Corporate Material Identification System—For Raw Materials, Purchased Components and Expense Items." *Group Technology Seminar Proceedings.* Arlington, 1979, pp. 53–58.

"Capital Equipment—Machining." *Production Engineering* 32, no. 1 (January 1985) pp. 6–18.§

Carlson, Mark. "Table-Top Robots." *Production Engineering* 32, no. 4 (April 1985), pp. 86–88.§

Carrie, A.S. "Numerical Taxonomy Applied to Group Technology and Plant Layout." *International Journal of Production Research* 11, no. 4 (1973), pp. 399–416.

Chamberlain, Robert G. "The Shifting Balance between...Centralized and Distributed Systems." *Production Engineering* 27, no. 11 (November 1980), pp. 58–62.§

Chapman, J.R. "Integrating Engineering and Manufacturing Systems." *American Production and Inventory Control Society 25th Annual Conference Proceedings*. Chicago, 1982, pp. 570–574.

Chasen, S.H. "Principles of Geometric Modeling." *CIM Technology* 5, no. 2 (Summer 1986), pp. 15–18.§

Cherubino, P.M. "Machine Monitors Maintain Downtime Control." *Industrial Engineering* 3, no. 10 (October 1971), pp. 22–24.

Chevalier, P.M. "New Developments at OIR." *The 1981 OIR Users' Meeting Summary*. Palo Alto, 1981, pp. 7–15.

Chong, Philip S. "Manufacturing Resource Planning Games." *P&IM Review and APICS News* 5, no. 8 (August 1985), pp. 26, 56.§

"CIM: The View from Manufacturing." *P&IM Review and APICS News* 5, no. 3 (March 1985), pp. 54–55.§

Claire, Frank V. "The Weakest Link?" *P&IM Review and APICS News* 6, no. 2 (February 1986), pp. 36, 40, 44, 65.§

Clark, M. "Computer Simulation Reduces Guesswork in Production." *Production Engineering* 32, no. 6 (June 1985), pp. 78–79.§

Cohen, Stephen S., and Zysman, John. *Manufacturing Matters*. New York: Basic Books, 1987.

Cole, Patrick E. "How Calcomp Streamlined Itself." *Business Week* (June 6, 1988), p. 106.§

"Computer Directs Picking from Racks and Carousels." *Modern Material Handling* (January 9, 1984), pp. 54–55.§

"Computer-Aided Maintenance: Instant Troubleshooting at the Tip of Your Finger." *Modern Material Handling* 38, no. 3 (February 7, 1983), pp. 62–65.

"Computer-Integrated Manufacturing—From Vision to Reality." *Production Engineering* 27, no. 11 (November 1983), pp. 46–53.§

Copeman, R. "Profit Improvement from Improved Through-put." *American Production and Inventory Control Society 21st Annual Conference Proceedings*. Hollywood, Fla., 1979, pp. 142–147.

Corbet, J. Terry. "Local Area Networks and Shop Floor Data Collection." *P&IM Review and APICS News* 5, no. 12 (December 1985), pp. 40–48, 72.§

Crawford, C. Merle. "Defining the Charter for Product Innovation." *Sloan Management Review* (Fall 1980), pp. 3–12.§

Crawford, Karlene M., and Cox, James F. *Management and Technological Changes to Improve Competitiveness: Case Studies of Small- to Medium-Sized Apparel Industries*. Falls Church, Va.: American Production and Inventory Control Society, 1988.§

Dalrymple, R. "LAN Standards Efforts Begin to Pay Off." *Mini-Micro Systems* 19, no. 4 (March 1986), pp. 93–94, 96.§

"Data Collection to Go On-Line." *Industrial Engineering* 11, no. 9 (November 1973), pp. 33–35.

Davis, David. "SMR Forum: Computers and Top Management." *Sloan Management Review* (Spring 1984), pp. 63–67.§

Davis, Stanley M. *Future Perfect*. Reading, Mass.: Addison-Wesley, 1987.

Dean, James W., Jr., and Susman, Gerald I. "Organizing for Manufacturable Design." *Harvard Business Review* 67, no. 1 (January–February 1989), pp. 28–36.§

Dearden, John. "SMR Forum: Will the Computer Change the Job of Top Management?" *Sloan Management Review* (Fall 1983), pp. 57–60.§

Dertouzos, Michael L., Lester, Richard K., and Solow, Robert M. *Made In America: Regaining the Productive Edge*. Cambridge, Mass.: The Massachusetts Institute of Technology, 1989.

Derus, Kenneth G. "Systems Planning/QualityAssurance: A Vital Component for the Integration and Coordination Needed to Implement CIM/FMS." *Computer Integrated Manufacturing and Flexible Manufacturing Systems Seminar Proceedings*. New Orleans, April 24–26, 1985, pp. 171–178.§

Dilworth, James B., and McAlister, M. Khris. "Users Perceptions of FMS Characteristics." *Computer Integrated Manufacturing and Flexible Manufacturing Systems Seminar Proceedings*. New Orleans, April 24–26, 1985, pp. 201–208.§

Dolbey, Bradford I. "Stock Preparation in a Job Shop Using CIM." *Computer Integrated Manufacturing and Flexible Manufacturing Systems Seminar Proceedings*. New Orleans, April 24–26, 1985, pp. 129–132.§

Dollhoff, T.L. "Manufacturing Communications: Local Area Networks for the Integrated Factory." *Production Engineering* (February 1985), pp. 68–78.§

———. "Evaluating Manufacturing Software." *Production Engineering* 32, no. 6 (June 1985), pp. 68–73.§

Drodza,Thomas J. "Flexible Assembly System Features Automatic Setup." *Manufacturing Engineering* (December 1984), pp.75–76.§

Dunn, Alan G. "Robots and Flexible Manufacturing Systems—A Primer." *Computer Integrated Manufacturing and Flexible Manufacturing Systems Seminar Proceedings*. New Orleans, April 24–26, 1985, pp. 6–29.§

Edwards, G.A.B. "The Family Grouping Philosophy." *International Journal of Production Research* 9, no. 3 (1971), pp. 337–352.

Ellis, R. Jeffrey. "Improving Management Response in Turbulent Times." *Sloan Management Review* (Winter 1982), pp. 3–12.§

Ettlie, John E. *Taking Charge of Manufacturing*. San Francisco: Jossey-Bass 1988.§

Evans, Albert E. "On the Right Track with Transit America, Inc." *P&IM Review and APICS News* 5, no. 8 (August 1985), pp. 28–31, 55.§

"Existing Facilities, Flexible Handling and Modular Components." *Production Engineering* 30, no. 12 (December 1983), pp. 46–51.§

Farnum, Gregory T. "Industrial Robots—The Next 10 Years." *Manufacturing Engineering* (December 1985), pp. 55–56.§

Fazakerley, G.M. "A Research Report on the Human Aspects of Group Technology and Cellular Manufacture." *International Journal of Production Research* 14, no. 1(1976), pp. 123–134.

Fiderio, Janet. "Steps to a Smoother Lager." *Computerworld* 22, no. 22 (May 30, 1988), p. 46.§

Fitzgerald, Tim, and Richards, Jim. "Managing Materials Logistics at Check Technology." *P&IM Review and APICS News* 6, no. 1 (January 1986), pp. 46–48.§

Flesher, H.K. "CIM—The Short-Lived Advantage (Shrinking World Position—An Opportunity to Regain Lost Role)." *Computer Integrated Manufacturing and Flexible Manufacturing Systems Seminar Proceedings*. New Orleans, April 24–26, 1985, pp. 167–170.§

"Flexible Assembly: A Search for Economies of Scope." *Production Engineering* 32, no. 12 (December 1985), pp. 40–47.§

"Flexible Manufacturing Systems—Seven of the World's Best." *Modern Material Handling* 37, no. 13 (September 7, 1982), pp. 52–62.

Ford, David, and Ryan, Chris. "Taking Technology to Market." *Harvard Business Review* (March–April 1981), pp. 117–126.§

French, J.W. "Maintenance: Profit Drain or Catalyst." *Production Engineering* (February 1985), pp. 84–88.§

Frontini, Gian F., and Richardson, Peter R. "Design and Demonstration: The Key to Industrial Innovation." *Sloan Management Review* (Summer 1984), pp. 39–49.§

Frost, Chad. "Integrated Automation Program at Frost, Inc." *P&IM Review and APICS News* 4, no. 4 (April 1984), pp. 32–36.§

Fuller, J. "CAD for P/M Parts Manufacturing." *Manufacturing Engineering* (August 1984), pp. 149–150.§

Gallagher, C.C., Banerjee, S.K., and Southerm, G. "Group Technology in the Shipbuilding Industry." *International Journal of Production Research* 12, no. 1(1974), pp. 45–54.

Gallagher, C.C., Grayson, T.J., and Phillips, A. "A Review of the Development and Application of Group Technology at the University of Birmingham." *International Journal of Production Research* 9, no. 2 (1971), pp. 229–237.

Garwood, R. Dave. "Explaining JIT, MRP II, Kanban." *P&IM Review and APICS News* 4, no. 10 (October 1984), pp. 66–69, 88.§

Gerelle, Eric G.R., and Stark, John. *Integrated Manufacturing Strategy: Planning and Implementation.* New York: McGraw-Hill, 1988.

Gerwin, Donald. "Do's and Don'ts of Computerized Manufacturing." *Harvard Business Review* (March–April 1982), pp. 107–116.§

———. "A Theory of Innovation Processes for Computer-Aided Manufacturing Technology." *IEEE Transactions on Engineering Management* 35, no. 2 (May 1988), pp. 90–100.§

Gitlow, Howard S., and Hertz, Paul T. "Product Defects and Productivity." *Harvard Business Review* (September–October 1983), pp. 131–141.§

Glenney, Neil, and Rylander, Robert. "Real Time Shop Floor Control." *Production and Inventory Management* (Fourth Quarter 1982), pp. 121–138.§

Gralser, Harlan N. "CAD/Robotics Capability Helps Realize CIM." *Industrial Engineering* (August 1983), pp. 26–30.§

Gray, Larry. "The Automated Factory: An Emphasis on Controls." *P&IM Review and APICS News* (February 1983), pp. 20–25.§

Grayson, T.J. "Some Research Findings on Evaluating the Effectiveness of Group Technology." *International Journal of Production Research* 16, no. 2 (1978), pp. 89–102.

Gremillion, Lee L., and Pyburn, Philip. "Breaking the Systems Development Bottleneck." *Harvard Business Review* (March–April 1983), pp. 130–137.§

Grieco, Peter L., Jr. "The Just-in-Time Solution." *Computer Integrated Manufacturing and Flexible Manufacturing Systems Seminar Proceedings.* New Orleans, April 24–26,1985, pp. 133–140.§

Grieco, Peter L., Jr., and Gozzo, Michael W. "JIT and TQC—Is It for You?" *P&IM Review And APICS News* 5, no. 9 (September 1985), pp. 50–55.§

______. "The Just-in-Time Solution." *Computer Integrated Manufacturing and Flexible Manufacturing Systems Seminar Proceedings.* New Orleans, April 24–26, 1985, pp.133–140.§

Groover, Mikell P. "The Role of Robotics and Flexible Manufacturing in Production Automation." *Computer Integrated Manufacturing Systems: Selected Readings.* Ed. J.W. Nazemetz, W.W. Hammer, Jr., and R. Sadowski. Norcross, Ga.: Industrial Engineering and Management Press, 1985, pp. 8–15.§

Groover, Mikell P., and Zimmers, Emory W., Jr. *CAD/CAM: Computer-Aided Design and Manufacturing.* Englewood Cliffs, N.J.: Prentice-Hall, 1984.§

Groover, Mikell P., Hughes, J.E., Jr., and Odrey, N.G. "Productivity Benefits of Automation Should Offset Work Force Dislocation Problems." *Computer Integrated Manufacturing Systems: Selected Readings.* Ed. J.W. Nazemetz, W.W. Hammer, Jr., and R. Sadowski. Norcross, Ga.: Industrial Engineering and Management Press, 1985, pp. 47–52.§

Gross, J.L. "Components Can Be Added Gradually by Logically Mapping Out Present, Future Uses." *Industrial Engineering* (June 1984), pp. 28–37.§

Gunn, Thomas G. *Computer Applications in Manufacturing.* New York: Industrial Press, 1981.§

______. *Manufacturing for Competitive Advantage: Becoming a World Class Manufacturer.* Cambridge, Mass.: Ballinger, 1987.

Haber, L. "Cooperative 'MAP' Standards Effort Nets Results." *Mini-Micro Systems* 18, no. 14 (November 1985), pp. 50–54.§

Hales, H. Lee. "Computerized Facilities Planning and Design: Sorting Out the Options Available Now." *Computer Integrated Manufacturing Systems: Selected Readings.* Ed. J.W. Nazemetz, W.W. Hammer, Jr., and R. Sadowski. Norcross, Ga.: Industrial Engineering and Management Press, 1985, pp. 57–62.§

______. "How Small Firms Can Benefit from Computer-Integrated Manufacturing Systems." *Industrial Engineering* (June 1984), pp. 42–51.§

Hamilton, R. "Factory Systems Stuck in Low Gear." *Computerworld.* 1986.§

______. "MAP Enthusiasts Settle Down, Take Long-Range Positions: Evolving Status Dampens Ardor." *Computerworld* 20, no. 38 (September 22, 1986), p. 12.§

______. "TOP Faces SNA, User Apathy in Struggle to Become Standard." *Computerworld* 20, no. 40 (October 6, 1986), p. 6.§

Harley, Kenneth. "How to Plan and Organize an MRP Project." *Production and Inventory Management* (First Quarter 1983), pp. 53–65.§

Harmon, Paul, and King, David. *Expert Systems*. New York: John Wiley & Sons, 1985.§

Harrington, J., Jr. *Computer Integrated Manufacturing*. New York: Industrial Press, 1973.§

Harris, Marilyn A. "For IBM, Automation Means 'Made in the U.S.A.'" *Business Week* (June 16, 1986), pp. 104, 106.§

Hayes, Robert H., and Jaikumar, Ramchandran. "Manufacturing's Crisis: New Technologies, Obsolete Organizations." *Harvard Business Review* 66, no. 5 (September–October 1988), pp. 77–85.§

Hegland, Donald E. "CAD/CAM Integration —Key to the Automatic Factory." *Production Engineering* 28, no. 8 (August 1981), pp. 30–35.§

______. "CIM—The Foundation for Factory Automation." *Production Engineering* (May 1985), pp. 36–42.§

______. "Flexible Manufacturing." *Production Engineering* 28 no. 5 (May 1981), pp. 38–43.§

______. "Flexible Manufacturing—A Strategy for Winners." *Production Engineering* 25, no. 9 (September 1982), pp. 40–46.§

______. "Out in Front with CAD/CAM at Lockheed-Georgia." *Production Engineering* 28, no. 11 (November 1981), pp. 45–48.§

Heldreth, N.E. "Invest in Tomorrow's Workforce." *Production Engineering* (June 1985), pp. 38–40.§

Hill, T. *Manufacturing Strategy*. London: MacMillan Education, 1987.

Hocken, R.J. "NBS Building Facility for Research into Problem of Automating Small Machine Shops." *Industrial Engineering* (April 1982), pp. 82–84.§

Hoffman, William H. "Computer-Aided Process Planning." *Computer Integrated Manufacturing and Flexible Manufacturing Systems Seminar Proceedings*. New Orleans, April 24–26, 1985, pp. 141–148.§

Hohenstein, C. Louis. "Computers in Manufacturing: A Judgement Call." *P&IM Review and APICS News* 3, no. 1 (January 1983), pp. 20–26.§

Hohman, T., and West, J. "I/O Modules Add Power to the PC." *Production Engineering* (April 1985), pp. 62–64.§

Imai, Masaaki. *Kaizen*: *The Key to Japan's Competitive Success*. New York: Random House Business Division, 1986.§

Ingersoll Engineers. "'The Factory of the Future' Is a Myth." Rockford, Ill.: Ingersoll Engineers, Publication No. IEI 713-687-5M-TR, 1988.

Ingram, Frederick B. "Group Technology." *Production and Inventory Management* (Fourth Quarter 1982), pp. 19–33.§

Internicola, Philip R. "MRP II and the First-Time User: Avoiding the Pitfalls." *P&IM Review and APICS News* 5, no. 9 (September 1985), pp. 58–64.§

Ireland, T. "Paving the Way to CIM." *Manufacturing Systems* 5, no. 4 (April 1987), pp. 65–66.

Jacobs, James A. "Excellence Through Quality Management: The Secret to Business Growth and Profits." *Computer Integrated Manufacturing and Flexible Manufacturing Systems Seminar Proceedings.* New Orleans, April 24–26, 1985, pp. 189–194.§

Jelinek, Mariann, and Goldhar, Joel D. "The Strategic Implications of the Factory of the Future." *Sloan Management Review* (Summer 1984), pp. 29–37.§

Jenkins, Kenneth M., and Raedels, Alan R. "The Robot Revolution: Strategic Considerations for Managers." *Production and Inventory Management* (Third Quarter 1982), pp. 107–116.§

Johnson, H. Thomas, and Kaplan, Robert S. *Relevance Lost: The Rise and Fall of Management Accounting*. Boston: Harvard Business School Press, 1987.

Johnson, Steven B. "Comparing R&D Strategies of Japanese and U.S. Firms." *Sloan Management Review* (Spring 1984), pp. 25–34.§

Judson, David L. "Integrated Information Support Systems." *CIM Technology* 5, no. 2 (Summer 1986), pp. 23–30.§

"Just-In-Time (JIT)—Modern Materials Handling, 1985." *Manufacturing Guidebook*, p. 29.§

Kannewurf, Adolph. "Automated Material Handling Systems in a Computer-Integrated Manufacturing Plant." *Computer Integrated Manufacturing and Flexible Manufacturing Systems Seminar Proceedings.* New Orleans, April 24–26, 1985, pp. 44–47.§

Kasarda, J.B. "Shop Floor Control Must Be Able to Provide Real-Time Status and Control." *Industrial Engineering* (November 1980), pp. 74–78, 96.§

Kehrl, Howard H. "Engineering's Role in National Policy." *Production Engineering* 28, no. 8 (August 1981), pp. 24–25.§

Kelly, Kevin. "That Old-Time Accounting Isn't Good Enough Anymore." *Business Week* (June 6, 1988), p. 112.§

Kester, William A., Lannon, Robert P., and Bellandi, Robert. "Robot Users Need to Have Program for Maintenance." *Industrial Engineering* (January 1985), pp. 28–32.§

Kimemia, Joseph, and Gershwin, Stanley B. "Flow Optimization in Flexible Manufacturing Systems." *International Journal of Production Research* 23, no. 1 (1985), 81–96.§

Kloepper, Guenter. "Using Bar Codes in Shop Floor Management Systems." *Manufacturing Engineering* (November 1984), p. 71.§

Knox, C. "CAD/CAM and Group Technology: The Answer to Systems Integration?" *Industrial Engineering* (November 1980), pp. 66–73.§

Koelsch, J.R., and Hegland, D.E. "Robots—Catalysts for Progress in Manufacturing." *Production Engineering* (June 1985), pp. 46–51.§

Kolodziej, S. "Will TOP Take Off?" *Computerworld* 20, no. 19a (May 14, 1986), pp. 31–33.§

Korzeniowski, P. "GM and Boeing Give Protocol a Hard Sell at Autofact '85." *Computerworld* 19, no. 46 (November 18, 1985), pp. 24–25.§

Kosanchick, Mel, and Maul, Gary P. "Programmable Feeding Systems: The Right Step for Flexible Systems." *Computer Integrated Manufacturing and Flexible Manufacturing Systems Seminar*. New Orleans, April 24–26, 1985, pp. 209–212.§

Kramer, M. "Blue-Collar PCs: PCs Serve as the Electronic Partners of Factory Managers." *PC Week* 3, no. 30 (May 20, 1986), pp. 101–104.§

______. "Corporations Cope with All Sorts of E-Mailing." *PC Week* 3, no. 25 (June 24, 1986), pp. 81–82.§

______. "18 Firms Join Forces to Push OSI Standards." *PC Week* 3, no. 2 (January 14, 1986), p. 5.§

Krauskopf, Bruce. "Flexible Manufacturing of Machine Tools." *Manufacturing Engineering* (August 1984), pp. 141–142.§

______. "Robot Assisted Fabrication." *Manufacturing Engineering* (September 1984), pp. 114–115.§

Kuttner, B.C. "Computer-Aided Plastics Engineering." *Manufacturing Engineering* (March 1985), pp. 90–91.§

Lamb, J. "Standards: Standing on the Sidelines." *Datamation* 32, no. 3 (February 1, 1986), pp. 39–40.§

Layne, R. "Users Forced the Vendors to MAP Out the Standard." *Information Week* (May 12, 1986), p. 60.§

Lee, Bodil C., Kennedy, Bryan F., and Brecker, James N. "Software Application in Sheet Metal FMS." *Computer Integrated Manufacturing and Flexible Manufacturing Systems Seminar Proceedings*. New Orleans, April 24–26, 1985, pp. 87–96.§

Lee, L.C. "A Study of System Characteristics in a Manufacturing Cell." *International Journal of Production Research* 23, no. 6 (1985), pp. 1101–1114.§

Lee, Thomas H., Fisher, John C., and Yau, Timothy S. "Is Your R&D on Track?" *Harvard Business Review* (January–February 1986), pp. 34–44.§

Leffin, John W. "Improving Machining Process Flow: The Application of Group Technology to Manufacturing Cell Design." *Computer Integrated Manufacturing and Flexible Manufacturing Systems Seminar Proceedings.* New Orleans, April 24–26, 1985, pp. 97–100.§

Levulis, R.J. "Group Technology Strategies in the U.S." *Spring Annual Conference Proceedings—AIEE.* Atlanta, 1980, pp. 267–273.

———. "The Impact of Group Technology on Production and Inventory Control." *American Production and Inventory Control Society 25th Annual International Conference Proceedings.* Chicago, 1982, pp. 183–184.

Lingenfeiter, Bill. "Case Book 1: Harris Corp." *P&IM Review and APICS News* 3, no. 9 (September 1983), pp. 34–37.§

Link, Pete. "CAD/CAM: A Much Needed Overview." *P&IM Review and APICS News* 2, no. 10 (October 1982), pp. 40–44, 77–78).§

Lippolt, Thomas. "Collecting Manufacturing Data with Micro Networks." *P&IM Review and APICS News* (February 1983), pp. 35–36.§

Lofgren, Gunnar K. "Automatic Guide Vehicle Systems." *P&IM Review and APICS News* (February 1983), pp. 28–29.§

Longmire, Robert C. "Robots Are Joining the Work Force." *P&IM Review and APICS News* 2, no. 10 (October 1982), pp. 54–58.§

Loop, Jo An. "Converging on CIM: Megatest Corporation." *P&IM Review and APICS News* 5, no. 3 (March 1985), pp. 32–37, 66.§

Lotenschten, Sergio. "Just-in-Time in the MRP II Environment." *P&IM Review and APlCS News* 6, no. 2 (February 1986), pp. 26, 29, 52–54.§

Lott, Charles E. "The Case for a Value-Centered Manufacturing Strategy." *Computer Integrated Manufacturing and Flexible Manufacturing Systems Seminar.* New Orleans, April 24–26, 1985, pp. 213–219.§

Lowe, P.H., and Walshe, K.B.A. "Computer-Aided Tool Estimating: An Evaluation of the Labour Content of Injection Moulds." *International Journal of Production Research* 23, no. 2 (1985), pp. 371–380.§

Lunn, Terry. "Component Parts Control." *Computer Integrated Manufacturing and Flexible Manufacturing Systems Seminar Proceedings.* New Orleans, April 24–26, 1985.§

Macfarlane, M.L. "Plan for Today—And Tomorrow, Too." *Production Engineering* (April 1985), pp. 34–36.§

Maidique, Modesto A., and Hayes, Robert H. "The Art of High-Technology Management." *Sloan Management Review* (Winter 1984), pp. 17–31.§

Majchrazak, Ann. *The Human Size of Factory Automation.* San Francisco: Jossey-Bass, 1988.§

Makepeace, Darryl. "NPS Grows/NPS Slows a Solid Manufacturing Management System." *P&IM Review and APICS News* 2, no. 9 (September 1982), pp. 43–44, 50.§

Manna, Michael A. "CIM: Will It Work in a Small Company? A Case Study." *Computer Integrated Manufacturing and Flexible Manufacturing Systems Seminar Proceedings.* New Orleans, April 24–26, 1985, pp. 121–128.§

"Manufacturing Control Systems: The Sperry Univac UNIS 90." *P&IM Review and APICS News* 1, no. 5 (May 1981), pp. 27–29.§

Manufacturing Studies Board. *Toward a New Era in U.S. Manufacturing: The Need For a National Vision.* Washington, D.C.: National Academy Press, 1986.§

"Marrying JIT and MRP." *CTM Technology* 5, no. 1 (Spring 1986), p. 12.§

"Material Handling Equipment." *Production Engineering* 32, no. 1 (January 1985), pp. 46–52.§

"Materials Handling—Modern Challenges and a Mature Technology." *Production Engineering* (February 1985), pp. 38–46.§

Mather, H. "Change the System or Change the Environment." *Computer Integrated Manufacturing Systems: Selected Readings.* Ed. J.W. Nazemetz, W.W. Hammer, Jr., and R. Sadowski. Norcross, Ga.: Industrial Engineering and Management Press, 1985, pp. 27–31.§

McDonald, Richard A. "The Symbol Makes Symbology Work!" *P&IM Review and APICS News* 3, no. 9 (September 1983), pp. 36–37.§

McFarlan, F. Warren, and McKenney, James L. "The Information Archipelago—Governing the New World." *Harvard Business Review* (July–August 1983), pp. 91–99.§

McFarlan, F. Warren, McKenney, James L., and Pyburn, Philip. "The Information Archipelago—Plotting a Course." *Harvard Business Review* (January–February 1983), pp. 145–156.§

McKenney, James, and McFarlan, F. Warren. "The Information Archipelago—Maps and Bridges." *Harvard Business Review* (September–October 1982), pp. 109–119.§

Melnyk, Steven A., and Jacob, Satish. "Evaluating the Layout Decision in a Group Technology Setting." *Proceedings of the American Institute of Decision Sciences Conference.* San Antonio, Texas, 1983.

Melnyk, Steven A., and Narasimhan, Ram. "The Successful Implementation of CIM." *American Production and Inventory Control Society 31st Annual International Conference Proceedings.* Las Vegas, October 17–21, 1988, pp. 417–422.§

Melnyk, Steven A., Carter, Phillip L., Dilts, David M., and Lyth, David M. *Shop Floor Control*. Homewood, Ill.: Dow Jones-Irwin, 1985.§

Meredith, Jack R. "Automating the Factory: Theory Versus Practice." *International Journal of Production Research* 25, no. 10 (1987), pp. 1493–1510.§

———. "A Champion for Automation." *Industrial Engineering* (May 1985), pp. 25–26.§

———. "The Economics of Computer Integrated Manufacturing." *Computer Integrated Manufacturing Systems: Selected Readings*. Ed. J.W. Nazemetz, W.W. Hammer, Jr., and R. Sadowski. Norcross, Ga.: Industrial Engineering and Management Press, 1985, pp. 42–46.§

Meredith, Jack R., Hyer, Nancy L., Gerwin, Donald, Rosenthal, Stephen R., and Wemmerlow, Urban. "Research Needs in Managing Factory Automation." *Journal of Operations Management* 6, no. 2 (February 1986), pp. 203–218.§

Miles, Gregory L. "Selling Rockwell on Automation." *Business Week* (June 6, 1988), p. 104.§

Miller, R. "Factory with a Future...No Pushover!" *Computer Integrated Manufacturing and Flexible Manufacturing Systems Seminar Proceedings*. New Orleans, April 24-26, 1985, pp. 149–161.§

Miller, Robert D. "JIT and CIM: An Integrative Look at the Technologies." *Computer Integrated Manufacturing and Flexible Manufacturing Systems Seminar Proceedings*. New Orleans, April 24–26, 1985, pp. 30–40.§

Mishkoff, Henry C. *Understanding Artificial Intelligence*. Dallas: Texas Instrument Corporation, 1986.

Mitchell, Russell. "Detroit Stumbles on Its Way to the Future." *Business Week* (June 16, 1986), pp.103–104.§

Mitrofanow, S.P. *Scientific Fundamentals of Group Technology*. USSR, 1959.

"Modern Materials Handling, 1985." *Manufacturing Guidebook*, pp. 66–67.§

Monden, Y. "What Makes the Toyota Production System Really Tick?" *Industrial Engineering* (January 1981), pp. 36–46.§

Montag, Alfred C. "Flexible Automation for High Volume Production." *Manufacturing Engineering* (November 1984), pp. 79–80.§

Moore, J.M. "Computer Methods in Facilities Layout." *Industrial Engineering* (September 1980), pp. 83–93.§

Morrison, Dave. "MRP/Case-in-Point: Gilford Instrument Laboratories." *P&IM Review and APICS News* 4, no. 9 (September 1984), pp. 56–57.§

Mottern, Jim. "Executing the CIM Plan: Managing Technology Change." *Computer Integrated Manufacturing and Flexible Manufacturing Systems Seminar Proceedings*. New Orleans, April 24–26, 1985, pp. 1–5.§

Muir, William T. "Technology Management and Factory Automation." *Computer Integrated Manufacturing and Flexible Manufacturing Systems Seminar Proceedings*. New Orleans, April 24–26, 1985, pp. 101–110.§

Munro, H., and Noori, H. "Measuring Commitment to New Manufacturing Technology: Integrating Technological Push and Marketing Pull Concepts." *IEEE Transactions on Engineering Management* 35, no. 2 (May 1988), pp. 63–70.§

Muramatsu, Rintaro, Ishii, Kazuyoshi, and Takahashi, Katsuhiko. "Some Ways to Increase Flexibility in Manufacturing Systems." *International Journal of Production Research* 23, no. 4 (1985), pp. 694–703.§

Musselman, Kenneth J. "Computer Simulation: A Design Tool for FMS." *Manufacturing Engineering* (September 1984), pp. 117–120.§

Nakane, Jinichiro, and Hall, Robert W. "Management Specs for Stockless Production." *Harvard Business Review* (May–June 1983), pp. 84–91.§

New, C.C. "MRP and GT: A New Strategy for Component Production." *Production and Inventory Management* 18, no. 3 (1977), pp. 50–56.

Norman, R. Henry, and Blair, Patricia. "The Coming Growth in Appropriate Technology." *Harvard Business Review* (November–December 1982), pp. 62–64.§

Ogorek, M. "Interactive Graphics and Conversational Programming." *Manufacturing Engineering* (January 1985), pp. 47–48.§

"One Step in Managing Growth—Automate Manufacturing." *Production Engineering* (February 1985), pp. 92–93.§

"Our Machine Tools Never Wait for Materials." *Modern Material Handling* 37, no. 14 (September 21, 1982), pp. 46–50.

Parsons, Gregory L. "Information Technology: A New Competitive Weapon." *Sloan Management Review* (Fall 1983), pp. 3–14.§

"Part I—Meeting the Economic Challenge." *Modern Material Handling* 41, no. 9 (August 1986), pp. 64–69.§

"PC Automates Cutting Machine for Less." *Production Engineering* (April 1985), pp. 106–107.§

Pease, Richard L. "Automation and Electric Drives." *Production Engineering* 35, no. 5 (May 1987), pp. 36–39.

Pennar, Karen. "The Productivity Paradox." *Business Week* (June 6, 1988), pp. 100–102.§

Pereira, C.L. "Is Top Management Really Committed to Innovation." *Industrial Engineering* (June 1985), pp. 12–14.§

Petterson, Wilbur A. "CAM Interface with Shop Scheduling." *P&IM Review and APICS News* 3, no. 11 (November 1983), pp. 62–63.§

Pflanzer, Dale. "Robots Automate Gear-Cutting Operation." *Production Engineering* 32, no. 11 (November 1985), pp. 16–19.§

Pfleger, D. "Microprocessors: What They Are, How They're Used, and in What Applications." *Industrial Engineering* (November 1980), pp. 80–83.§

Phelps, R.I. "Artificial Intelligence—An Overview of Similarities with O.R." *Journal of the Operational Research Society* 37, no. 1 (1986), pp. 13–20.§

Pittiglio, Ted, and Moore, Jack. "Computer Integrated Assembly and Test of Printed Circuit Boards." *Computer Integrated Manufacturing and Flexible Manufacturing Systems Seminar Proceedings*. New Orleans, April 24–26, 1985, pp. 162–166.§

Pittiglio, Rabin, Todd, and McGrath. "Inventory Performance for High Technology Industries, 1982." *P&IM Review and APICS News* (June 1983), pp. 27–30, 58.§

"Plant-Floor/Process Software: Next Manufacturing Challenge." *MIS Week* 9, no. 44 (October 31, 1988), p. 46.§

Plossl, George W., and Lankford, Raymond L. "The Redirection of U.S. Manufacturing: Part 1—The Ending Era." *P&IM Review and APICS News* 4, no. 10 (October 1984), pp. 74–78; 88.§

Polcyn, Stanley. "Robots and American Industrialization." *P&IM Review and APICS News* 1, no. 11 (November 1981), pp. 18–20.§

Port, Otis. "High Tech to the Rescue." *Business Week* No. 2951 (June 16, 1986), pp. 100–103.§

Port, Otis, King, Resa, and Hampton, William J. "How the New Math of Productivity Adds Up." *Business Week*. No. 3055 (June 6, 1988), pp. 103–114.§

Porter, Robert W. "Classification and Coding for Today's Manufacturing." *Computer Integrated Manufacturing and Flexible Manufacturing Systems Seminar Proceedings*. New Orleans, April 24–26, 1985, pp. 48–57.§

Poza, Ernesto J. "Twelve Actions to Build Strong U.S. Factories." *Sloan Management Review* (Fall 1983), pp. 27–38.§

"Productivity Increases at Buick Through AS/RS." *P&IM Review and APICS News* 2, no. 4 (April 1982), pp. 30–31.§

Pukanic, R. "Flexibility and Integration Are Key Material Handling Concepts in Electronic Assembly Environment." *Industrial Engineering* (April 1985), pp. 40–43, 115.§

Pullen, R.D. "A Survey of Cellular Manufacturing Cells." *The Production Engineer* (September 1976), pp. 451–454.

Pun, L., Doumeingts, G., and Bourely, A. "The GRAI Approach to the Structural Design of Flexible Manufacturing Systems." *International Journal of Production Research* 23, no. 6 (1985), pp. 1197–1215.§

Purcheck, Gunter. "Computer-Aided Organization for Manufacture." *International Journal of Production Research* 23, no. 5 (1985), pp. 887–910.§

______. "Machine-Component Group Formation: An Heuristic Method for Flexible Production Cells and Flexible Manufacturing Systems." *International Journal of Production Research* 23, no. 5, (1985), pp. 911–943.§

Quinn, James Brian. "U.S. Industrial Strategy: What Directions Should It Take?" *Sloan Management Review* (Summer 1983), pp. 3–21.§

Ranky, P.G. *Computer Integrated Manufacturing*. Englewood Cliffs, N.J.: Prentice-Hall International, 1986.§

Rasmus, Daniel W. "Orchestrating Change—Progressing Toward CIM." *Manufacturing Systems* 5, no. 4 (April 1987), pp. 30–32.

Reddy, Jack, and Berger, Abe. "Three Essentials of Product Quality." *Harvard Business Review* (July–August 1983), pp. 153–159.§

Roberts, Edward B., and Alan R. Fusfeld. "Staffing the Innovative Technology-Based Organization." *Sloan Management Review* (Spring 1981), pp. 19–34.§

"Robots: Combining Technologies for Added Benefits." *Production Engineering* 31, no. 6 (June 1984), pp. 80–84.§

Rockart, John F., and Crescenzi, Adam D. "Engaging Top Management in Information Technology." *Sloan Management Review* (Summer 1984), pp. 3–16.§

Roman, Anthony W. "How to Tune In on Company Goals and Objectives." *Production Engineering* 28, no. 5 (May 1981), pp. 30–31.§

Ronstadt, Robert, and Kramer, Robert. "Getting the Most Out of Innovation Abroad." *Harvard Business Review* (March–April 1982), pp. 94–99.§

Rubinger, Bruce. "Industrial Innovation: Implementing the Policy Agenda." *Sloan Management Review* (Spring 1983), p. 43–57.§

Sadowski, R.P. "History of Computer Use in Manufacturing Shows Major Need Now Is for Integration." *Computer Integrated Manufacturing Systems: Selected Readings*. Ed. J.W. Nazemetz, W.W. Hammer, Jr., and R. Sadowski. Norcross, Ga.: Industrial Engineering and Management Press, 1985, pp. 3–7.§

Sampson, R.A. "Integration of MRP with the Flexible Manufacturing System." *Computer Integrated Manufacturing and Flexible Manufacturing Systems Seminar Proceedings*. New Orleans, April 24–26, 1985, pp. 111–119.§

Sanders, I.F. "A Step Toward Tomorrow: AS/RS at Lockheed." *P&IM Review and APICS News* 2, no. 4 (April 1982), pp. 24–28.§

Sartori, Luca G. *Manufacturing Information Systems.* New York: Addison-Wesley, 1988.§

Savage, Edward, and Mikurak, Michael. "Finite Scheduling: Staging a Comeback?" *CIM Technology* 5, no. 1 (Spring 1986), pp. 26–31.§

Schaefer, Thomas J. "Modular Approach to CIM: Integrate Your Factory in Pieces." *Production Engineering* (April 1985), pp. 102–104.§

Schiller, Zachary. "Numbers Don't Tell the Story." *Business Week.* No. 3055 (June 6, 1988), p. 105.§

Schlegel, G.L. "Computer Integrated Manufacturing (CIM): One Company's Strategy Towards Managing the New 'Fifth Factor of Production.'" *Computer Integrated Manufacturing and Flexible Manufacturing Systems Seminar Proceedings.* New Orleans, April 24–26, 1987, pp. 67–75.§

Schreiber, R.R. "CNC Programming: An Occupation in Transition." *Manufacturing Engineering* (June 1985), pp. 75–76.§

Schultz, Hilbert K. "Microcomputers and Manufacturing, Distribution, and Service Organizations." *Production and Inventory Management* no. 4 (1983), pp. 87–93.§

Schultz, K.B. "A Machine Tool Manufacturer's View of the Factory Environment of the Future." *Computer Integrated Manufacturing Systems: Selected Readings.* Ed. J.W. Nazemetz, W.W. Hammer, Jr., and R. Sadowski. Norcross, Ga.: Industrial Engineering and Management Press, 1985, pp. 16–21.§

Schwendinger, James R. "Do Your Homework Before You Modernize the Factory." *Computer Integrated Manufacturing and Flexible Manufacturing Systems Seminar.* New Orleans, April 24–26, 1985, pp. 228–239.§

Shank, D.L. "Group Technology Provides Organized Approach to Realizing Benefits of CIMS." *Industrial Engineering* (April 1985), pp. 74–80.§

Sibbald, G.W. *CIM Today.* Menlo Park, Calif.: SRI International, March 1986.§

Smith, J.L. "CIM—Target or Philosophy By-Product." *Computer Integrated Manufacturing Systems: Selected Readings.* Ed. J.W. Nazemetz, W.W. Hammer, Jr., and R. Sadowski. Norcross, Ga.: Industrial Engineering and Management Press, 1985, pp. 32–34.§

Smith, R.E. "Robotic Vehicles Will Perform Tasks Ranging from Product Retrieval to Sub-assembly Work in Factory of Future." *Industrial Engineering* (September 1983), pp. 62–72.§

Smith, Spencer B. *Computer Based Production and Inventory Control.* Englewood Cliffs, N.J.: Prentice-Hall, 1989.

Smith, Steven R. "MRP II and JIT: Evolution of a Competitor." *P&IM Review and APICS News* (August 1985), pp. 32–36, 60.§

Spencer, Michael S. "Scheduling Components for Group Technology Lines (A New Application for MRP)." *Production and Inventory Management* (Fourth Quarter 1980), pp. 43–49.§

Sprague, L.G. "The Future Factory: Is the Push for Automation Coming from Below?" *Purchasing Management* (December 1987), pp. 10–12, 20.§

Staeheli, Michael. "Manufacturing Change: A Work in Process." *P&IM Review and APICS News* 6, no. 2 (February 1986), pp. 46–51.§

Stauffer, R. N. "The Pontiac Fiero: Advent of a New Manufacturing Philosophy." *Manufacturing Engineering* (November 1983), pp. 65–68.§

______. "Robots Provide Efficiency and Quality in New Production Welding Lines at Jeep." *Manufacturing Engineering* (October 1983), pp. 57–60.§

Stevens, Larry. "The CIM Connection: Reaching Out to the Shop Floor." *Computerworld* 22, no. 22 (May 30, 1988), pp. 45–52.§

Stobaugh, Robert, and Telesio, Piero. "Match Manufacturing Policies and Product Strategy." *Harvard Business Review* (March–April 1983), pp. 113–120.§

Sunderland, F.O., and Fox, R.E. "A Synchronized Manufacturing Challenge." *P&IM Review and APICS News* 6, no. 3 (March 1986), pp. 36–44.§

Swamidass, P.M. *Manufacturing Flexibility*. Monograph. Waco, Tex.: Operations Management Association, January 1988.§

Swartley-Loush, J. "Just-in-Time: Is It Right for You?" *Production Engineering* (June 1985), pp. 60–63.§

Takeuchi, Hirotaka, and Quelch, John A. "Quality Is More Than Making a Good Product." *Harvard Business Review* (July–August 1983), pp. 139–145.§

"Taking the Step from CAD/CAM to CIM." *Production Engineering* 30, no. 11 (November 1983), pp. 46–57.§

Tanner, John P. "The Learning Curve: A Line on Labor Cost." *Production Engineering* (May 1985), pp. 72–78.§

"Test and Inspection Equipment." *Production Engineering* 32, no. 1 (January 1985), pp. 38–42.§

Thompson, James D. *Organizations in Action*. New York: McGraw-Hill, 1967.

Thompson, Olin W. "Practitioner—Know Thy Self." *Computer Integrated Manufacturing and Flexible Manufacturing Systems Seminar*. New Orleans, April 24–26, 1985, pp. 240–247.§

Thorton, Jack. "Strategic Management: New Marketing Muscle (Part 5)." *Industry Week* 233, no. 3 (May 4, 1987), pp. 37–40.

Thurwachter, William A. "FM$ vs. FMc." *Computer Integrated Manufacturing and Flexible Manufacturing Systems Seminar Proceedings*. New Orleans, April 24–26, 1985, pp. 179–188.§

Tombak, Mihkel, and De Meyer, Arnoud. "Flexibility and FMS: An Empirical Analysis." *IEEE Transactions on Engineering Management* 35, no. 2 (May 1988), pp. 101–107.§

Tompkins, James A. "CIM is Great: But What Do You Do?" *Modern Materials Handling* 41, no. 9 (August 1986), p. 45.§

———. "Successful Facilities Planner Must Fulfill Role of Integrator in the Automated Environment." *Industrial Engineering* (May 1984), pp. 54–58.§

"Transfer the Heads, Hold the Workpieces." *Industrial Engineering* 9, no. 11 (1977), pp. 26–30.

Tulkoff, J. "Implementing a Computer Process Planning System Based on a Group Technology Classification and Coding Scheme." *Computer Integrated Manufacturing Systems: Selected Readings.* Ed. J.W. Nazemetz, W.W. Hammer, Jr., and R. Sadowski. Norcross, Ga.: Industrial Engineering and Management Press, 1985, pp. 72–75.§

Turban, E. "Review of Expert Systems Technology." *IEEE Transactions on Engineering Management* 35, no. 2 (May 1988), pp. 71–81.§

"Turning Cells Boost Gear Blank Output." *Production Engineering* 31, no. 8 (August 1984), pp. 60C–60F.§

Ullman, Allan. "Streamlining MiroFlector's Manufacturing Operation." *P&IM Review and APICS News* 3, no. 1 (January 1983), pp. 17, 26.§

"Unexpected Benefits Through Automation." *P&IM Review and APICS News* 1, no. 5 (May 1981), pp. 23-24.§

Voss, C A. *Managing New Managing Technologies.* Monograph. East Lansing, Mich.: Operations Management Association, 1986.§

Webster, W.B. "Production and Order Planning in the FMS Environment." *Computer Integrated Manufacturing Systems: Selected Readings.* Ed. J.W. Nazemetz, W.W. Hammer, Jr., and R. Sadowski. Norcross, Ga.: Industrial Engineering and Management Press, 1985, pp. 63–71.§

Weimer, G. "Autofact '85: Integrated Manufacturing Comes of Age." *Production Engineering* 32, no. 12 (December 1985), pp. 48–54.§

Weisel, Walter K. "The Robot's Role in Productivity." *Production Engineering* 24, no. 12 (December 1981), pp. 18–19.§

Welter, W.R. "Factory of the Future/IE of the Future." *Computer Integrated Manufacturing Systems: Selected Readings.* Ed. J.W. Nazemetz, W.W. Hammer, Jr., and R. Sadowski. Norcross, Ga.: Industrial Engineering and Management Press, 1985, pp. 22–26.§

———. "Manufacturing Strategy—The Why and How." *Computer Integrated Manufacturing and Flexible Manufacturing Systems Seminar.* New Orleans, April 24–26, 1985, pp. 220–227.§

"Why We're Enthusiastic about Industrial Robots." *Modern Material Handling* (September 20, 1983), pp. 54–57.§

Wick, C. "Turning Centers Update." *Manufacturing Engineering* 99, no. 1 (July 1987), pp. 41–48.§

Wild, William J., Jr., and Port, Otis. "This Video 'Game' Is Saving Manufacturing Millions." *Business Week* no. 3012 (August 17, 1987), pp. 82–84.§

Wilhelm, W.E., and Shin, Hyun Myung. "Effectiveness of Alternative Operations in a Flexible Manufacturing System." *International Journal of Production Research* 23, no. 1 (1985), pp. 65–79.§

Williams, Joe L. "Decision Support Systems in Production Planning." *P&IM Review and APICS News* 3, no. 11 (November 1983), pp. 82–85.§

Willams, Theodore J. "Recent Developments in the Application of Plant-Wide Computer Control." *Computers in Industry* 8, no. 2/3 (April 1987), pp. 233–254.

Willis, R.G., and Sullivan, K.H. "CIMS in Perspective: Costs, Benefits, Timing, Payback Periods are Outlined." *Computer Integrated Manufacturing Systems: Selected Readings*. Ed. J.W. Nazemetz, W.W. Hammer, Jr., and R. Sadowski. Norcross, Ga.: Industrial Engineering and Management Press, 1985, pp. 35–41.§

Willoughby, Bert R. "Automatic Real Time Tracking and Control in the Manufacturing Plant." *Computer Integrated Manufacturing and Flexible Manufacturing Systems Seminar Proceedings*. New Orleans, April 24–26, 1985, pp. 76–83.§

Wiltsie, Steve. "Microcomputers in Manufacturing." *P&IM Review and APICS News* (January 1983), pp. 27–29.§

Winston, P.H. "The AI Business: A Perspective." *Manufacturing Engineering* (March 1985), pp. 75–78.§

Yao, David D., and Buzacott, J A. "Modelling the Performance of Flexible Manufacturing Systems." *International Journal of Production Research* 23, no. 4 (1985), pp. 945–959.§

Yoshikawa, Hiroyuki. "Computer Integrated Manufacturing: Visions and Realities About the Factory of the Future." *Computers in Industry* 8, no. 2/3 (April 1987), pp. 181–196.

Zimmerman, S.M., and Glover, D. "Safety Stocks: Balancing Outages and Inventory Costs." *Industrial Engineering* (May 1982), pp. 38–42.§

Zuboff, Shoshana. "New Worlds of Computer-Mediated Work." *Harvard Business Review* (September–October 1982), pp. 142–152.§

Zygmont, J. "Guided Vehicles Set Manufacturing in Motion." *High Technology* 6, no. 12 (December 1986), pp. 16–21.§

INDEX

D

G

H

I

Q

R

T

U

V

W

Y

Z

MEASURING UP
Charting Pathways to Manufacturing Excellence
Robert W. Hall, H. Thomas Johnson, and Peter B. B. Turney
The Business One Irwin/APICS Series in Production Management

This is the ultimate guide for manufacturing executives and managers looking to improve their overall competitiveness. *Measuring Up* covers both operational and cost measurement so you can improve costing for better results, manage your organization for continual improvement, and measure the proper functions to ensure excellence.

ISBN: 1-55623-359-0 $42.50